# Advances in Sustainability Science and Technology

The book series aims at bringing together valuable and novel scientific contributions that address the critical issues of renewable energy, sustainable building, sustainable manufacturing, and other sustainability science and technology topics that have an impact in this diverse and fast-changing research community in academia and industry.

The areas to be covered are

- Climate change and mitigation, atmospheric carbon reduction, global warming
- Sustainability science, sustainability technologies
- Sustainable building technologies
- Intelligent buildings
- Sustainable energy generation
- Combined heat and power and district heating systems
- Control and optimization of renewable energy systems
- Smart grids and micro grids, local energy markets
- Smart cities, smart buildings, smart districts, smart countryside
- Energy and environmental assessment in buildings and cities
- Sustainable design, innovation and services
- Sustainable manufacturing processes and technology
- Sustainable manufacturing systems and enterprises
- Decision support for sustainability
- Micro/nanomachining, microelectromechanical machines (MEMS)
- Sustainable transport, smart vehicles and smart roads
- Information technology and artificial intelligence applied to sustainability
- Big data and data analytics applied to sustainability
- Sustainable food production, sustainable horticulture and agriculture
- Sustainability of air, water and other natural resources
- Sustainability policy, shaping the future, the triple bottom line, the circular economy

High quality content is an essential feature for all book proposals accepted for the series. It is expected that editors of all accepted volumes will ensure that contributions are subjected to an appropriate level of reviewing process and adhere to KES quality principles.

The series will include monographs, edited volumes, and selected proceedings.

More information about this series at https://link.springer.com/bookseries/16477

Anca Draghici · Larisa Ivascu
Editors

# Sustainability and Innovation in Manufacturing Enterprises

## Indicators, Models and Assessment for Industry 5.0

Springer

*Editors*
Anca Draghici
Faculty of Management in Production and Transportation
Politehnica University of Timisoara
Timisoara, Romania

Larisa Ivascu
Faculty of Management in Production and Transportation
Politehnica University of Timisoara
Timisoara, Romania

Academy of Romanian Scientists
Bucharest, Romania

ISSN 2662-6829 ISSN 2662-6837 (electronic)
Advances in Sustainability Science and Technology
ISBN 978-981-16-7364-1 ISBN 978-981-16-7365-8 (eBook)
https://doi.org/10.1007/978-981-16-7365-8



This Springer imprint is published by the registered company Springer Nature Singapore Pte Ltd.
The registered company address is: 152 Beach Road, #21-01/04 Gateway East, Singapore 189721, Singapore

# Preface

The book presents innovative and important approaches for organizational sustainability in the context of Industry 5.0; by following the book content, readers will discover a large debate on green manufacturing, Industry 5.0 particularities and perspectives, smart manufacturing, intelligent monitoring, warehouse ergonomics, innovative services and trends for the manufacturing processes considering the new and dynamic context of the Industrial Revolution. The purpose of this book is to inform and educate its public about sustainable development, approaches and applications related to the manufacturing processes and to present the trends to the next economic and social paradigms, the Industry 5.0 and Society 5.0. Furthermore, educational aspects, case studies from various companies, together with the analysis and synthesis of the literature have been included in the chapters. Thus, this book could be considered as a started point and foundation for researchers and practitioners interested in the evolution and the trends of the manufacturing systems.

The current economic conditions and the business environment dynamics have contributed to the development of a new production paradigm. Starting from the Industry 4.0 advances in robotics, Artificial Intelligence (AI) and Machine Learning (ML), there has been launched a new age of automation, as machines match or outperform human performance in a range of different work activities, including the ones requiring cognitive capabilities (McKinsey Company, 2017[1] and 2018[2]). Therefore, it is not new that manufacturing conditions and economic framework conditions change over time and thus, Industry 5.0 has been introduced and is considered as the next Industrial Revolution with an impact on industry and economy, civil society, governance, structures as well as human identity.

New and emerging technologies of Industry 5.0 should always be designed for sustainability, and they will support the management of sustainable development

[1] McKinsey & Company (2017) A future that works: automation, employment, and productivity, McKinsey Global Institute.

[2] McKinsey & Company (2018) Unlocking success in digital transformations, Retrieved from: https://www.mckinsey.com/business-functions/organization/our-insights/unlocking-success-in-digital-transformations.

(mandatory actions). An actual and realistic company's goal must be to conserve its resources and to continuously improve resource efficiency, and keep the environment clean and create added value for people (clients, customers and communities) through their products and services offered; then a company in Industry 5.0 will be successful. The research and development approaches together with collaborative innovation must be based on a theoretical analysis and best practice collection of the basic blocks of the domain or topic in manufacturing systems. The concept of Industry 4.0 and digitization are meant to create the foundation. A core component of the respective developments can be traced back to the technological innovations. The technologies such as AI or Big Data, Internet of (every) Things, Robots and Human-Computer Interaction are among the most emergent ones, during the current decade. These announce new changes in the design, manufacturing and innovation processes that should also be addressed with relationship management and sustainable development management. Changes affect product life cycle management and sustainable development awareness (referring to economic, social and environment awareness) by determining the transformation of traditional approaches mostly to circular, blue economy.

The triumvirate of Industry 5.0 (characterized by a deep and extended digitization of the processes) consists of Process, People and Technology but integrated by the sustainable development principles and values. Technology enables the automated execution of processes and thus creates added value, while people follow the process and use technology in a leveraged manner. It is important to point out the changes brought about by Industry 5.0 and create the necessary openness to change, because change is constantly taking place in many different facets.

Sustainable development principles and approaches are already recognized and implemented by many companies all over the world. It has become mandatory because of the pressure of climate change facts and by the numerous benefits that contribute to improving the attractiveness and the degree of implementation in the organizational framework. In recent years, there have been several concerns for the "health of individuals and the environment" which have arisen due to pollution, the natural resource scarcity and the increase in the amount of waste generated by manufacturing processes. Furthermore, all these concerns have been addressed through sustainability which can respond to these concerns and is an essential feature of actual design and exploitation of the manufacturing systems. It has been estimated that new business models will be developed in this new digital context of the economy.

Sustainable manufacturing contributes to the creation of healthy products through healthy processes. These processes have a reduced effect on society, streamline the number of resources used, are economically healthy and conserve energy. In addition, manufacturing sustainability must consider the aspects related to the social dimension of the organization or the human side of enterprise. Employees' safety and occupational risk management are still discussed in the literature because there is still human intervention in manufacturing processes. At the same time, companies are involved in social responsibility actions, and communities are supported to encourage healthy and equitable practices, responsible consuming of non-polluting products that have a low impact on the environment.

The development of manufacturing systems that generate zero waste is a challenge for the economic environment. Applying the principles of sustainable development to reduce the amount of waste disposed of is an important direction for businesses. Current trends in streamlining waste management are a real challenge for businesses. The needs and desires of customers are more numerous and diversified, so sustainable processes must be found in the design and approach of stakeholders and shareholders, too.

The book opens with a Preface and ends with a chapter dedicated to the debate on future trends in manufacturing.

Chapter 1 discusses the green production approach in the context of the circular economy and sustainable development. This chapter presents important aspects related to natural resources, greenhouse gas emissions, modelling emissions in production processes and the transition from waste management to resource management.

Chapter 2 presents a detail analysis of the main characteristics related to Industry 4.0, Industry 5.0 and Society 5.0 and identifies differences, implications for research and industrial/business practice. This chapter presents an exhaustive reference analysis and synthesis about the New Industrial Revolution, challenges and implications.

Chapter 3 evaluates smart manufacturing systems and presents their particularities throughout the organizational development. The study also presented the major software providers' prediction, before and during the COVID pandemic time, which greatly impacted the consumer behaviour.

Chapter 4 presents a synthesis about the production processes' automation and their intelligent monitoring. Various aspects of AI control systems are discussed, mainly in the case of machine tools. The authors identified gaps and opportunities to develop other blocks of intelligent monitoring of the manufacturing processes.

Chapter 5 debates over aspects related to the new challenges of reducing ergonomics risks in manufacturing warehouse logistics by using emerging technologies to achieve workplace wellbeing. After an extended literature review regarding the relevant ergonomics approaches in warehouse logistics, there are presented some warehouse ergonomics solutions (as examples of best practices) to be considered for the next generation of the logistics system.

Chapter 6 presents a model of the product development process including a technical configurator with an adaptable interface and a e-service web application (the authors' own development), which involves customers' direct involvement in the design process by exploiting concepts such as mass customization, personalization and individualization. The development of new products in the context of sustainable development and digitalization is adapted to specific customers' needs.

Chapter 7 provides a large scientific debate on complexity management in manufacturing. Based on the findings in the literature, there is proposed a model for managing complexity in large-scale manufacturing environments (considering the case of big projects dedicated to services' outsourcing). The focus is on defining an integrated approach of complexity, specific manufacturing-related complexity criteria, cyclical reporting, targeted measures to identified specific measures to optimize the management process.

Chapter 8 presents an overview regarding the “presence” of the Industry 4.0 elements in the European Union manufacturing companies using publicly available data from the European Statistical Office.

Chapter 9 brings some new insights into the way the components of the Industry 5.0 readiness model influences the value added reported by countries, for the period between 2017 and 2019, on four levels of analysis, namely (i) country gross value added; (ii) environmental activities’ value added; (iii) information and communication technology sector value added; (iv) labour productivity. The results prove that selected components of the Industry 5.0 readiness model determine a positive marginal effect on the value added reported by countries.

Chapter 10 addresses AI and its importance in the context of Industry 5.0. At the same time, the importance of AI at the university level is addressed.

Chapter 11, the last chapter of this book, concludes by presenting a series of organizational strategies that contribute to increasing organizational competitiveness and implications for innovation, evaluates the implications of Industry 4.0 and makes an inventory of the technologies used in this industrial era. To consolidate the importance of this revolution for the manufacturing industry, a presentation of the situation in Romania is made from the perspective of various important indicators for Industry 4.0 (showing a case of transition to Industry 5.0). In the end of the chapter is presented a market research and a debate on the proposed strategic roadmap for the future of manufacturing.

Timisoara, Bucharest, Romania
September 2021

Anca Draghici
Larisa Ivascu

# Contents

# Editors and Contributors

## About the Editors

**Anca Draghici** received the M.Sc. in Machine-Tools Design (1989, Transilvania University of Brasov, Romania) and B.Sc. in Business Management (2001, Babes-Bolyai University of Cluj-Napoca, Romania). In 2001, she achieves the Ph.D. degree in the research field of machine-tools ergonomics, at Transilvania University of Brasov, Romania. Now she is full professor (and Ph.D. supervisor) at Politehnica University Timisoara, Romania, Faculty of Management in Production and Transportation. Her teaching subjects are related to Industrial Ergonomics, Human Resources Management, Occupational Health and Safety, Investments Management and Knowledge Management. Her research fields of interest are linked with the impact of sustainability on organizational dynamics and business models. She regularly publishes and participates on international scientific conferences (she has more than 130 articles index in Clarivate Analytics database). In the last 10 years she was involved in different Erasmus+ educational projects dedicated to sustainable development skills development and education (for higher education and vocational training). Anca Draghici is president of the Society for Ergonomics and Work Environment Management in Romania (ErgoWork), member of the Member of Academic Management Society of Romania (SAMRO), former member of the European Manufacturing and Innovation Research Association a Cluster Leading Excellence (EMIRAcle) and member of the European Certification and Qualification Association (ECQA). As a recognition to her research achievements, Anca DRAGHICI has been guest editor at Human Systems Management journal (2020, 2021), Sustainability journal (2020) and has act as invited speaker in several international conferences; she is editor of the Scientific Bulletin of Politehnica University of Timisoara, Transaction on Engineering and Management.

**Larisa Ivascu** is currently professor at the Politehnica University of Timisoara, Romania. In 2007 she graduated Software Development Bachelor programme at the Faculty of Computer Science and in 2010 she graduated the MBA programme at the

Faculty of Management in Production and Transportation, both at Politehnica University of Timisoara, Romania. She has a Ph.D. degree in Engineering and Management from Politehnica University of Timisoara, Romania (2013). Her research focuses on corporate social responsibility, risk management, sustainability, strategic and business management. She has published over 190 papers in international journals in which 86 are Clarivate Analytics indexed, and 95 as first or sole author, and took part in many international conferences as an Invited Speaker. She is vice-president of the Society for Ergonomics and Work Environment Management in Romania (ErgoWork) and member of the World Economics Association (WEA), International Economics Development and Research Center (IEDRC). She is Director of the Engineering and Management Research Center (CCIM) of Politehnica University of Timisoara, Romania and member of Academic Management Society of Romania (SAMRO). As a recognition to her research achievements, Larisa IVASCU has been guest editor at Safety Journal and Resources Journal. In addition, she is Editorial Board Member of the International Journal of Social Computing and Cyber-Physical Systems, and an Editorial Office Member of Scientific Bulletin of Politehnica University of Timisoara, Transaction on Engineering and Management.

## Contributors

**Lidia Alexa** "Gheorghe Asachi" Technical University of Iasi, Engineering and Management Department, Iasi, Romania

**Ben-Oni Ardelean** University of Bucharest, Faculty of Baptist Theology, Bucharest, Romania

**Silvia Avasilcăi** "Gheorghe Asachi" Technical University of Iasi, Engineering and Management Department, Iasi, Romania

**Felicia Banciu** Politehnica University of Timisoara, Faculty of Mechanical Engineering, Timisoara, Romania

**Cristin Bigan** Ecological University of Bucharest, Faculty of Managerial Engineering, Bucharest, Romania

**Oana Bogdan** West University of Timisoara, Faculty of Economics and Business Administration, Timioara, Romania

**Valentin Burca** West University of Timisoara, Faculty of Economics and Business Administration, Timioara, Romania

**George Carutasu** Romanian-American University, Faculty of Computer Science for Business Management, Bucharest, Romania

**Nicoleta Luminita Carutasu** University Politehnica of Bucharest, Faculty of Industrial Engineering and Robotics, Bucharest, Romania

**Aura Domil** West University of Timisoara, Faculty of Economics and Business Administration, Timioara, Romania

**Florin Dragan** Politehnica University of Timisoara, Department of Automation and Applied Informatics, Faculty of Automation and Computers, Timisoara, Romania

**Anca Draghici** Faculty of Management in Production and Transportation, Politehnica University of Timisoara, Timisoara, Romania

**George Draghici** Politehnica University Timisoara, Faculty of Mechanical Engineering, Timisoara, Romania

**Anamaria Feier** Politehnica University of Timisoara, Faculty of Mechanical Engineering, Timisoara, Romania

**Adrian Ciprian Firu** Politehnica University of Timisoara, Faculty of Mechanical Engineering, Timisoara, Romania

**Alin Gaureanu** Faculty of Management in Production and Transportation, Politehnica University of Timisoara, Timisoara, Romania

**Larisa Ivascu** Faculty of Management in Production and Transportation, Politehnica University of Timisoara, Timisoara, Romania;
Academy of Romanian Scientists, Bucharest, Romania

**Caius-Tudor Luminosu** Faculty of Management in Production and Transportation, Politehnica University of Timisoara, Timisoara, Romania

**Mirabela Metzler** Faculty of Management in Production and Transportation, Politehnica University of Timisoara, Timisoara, Romania

**Anca Mocan** Amaris Consulting, Vernier, Switzerland

**Beata Mrugalska** Faculty of Engineering Management, Poznan University of Technology, Poznan, Poland

**Elif Ocakci** Faculty of Management in Production and Transportation, Politehnica University of Timisoara, Timisoara, Romania;
Department of Strategy and Business Development, Frankfurt am Main, Germany

**Daniel Paschek** ServiceNow, Frankfurt am Main, Germany

**Marius Pîslaru** “Gheorghe Asachi” Technical University of Iasi, Engineering and Management Department, Iasi, Romania

**Frank Rennung** Deutsche Telekom IT GmbH - Business Solution, Darmstadt, Hessen, Germany

**Dorian Stef** Politehnica University Timisoara, Faculty of Mechanical Engineering, Timisoara, Romania

**Gyula Szabó** Donát Bánki Faculty of Mechanical and Safety Engineering, Óbuda University, Budapest, Hungary

**Alin Ion Tapirdea** Politehnica University of Timisoara, Faculty of Mechanical Engineering, Timisoara, Romania

# Chapter 1
# Green Manufacturing in the Context of Circular Economy

**Anca Draghici and Larisa Ivascu**

**Abstract** The green manufacturing (GM) approach has been defined and described in the context of improving the level of companies' competitiveness, innovation, and efficiency. Engaging businesses in adopting the principles of the circular economy helps them respond to regulatory constraints and opportunities related to green manufacturing, too. In addition, a debate of the new concept family of the product life cycle management related to Industry 5.0 will be presented being considered another perspective of Green Manufacturing (related to organizations' environmental policies, highlighting the efforts for sustainable product life cycle by simultaneously considering stakeholders' management related to the value chain). Overall, the need to reduce organizations' impact on the environment and strengthen the awareness of this is emphasized by the current chapter.

**Keywords** Manufacturing · Innovation · Sustainability · Environment · Footprint · Financial benefits · Non-financial benefits

## 1 Introduction

Today's enterprises are increasingly applying the principles of sustainable development. Thus, sustainability has become a necessity and a condition for many business collaborations [13, 16, 32, 38]. The objectives of sustainable development target organizational processes, innovation, infrastructure, employees, solid partnerships, and many other adjacent elements [11, 31, 49, 48]. These objectives are set by the 2030 Agenda.

---

A. Draghici · L. Ivascu (✉)
Faculty of Management in Production and Transportation, Politehnica University of Timisoara, 300006 Timisoara, Romania
e-mail: larisa.ivascu@upt.ro

A. Draghici
e-mail: anca.draghici@upt.ro

L. Ivascu
Academy of Romanian Scientists, 3 Ilfov Street, 050044 Bucharest, Romania

 
A. Draghici and L. Ivascu (eds.), *Sustainability and Innovation in Manufacturing Enterprises*, Advances in Sustainability Science and Technology,
https://doi.org/10.1007/978-981-16-7365-8_1

In the context of sustainable development, the concept of Green Manufacturing (GM) is outlined. GM refers to those production processes that pollute less, have a lower impact on the environment, the amount of waste generated is lower, resource consumption is minimized, and clean technologies are used. At the same time, GM supports not only the recycling and reuse of waste but also the reduction of the amount of greenhouse gases [3, 23, 25]. This approach has also become interesting in terms of automation and computerization of the industry.

In addition, from the technical perspective of the GM approach (often done by engineers), it is defined as organizations' responses to the environmental policies, by highlighting the need to develop a sustainable product life cycle that includes stakeholders' management related to the value chain.

## 2 Current Market Requirements

The current requirements of the manufacturing industry refer to a series of business directions for automation, digitization, and sustainability. Automation and digitization define Industry 4.0 and prepare the industry for the next step, Industry 5.0. Sustainability incorporates a series of principles that refer to the reduction of the organizational impact on the environment through efficient use of organizational resources [17, 42, 44, 46]. Market requirements are developed by customer requirements and related needs. Among these market requirements are the following [2, 3, 23, 25, 31, 41, 49, 48]:

- Integrated approaches refer to the whole process produced in systems and subsystems;
- Transport—streamlining the distribution network to reduce the impact on the environment. Emphasis is placed on reducing the number of intermediaries in the distribution network, so that this network is more efficient;
- Location—production locations are important because they integrate several sustainability approaches. These locations must be lighted, have easy transportation, provide easy transportation for employees, and other details;
- Water efficiency is another aspect of production. Streamlining the amount of water and reducing the water footprint are steps for sustainability;
- Energy efficiency is a priority of GM. Different aspects of information technology are applied to reduce energy consumption;
- Materials and resources—the emphasis is placed on recycling, reuse, reconditioning, remanufacturing, repair, and other functions that reduce the quantity of raw materials. When the product is no longer used by the customer, it can return to the production process as a raw material, thus meeting the concept of reverse logistics. Resources must be used in efficient quantities to reduce the impact of manufacturing on the environment;

- Innovation is strongly connected with GM, and it is directly linked to digitization and automation. Globalization involves innovation and adaptation. These aspects are requirements of the market and customers;
- Automation is part of the characteristics of the present industrial revolution and is a necessity for current manufacturing;
- Digitization is another feature of the current industrial revolution, Industry 4.0, and thus represents a market requirement. Digitization is an important step being a support in the connection and communication between production machines and people;
- Improving air quality is an issue targeted by production plants. Often, air quality is low in these production processes and therefore there is a need to improve air quality in production;
- Prioritization—the activities of the production processes must be prioritized and approached according to the vision, mission, and organizational objectives.

## 3 Green Manufacturing: Benefits and Drivers

The GM process includes the six basic stages: design, procurement, manufacturing, packaging and distribution, customer use to end of life, and remanufacture [11, 36], Fig. 1.

This paradigm is a result of market factors, technology, digitization, and process automation. Improving the awareness of enterprises regarding environmental risks has contributed to new production directions by designing environmentally friendly and less polluting systems [1, 14, 24, 30]. There is currently a need to identify environmental improvements that can be made to operational activities, processes, and systems. Table 1 presents the benefits of GM. It can be observed that the benefits

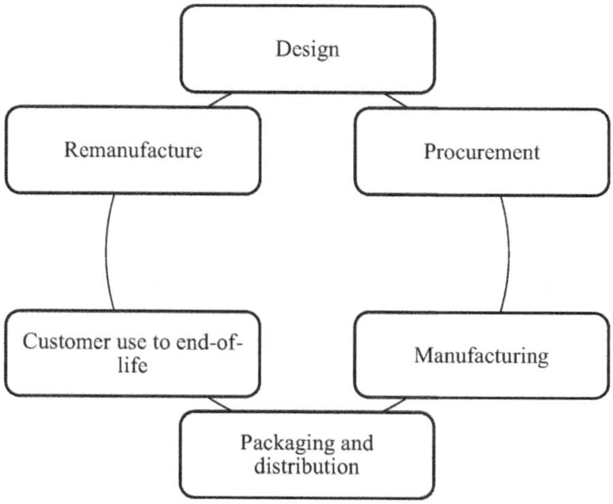


**Fig. 1** GM process cycle

**Table 1** The benefits of GM

| Benefit | Description |
|---|---|
| Cost reduction | Machines, equipment, and systems are becoming more energy efficient. At the same time, investments in modern technologies contribute to achieving organizational goals and are a way to reduce costs over a long period of time. Your activity regarding the reduction of polluting activities contributes to the development of a favourable, motivating work environment. Thus, productivity can increase |
| Efficiency of the quantity of raw materials | Many of the enterprises accept involvement in reverse logistics. This concept means that a product that reaches the end of its life cycle is returned to the manufacturer. The use of materials recovered from the product returned to the manufacturer can reduce the cost of raw materials to produce new products. This approach to reverse logistics has positive effects not only on the producer but also on the environment because of reducing the amount of waste |
| Reducing the carbon footprint | The implementation of green production processes contributes to reducing the amount of greenhouse gases generated. This can reduce the carbon footprint. A significant reduction in the amount of carbon dioxide will bring the company several government benefits, image, or partnerships. Currently, the focus is on an approach to reducing greenhouse gas emissions |
| Improving the image and the brand | In the business environment, there are a series of business networks that evolve partners according to certain criteria and give confidence according to certain implications for the protection of the environment. An enterprise that develops green production can improve its marketing effort. A manufacturer that invests in production equipment can improve both its image and brand |
| Improving production time | The use of green and innovative equipment can be designed to streamline time and reduce downtime lost in production processes. Reducing production time means cost efficiency and improving financial results |

are cost reduction, efficiency of the quantity of raw materials, reducing the carbon footprint, improving the image and the brand, and improving production time [7, 17, 33, 34, 42, 44].

The literature review has identified a series of barriers and drivers for GM [5, 20, 21, 27, 29]. There are also several drivers that facilitate the GM approach, Fig. 2 [15, 17, 39, 41]. Barriers are encountered in the application of manufacturing principles and aim at different directions which are presented in the map below, Fig. 3 [4, 7, 25, 41–44].

## 4 An Inventory of "Green" Concepts

It is obvious that a new approach related to "Green" concepts has been generalized in the organizations' practices, and extended use related to Industry 5.0 development and the Smart Society (next society that is because of the actual accelerated digital transformation of all economic activities). Taking into consideration the actual accepted definition of GM, already explained in Fig. 1, the strong relation with the general concept of the product life cycle that is seen from a "green" perspective,

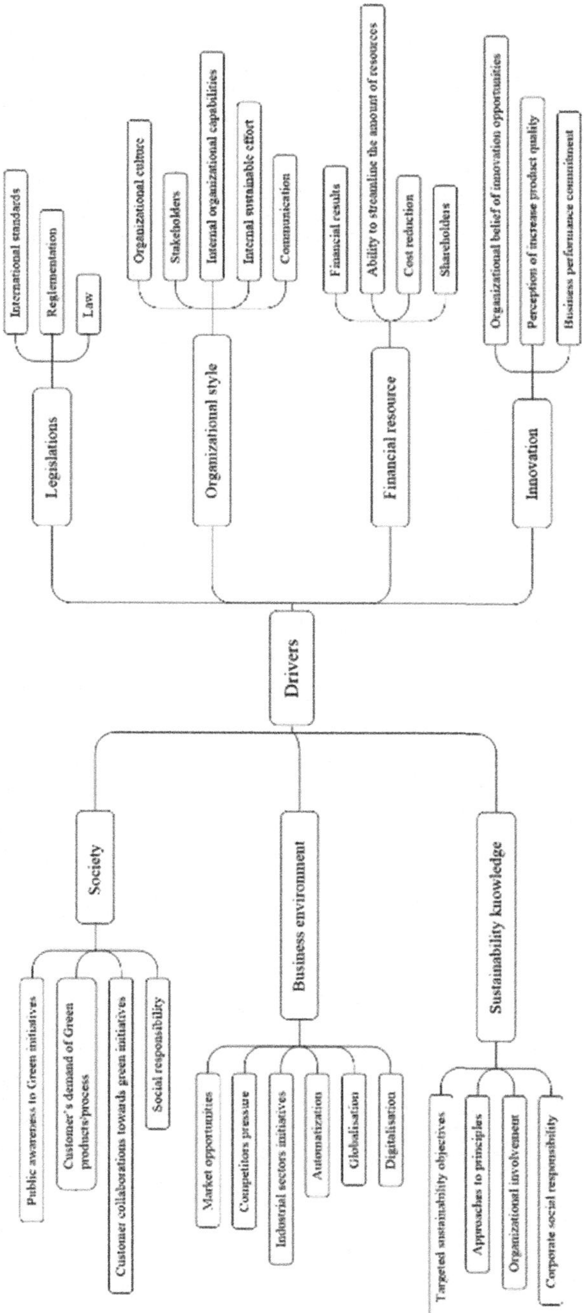


**Fig. 2** The drivers of GM

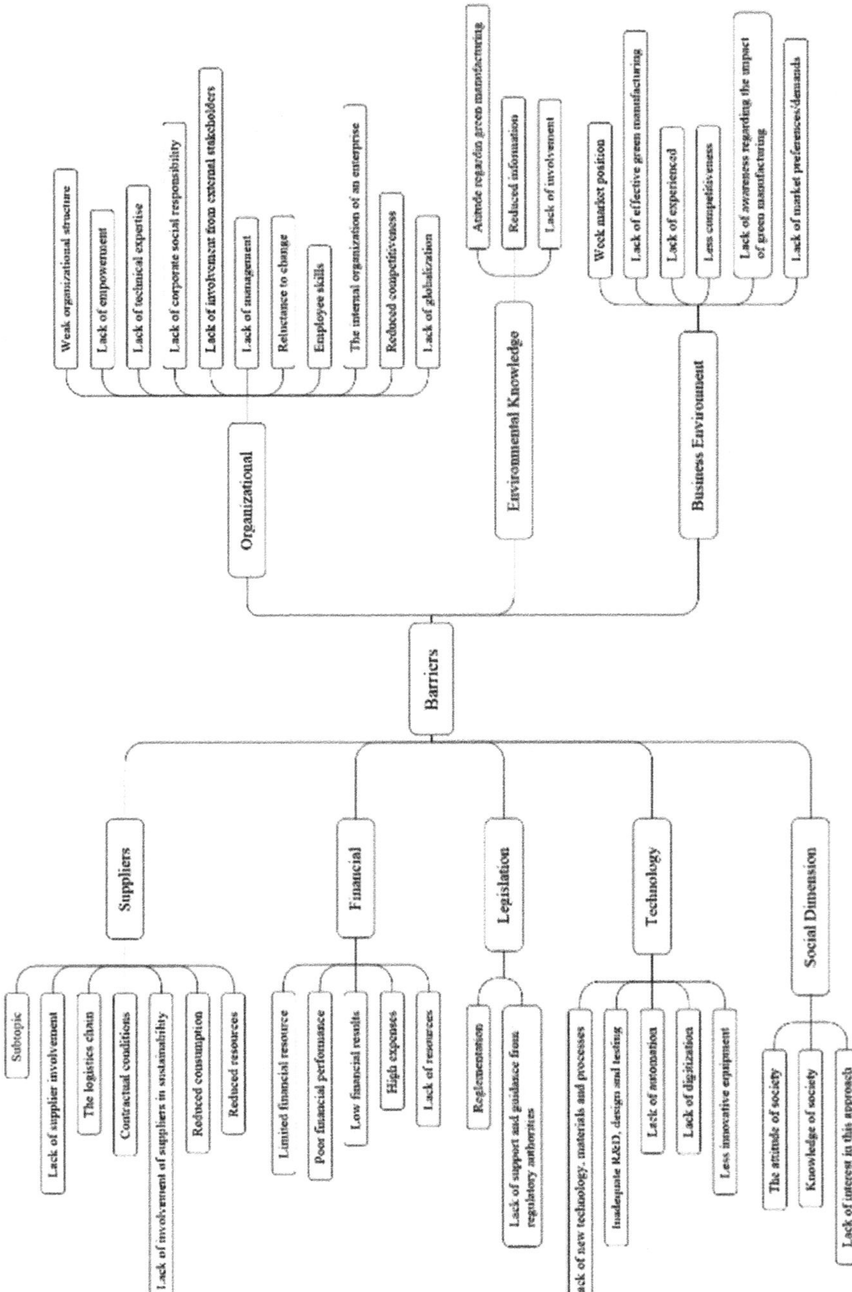


**Fig. 3** The barriers of GM

also known as the environment-based product, can be seen [22, 47]. So, the life cycle thinking has been stated as a general philosophy that impacts product life cycle activities for achieving a much responsible behaviour related to the awareness of the environment, and social and economic dimensions of sustainability. The approach strongly supports the principles and values of circular economy and demonstrates the current tendencies, too. Thus, Table 2 summarizes these ideas based on the literature of last years related to the "Green" trend. In addition, a terminological map is presented in Fig. 4.

## 5 The Future of Green Manufacturing

Future directions of GM will aim to outline the characteristics for Industry 5.0. This approach aims to reduce human action, complete automation, and customized mass production. In this framework, a series of activities can be carried out to contribute to the reduction of the organizational impact on the environment.

Organizations aim, first, to reduce the costs and to improve the production capacity. If these aspects are completed by the reduction of waste, good waste management, the efficiency of the quantity of water and energy used, and the involvement of the employees in the decision-making process, an image of the next stage of manufacturing is outlined [3, 19, 23, 25, 28, 45, 47].

Furthermore, Culot et al. [10] have recognized that manufacturing will be affected by "profound changes to the configuration" of the value proposition, value creation, or most of co-creation of value to their customers. Thus, it is supported by this study that the future belongs to the GM network which will be supported by Smart technologies (as the Internet of Everything) to interconnect diverse types of suppliers and customers [40]. This is because of the digital transformation that allows the Fifth Industrial Revolution to take place. The literature on the GM topic is still fragmented; the features of the Industry 5.0 emerging paradigm appear to be not recognized yet despite the progress and disruptions in innovation in the last years.

Culot et al. [10] research results have been filtered and confronted with the already discussed aspect of the literature in the field of GM, and they were filtered by the consideration of the identified benefits and drivers (taxonomies presented in Figs. 2 and 3). Based on this approach, there could be summarized the following trends of GM (based on the life cycle thinking paradigm, the circular economy perspective, and the actual digital transformations):

- The drivers of GM internalization are balanced by equally important drivers to outsourcing. Thus, the configuration will be segment-specific:
  - Manufacturers will focus on green production that will be internalized to support the adequate control for process flexibility and product customization and personalization (dedicated to market niche). This is the condition for the increasing of the GM quality control in terms of product functions and processes' low environmental impact;

**Table 2** An inventory of "Green" concepts related to GM in the transition to Industry 5.0 and the smart economy

| Concept | Brief description | References |
|---|---|---|
| Green design similar to design for the environment, *green product design*. Connected with: | Synonymous with eco-design or Green design strategy<br>It is based on the life cycle thinking and it is focused on lifetime environmental impact of *products*<br>It is considered as a case of "Design for X" where X is environment focus of designers<br>The approach for green product design takes into consideration the consumers' green awareness, the compliance to environmental regulations and quality | Hong et al. [22], Xu et al. [47] |
| Green procurement from a green supply chain, green logistics (including green warehouses, green transportation, green distribution, etc.), and reverse logistics (related to circular economy practices) | It refers to the eco-efficiency of the supply chain<br>The aim is to reduce all types of losses using different eco-approaches (e.g., reuse of materials, non-polluting technologies, inputs, efficient and less polluting production processes, legal compliance, and environmental awareness among all participants of the entire production chain)<br>It stimulates the adoption of green practices, mostly related to social and environmental sustainability dimensions, in the supply chain management | Cinar [9], Guo et al. [19], Liu et al. [28],Teixeira et al. [45], Xu et al. [47] |
| Green manufacturing and cleaner production, green technology | GM expresses a cumulus of actions in the manufacturing field which are oriented to diminish pollution, lower production processes that impact the environment, adequately manage waste in accordance with legal and normative compliance and with circular economy principles and values<br>It is strongly connected with clean technologies, cleaner production, greening the production processes, recycling, remanufacturing, and reuse of waste with the focus on reducing the amount of greenhouse gases<br>Cleaner production is "accounted for processes (saving energy and raw materials, reducing the use of toxic substances, diminishing the quantity of waste and emissions) and products (decreasing impacts along the whole life cycle), as well as for services" | Afum et al. [3], Huiling and Dan [23], Karuppiah et al. [25], Machado et al. [29], Rajput and Datta [38], Teixeira et al. [45] |

(continued)

**Table 2** (continued)

| Concept | Brief description | References |
|---|---|---|
| Green marketing (referring to the green marketing mix and including green consumer, green marketing strategies) | The marketing of environment-friendly products and/or services (Green Brands)<br>The attempts of marketers to develop strategies aiming to influence a Green behaviour of the consumer<br>It is mostly related to industrial ecology and the general concern for reducing the carbon footprint<br>GM is a monetarily determined framework to mitigate all types of waste streams related to the production processes (associated with the input transformation processes into output items) | Canavari and Coderoni [6], Dincer and Acar [12], Labrecque [26], Moravcikova et al. [35], Rajput and Datta [38] |
| End-of-life activities (reuse, recycling, recover, remanufacturing) reverse logistics | End-of-life activities are mandatory in the circular economy context, but not very often applied by organizations. "An advanced end of life system is one that uses approaches and technologies of the upstream processes and maintains a real-time data synergy with intelligent products"<br>Reverse logistics involves product returns from downstream members to the upstream, product reprocessing, and remanufacturing (e.g., recalls and returns of end-of-life products) | Cinar [9], Giannetti et al. [18], Guo et al. [19], Rahman et al. [37], Rajput and Datta [38] |
| Green competitiveness related to Green economy | Organizations successful, effective, and efficient initiatives, actions, projects (related to Green Investments) for global and local environmental problems, to reduce anthropogenic pressure on the environment to save resources, to form competitive development of the country and economic sectors. | Chygryn et al. [8] |

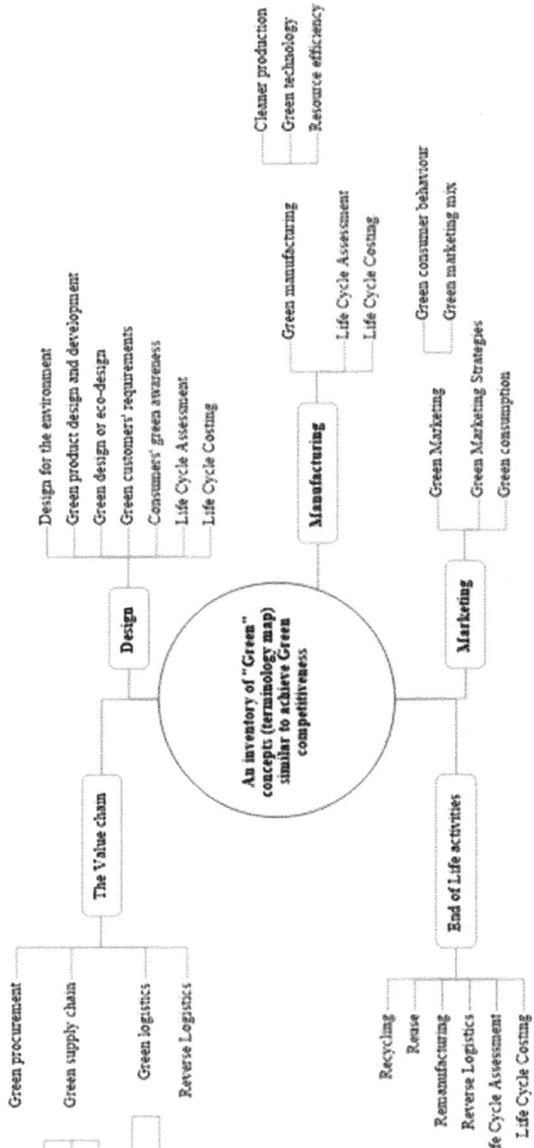


**Fig. 4** A terminological map of "Green" concepts

  - Extended application of Cloud technologies and interconnectivity solutions for GM will simplify access to outsourced manufacturing capabilities and will facilitate the identifications of the optimum business partners. In addition, it has been observed that data sharing, process integration, and digital coordination technologies could simplify outsourcing and green business process management;

- Organizations with GM will "internalize only digital data-driven services for smart products and those contributing to distinctive customer experiences. Traditional services requiring specialized capabilities will not be internalized":

  - GM will be based on green services that will be supported by local companies as the small and medium-size ones. This is supported by the spread of the Product Service Systems (PSS) or product-as-a-service (PaaS) approaches, all related to the smart product which is emerging;
  - Those eco-services creating a distinctive green customer experience will be internalized;
  - Additive manufacturing technologies will be of large scale used due to the extended market for these technologies and the knowledge achieved; the technologies will be considered mature in the next years in terms of costs, scope, performance, and accessibility;
  - Additive manufacturing technologies will impact the green maintenance development;
  - Small workshop related to the "small-scale production will be often established" to support the practical solutions of cleaner production with clean technologies (small and medium-size companies implementing GM);

- Resource efficiency aspects (related to the limited, scarce natural resources associated with the presence of a natural materials crisis) will be considered the core of GM:

  - New materials, synthetic substitutes will be developed urgently because they must "satisfy the exponentially increasing need for natural resources";
  - Most "recycling processes and circular economy practices will reintroduce natural resources into the process";

- Waste management value chain will be of great importance for green business and GM, too. Actor organizations in this chain will support activities as "disassembly and routing of components or materials back into production". The trends suggested that

  - Green practices are and will be driven not only by the green requirements of the green consumers but also for having green reputational advantages. Despite these, GM will still not be a generalized practice in many industrial areas of the world. Big companies will be leaders in this field and facilitators by promoting green supply chain;
  - Environmental standards and regulations will have a slow evolution (sustainable development legal compliance will be still of intense international debate),

  but mostly they will support the development and application of recycling technologies and the practices related to a circular economy;
  - Most GM will be confronted with difficulties to ensure "end-to-end supply chain collaboration as needed in circular economy";

- End-of-life activities will be still critical in the next period. Organizations will internalize "specific end-of-life product management activities in relation to the revenues and reputational opportunities":
  - GM will progress by supporting the end-of-life product management capabilities;
  - End-of-life activities will be subject of outsourcing, externalization, and companies will support these actions because of social responsibility behaviour and reputational reasons;
- Excessive digital transformation of manufacturing will be supported by software providers;
  - GM will include the problematics of green information technology and communication;
  - There will be a parallel development of smart products and digital services. Thus, digital actors of the market will support the replacement or substitute of the product ownership by product-as-a-service approaches;
  - The digital actors of the market will be enabled to develop dedicated software solutions substituting traditional services;
  - As a general estimation, other business services such as financial-accounting, legal consultancy, call centres as customer support services, and design will be provided via the Internet (digital services) and will be supported by extensive digital platforms.

## 6 Conclusions

Industry is an important branch for most countries. Industries contribute to the pollution of the environment through the amount of greenhouse gases that are generated. The ability to reduce this level of pollution is a priority for industries. However, the implications in this approach are conditioned by a series of benefits that the stakeholders, mainly the shareholders, see in the whole approach.

The implications of the enterprises in different approaches are realized if a financial benefit can be registered. The financial benefits of GM are long-term and a priority for new technologies developed.

However, it is difficult to estimate how long this pandemic crisis will last and especially its real effects and the impact on the global economy, in general, and on manufacturing, particularly. We do not know who the winners or losers of this moment of turbulence in the development of mankind are, but it is observed that individuals endowed with creativity and digital skills (open, flexible, and tolerant) have been more agile in finding a place in society. Likewise, manufacturing organizations that knew how to innovate quickly and comply with the new requirements imposed on the health and safety of workers were successful, continued their activity, and overcame the "shock" in a short term.

The trends in the field of GM presented in this chapter considered aspects found based on the current bibliographic analysis and synthesis and those reported by Culot et al. [10]. As a long-term consequence, structural measures will be needed in the field of GM, which implies a simultaneous action for the development of the green, circular economy "in the face of bleak prospects of recession and unemployment". Among these structural measures, we expect that investments in innovation, more likely supported by public–private partnerships and by non-government organizations (NGOs), will return from the spotlight on the trajectory of the transition from Industry 4.0 to Industry 5.0.

In many ways, the current perception and correct characterization of this transition to Industry 5.0 are still unclear (lack of knowledge and understanding, weak change acceptance), but certainties come from the manifestation of GM drivers, such as the evolution of applications in relation to data analytics and management, the emergent digital technologies, and the changes of the market demand.

## References

1. Abdul-Rashid SH, Sakundarini N, Raja Ghazilla RA, Thurasamy R (2017) The impact of sustainable manufacturing practices on sustainability performance: empirical evidence from Malaysia. Int J Oper Prod Manag. https://doi.org/10.1108/IJOPM-04-2015-0223
2. Abualfaraa W, Salonitis K, Al-Ashaab A, Ala'raj M (2020) Lean-green manufacturing practices and their link with sustainability: a critical review. Sustainability (Switzerland). https://doi.org/10.3390/su12030981
3. Afum E, Osei-Ahenkan VY, Agyabeng-Mensah Y, Amponsah Owusu J, Kusi LY, Ankomah J (2020) Green manufacturing practices and sustainable performance among Ghanaian manufacturing SMEs: the explanatory link of green supply chain integration. Manag Environ Qual: Int J. https://doi.org/10.1108/MEQ-01-2020-0019
4. Bai C, Satir A (2020) Barriers for green supplier development programs in manufacturing industry. Resour Conserv Recycl. https://doi.org/10.1016/j.resconrec.2020.104756
5. Ben Ruben R, Vinodh S, Asokan P (2019) State of art perspectives of lean and sustainable manufacturing. Int J Lean Six Sigma. https://doi.org/10.1108/IJLSS-11-2016-0070
6. Canavari M, Coderoni S (2019) Green marketing strategies in the dairy sector: consumer-stated preferences for carbon footprint labels. Strateg Chang. https://doi.org/10.1002/jsc.2264
7. Chan SW, Tiwari STS, Ahmad MF, Zaman I, Sia WL (2018) Green procurement practices and barriers in furniture manufacturing companies. Int J Supply Chain Manag
8. Chygryn O, Rosokhata A, Rybina O, Stoyanets N (2021) Green competitiveness: the evolution of concept formation. In: E3S web of conferences. https://doi.org/10.1051/e3sconf/202123400004
9. Cinar S (2021) Sustainable reverse logistic network design for end-of-life use-case study. RAIRO—Oper Res. https://doi.org/10.1051/ro/2019069
10. Culot et al. (2020) Behind the definition of Industry 4.0: Analysis and open questions
11. Deif AM (2011) A system model for green manufacturing. J Clean Prod. https://doi.org/10.1016/j.jclepro.2011.05.022

12. Dincer I, Acar C (2017) Innovation in hydrogen production. Int J Hydrog Energy. https://doi.org/10.1016/j.ijhydene.2017.04.107
13. Dornfeld DA (2014) Moving towards green and sustainable manufacturing. Int J Precis Eng Manuf—Green Technol. https://doi.org/10.1007/s40684-014-0010-7
14. Dubey R, Gunasekaran A, Childe SJ, Wamba SF, Papadopoulos T (2016) The impact of big data on world-class sustainable manufacturing. Int J Adv Manuf Technol. https://doi.org/10.1007/s00170-015-7674-1
15. Gandhi NS, Thanki SJ, Thakkar JJ (2018) Ranking of drivers for integrated lean-green manufacturing for Indian manufacturing SMEs. J Clean Prod. https://doi.org/10.1016/j.jclepro.2017.10.041
16. Gedam VV, Raut RD, de Sousa L, Jabbour AB, Narkhede BE, Grebinevych O (2021) Sustainable manufacturing and green human resources: critical success factors in the automotive sector. Bus Strateg Environ. https://doi.org/10.1002/bse.2685
17. Ghazilla RAR, Sakundarini N, Abdul-Rashid SH, Ayub NS, Olugu EU, Musa SN (2015) Drivers and barriers analysis for green manufacturing practices in Malaysian smes: a preliminary findings. Procedia CIRP. https://doi.org/10.1016/j.procir.2015.02.085
18. Giannetti BF, Agostinho F, Eras JJC, Yang Z, Almeida CMVB (2020) Cleaner production for achieving the sustainable development goals. J Clean Prod. https://doi.org/10.1016/j.jclepro.2020.122127
19. Guo S, Shen B, Choi TM, Jung S (2017) A review on supply chain contracts in reverse logistics: supply chain structures and channel leaderships. J Clean Prod. https://doi.org/10.1016/j.jclepro.2016.12.112
20. Hartini S, Ciptomulyono U (2015) The Relationship between lean and sustainable manufacturing on performance: literature review. Procedia Manuf. https://doi.org/10.1016/j.promfg.2015.11.012
21. He B, Bai KJ (2021) Digital twin-based sustainable intelligent manufacturing: a review. Adv Manuf. https://doi.org/10.1007/s40436-020-00302-5
22. Hong Z, Wang H, Gong Y (2019) Green product design considering functional-product reference. Int J Prod Econ. https://doi.org/10.1016/j.ijpe.2019.01.008
23. Huiling L, Dan L (2020) Value chain reconstruction and sustainable development of green manufacturing industry. Sustain Comput: Inform Syst. https://doi.org/10.1016/j.suscom.2020.100418
24. Hussain S, Jahanzaib M (2018) Sustainable manufacturing-an overview and a conceptual framework for continuous transformation and competitiveness. Adv Prod Eng Manag. https://doi.org/10.14743/apem2018.3.287
25. Karuppiah K, Sankaranarayanan B, Ali SM, Chowdhury P, Paul SK (2020) An integrated approach to modeling the barriers in implementing green manufacturing practices in SMEs. J Clean Prod. https://doi.org/10.1016/j.jclepro.2020.121737
26. Labrecque LI (2020) Color research in marketing: theoretical and technical considerations for conducting rigorous and impactful color research. Psychol Mark. https://doi.org/10.1002/mar.21359
27. Leng J, Ruan G, Jiang P, Xu K, Liu Q, Zhou X, Liu C (2020) Blockchain-empowered sustainable manufacturing and product lifecycle management in industry 4.0: a survey. Renew Sustain Energy Rev. https://doi.org/10.1016/j.rser.2020.110112
28. Liu Y, Blome C, Sanderson J, Paulraj A (2018) Supply chain integration capabilities, green design strategy and performance: a comparative study in the auto industry. Supply Chain Manag. https://doi.org/10.1108/SCM-03-2018-0095
29. Machado CG, Winroth MP, Ribeiro da Silva EHD (2020) Sustainable manufacturing in industry 4.0: an emerging research agenda. Int J Prod Res. https://doi.org/10.1080/00207543.2019.1652777
30. Madan Shankar K, Kannan D, Udhaya Kumar P (2017) Analyzing sustainable manufacturing practices—A case study in Indian context. J Clean Prod. https://doi.org/10.1016/j.jclepro.2017.05.097

31. Mao S, Wang B, Tang Y, Qian F (2019) Opportunities and challenges of artificial intelligence for green manufacturing in the process industry. Engineering. https://doi.org/10.1016/j.eng.2019.08.013
32. Mishra V, Mittal A (2019) Policies for sustainable manufacturing and extended producer responsibility for green and sustainable manufacturing in India—A review. Int J Mech Prod Eng Res Dev. https://doi.org/10.24247/ijmperdapr201922
33. Mittal VK, Sangwan KS (2014) Prioritizing barriers to green manufacturing: environmental, social and economic perspectives. Procedia CIRP. https://doi.org/10.1016/j.procir.2014.01.075
34. Modha M, Sumant M (2018) A literature review on drivers and Barriers of green manufacturing in Indian/global SME's. In: Procedings of ICIE 2017
35. Moravcikova D, Krizanova A, Kliestikova J, Rypakova M (2017) Green marketing as the source of the competitive advantage of the business. Sustainability (Switzerland). https://doi.org/10.3390/su9122218
36. Pang R, Zhang X (2019) Achieving environmental sustainability in manufacture: a 28-year bibliometric cartography of green manufacturing research. J Clean Prod. https://doi.org/10.1016/j.jclepro.2019.05.303
37. Rahman SM, Perry N, Müller JM, Kim J, Laratte B (2020) End-of-life in industry 4.0: ignored as before? Resour Conserv Recycl. https://doi.org/10.1016/j.resconrec.2019.104539
38. Rajput SPS, Datta S (2019) Sustainable and green manufacturing—A narrative literature review. Mater Today: Proc. https://doi.org/10.1016/j.matpr.2020.02.535
39. Rehman MAA, Shrivastava RR, Shrivastava RL (2014) Evaluating green manufacturing drivers: an interpretive structural modelling approach. Int J Prod Qual Manag. https://doi.org/10.1504/IJPQM.2014.062223
40. Saleem F, Khattak A, Ur Rehman S, Ashiq M (2021) Bibliometric analysis of green marketing research from 1977 to 2020. Publications. https://doi.org/10.3390/publications9010001
41. Seth D, Rehman MAA, Shrivastava RL (2018) Green manufacturing drivers and their relationships for small and medium(SME) and large industries. J Clean Prod. https://doi.org/10.1016/j.jclepro.2018.07.106
42. Singh C, Singh D, Khamba JS (2020) Analyzing barriers of green lean practices in manufacturing industries by DEMATEL approach. J Manuf Technol Manag. https://doi.org/10.1108/JMTM-02-2020-0053
43. Singh M, Singh K, Sethi A (2019) An empirical investigation and prioritizing critical barriers of green manufacturing implementation practices through VIKOR approach. World J Sci Technol Sustain Dev. https://doi.org/10.1108/wjstsd-08-2019-0060
44. Singh M, Singh K, Sethi AS (2020) Analytical hierarchy process and TOPSIS for selecting best parameters of green manufacturing. Meas Bus Excell. https://doi.org/10.1108/MBE-08-2019-0076
45. Teixeira CRB, Assumpção AL, Correa AL, Savi AF, Prates GA (2018) The contribution of green logistics and sustainable purchasing for green supply chain management. Indep J Manag Prod. https://doi.org/10.14807/ijmp.v9i3.789
46. Wang X, Chen L, Dan B, Wang F (2018) Evaluation of EDM process for green manufacturing. Int J Adv Manuf Technol. https://doi.org/10.1007/s00170-017-0892-y
47. Xu X, He P, Xu H, Zhang Q (2017) Supply chain coordination with green technology under cap-and-trade regulation. Int J Prod Econ. https://doi.org/10.1016/j.ijpe.2016.08.029
48. Zhang X, Ming X, Liu Z, Qu Y, Yin D (2019) General reference model and overall frameworks for green manufacturing. J Clean Prod. https://doi.org/10.1016/j.jclepro.2019.117757
49. Zhang H, Zhang Z, Pu X, Li Y (2019) Green manufacturing strategy considering retailers' fairness concerns. Sustainability (Switzerland). https://doi.org/10.3390/su11174646

# Chapter 2
# Industry 5.0 Challenges and Perspectives for Manufacturing Systems in the Society 5.0

**Daniel Paschek, Caius-Tudor Luminosu, and Elif Ocakci**

**Abstract** This chapter presents an exhaustive reference analysis and synthesis about the Industry 5.0 development, challenges, and implications. First, a state of the art on Industry 4.0 and digitalization will be presented considering the Industrial Revolution stages. Currently, manufacturing and business processes are increasingly using Cloud and mobile technologies, Internet of Things (IoT), Artificial Intelligence (AI), and Data Analytics, which heralds the fifth Industrial Revolution (Industry 5.0). Consequently, in the second part of the chapter, there will be presented synthesis research drawing the state-of-the-art overview on Industry 5.0, as a core idea of this chapter. Furthermore, a detailed characterization will be done to identify the differences and similarities from Industry 4.0 to Industry 5.0 and Society 5.0 by gaining benefits, advantages, and disadvantages as well as expected business value and ethical implications.

**Keywords** Industry 4.0 · Industry 5.0 · Digitalization · Digital transformation · Society 5.0 · Leadership 5.0

---

D. Paschek
ServiceNow, Frankfurt am Main, Germany
e-mail: daniel.paschek@servicenow.com

C.-T. Luminosu (✉) · E. Ocakci
Faculty of Management in Production and Transportation, Politehnica University of Timisoara, Timisoara, Romania
e-mail: caius.luminosu@upt.ro

E. Ocakci
e-mail: elif.ocakci@continental-corporation.com; elif.ocakci@student.upt.ro

E. Ocakci
Department of Strategy and Business Development, Frankfurt am Main, Germany

 
A. Draghici and L. Ivascu (eds.), *Sustainability and Innovation in Manufacturing Enterprises*, Advances in Sustainability Science and Technology,
https://doi.org/10.1007/978-981-16-7365-8_2

# 1 Introduction

The actual challenges and the dynamics of the manufacturing world (defined by different types of organizations, the market actors, and policymakers) are focused on the transformation and digitalization of their business processes and models to achieve a sustainable competitive advantage [27, 36]. Thus, manufacturing organizations need the ability and agility to recognize macroeconomic trends, customer requirements, and innovation dynamics to create new product-services systems and solutions for client satisfaction.

Furthermore, different actors from manufacturing supply chains are confronted and need to manage a large quantity of data which drive strategic and operational levels of decisions and determine the high complexity of the business processes [90, 120].

New technologies like Artificial Intelligence (AI) are claimed to be in the centre of the new Industrial Revolution having already a big impact on several business processes, changing customer requirements which are reflected by the Business Process Reengineering (BPR) and Business Intelligence (BI) processes which reshape the frameworks of Business-to-Business (B2B) and Business-to-Customers (B2C) models. The new model and process transformations are impacted by huge interconnections phenomena. Therefore, there is a theoretical need to analyse the present dependencies and relations to obtain an implementation approach [27, 90].

In the last years, scientific and academic representatives have recognized that digitalization automation solutions could better support employees and could offer new opportunities to generate new added value to the business management paradigm. Especially, AI holds significant power to improve business processes as well as enhance forecasting and decision-making capabilities within the next years [47].

Policymakers have also recognized the changes in the manufacturing world, and they developed studies and programs to better support the new skills needed and the implementation of new technologies. For example, Elżbieta Bieńkowska, the European Commissioner for Internal Market, Industry, Entrepreneurship and SMEs (2014–2019), underlined the statement for the "Digital Transformation Scorebook 2018" [92]:

> The future of industry is digital. Progress in technologies such as big data, AI and robotics, the IoT and high-performance computing is already transforming the very nature of work and society as a whole … AI is opening massive business opportunities and transforming value chains. It is therefore at the core of the renewed EU Industrial Policy, our work on SMEs, and the Digital Singe Market strategy.

Recently, Mariya Gabriel, Commissioner for Innovation, Research, Culture, Education and Youth, has the following statement (when the new report on Industry 5.0 has been launched [22]):

> The new concept of Industry 5.0 could not come at a better time. Many European industries are reinventing themselves, adapting to the new COVID reality, increasingly embracing digital and green technologies to remain the solution provider for all Europeans. Now is the time to make workplaces more inclusive, build more resilient supply chains and adopt more sustainable ways of production.

These declarations supported the need for rapid and agile actions, and the importance of AI as well as digitalization for the business and the effects on the society, too. Recently, it has been recognized that technologies based on AI are emerging; AI for process modelling and optimization will lead to seamless business processes and efficient and personalized process flows to achieve more customer satisfaction (e. g., by next best offer or next best action activities) [14].

In the context of this chapter, we shall describe the ongoing transformation processes by digitalization which announce the new Industrial Revolution: Industry 5.0 in the nearest future (already present in some high-tech industries). The state of the Revolution is described based on the literature review, and by a collection of relevant points expressed by opinion leaders as consultants or CEOs, too. The future industrial context of Industry 5.0 is announced by the accelerated technological developments of Industry 4.0, so fast that Industry 5.0 is already becoming part of the business landscape [6, 90].

The objective of the scientific debate in this chapter is to create a general overview of the transition from Industry 4.0 to Industry 5.0 by creating an inventory of the challenges and opportunities for the manufacturing systems in the Society 5.0. Thus, this chapter presents a discussion on a reference analysis and synthesis about the Industry 5.0 development, challenges, and implications. The chapter structure consists of the following parts:

(1) Presentation of a research synthesis for building a bibliographic and practical state-of-the-art overview on Industry 4.0 Revolution with business and social relation as the foundation for understanding manufacturing evolution. Furthermore, there will be compiled the main ideas related to the analysis and assessment of digitalization and important digital technologies that could support the digital transformation business status;
(2) A scientific debate on the synthesis for building a bibliographic and practical state-of-the-art overview on the Industry 5.0 Revolution. The debate will be concluded by a comparative study of Industry 4.0, Industry 5.0, and Society 5.0 (in terms of benefits, advantages and disadvantages, business value, and ethical implications).

## 2 Industry 4.0 and the History of Industrial Revolutions

A revolution process is accompanied by fundamental innovation or changes which occurred in time because new technologies and new triggers changed economic systems and/or social structures. A synthesis overview of the Industrial Revolution history is shown in Fig. 1. As can be seen, the First Industrial Revolution (Industry 1.0) was developed from 1760 to 1840 and was defined by the introduction and development (extensive use) of mechanical production equipment like the invention of the steam engine or the construction of the railroads. The Second Industrial Revolution (Industry 2.0) was characterized by the extensive use of the assembly line and the

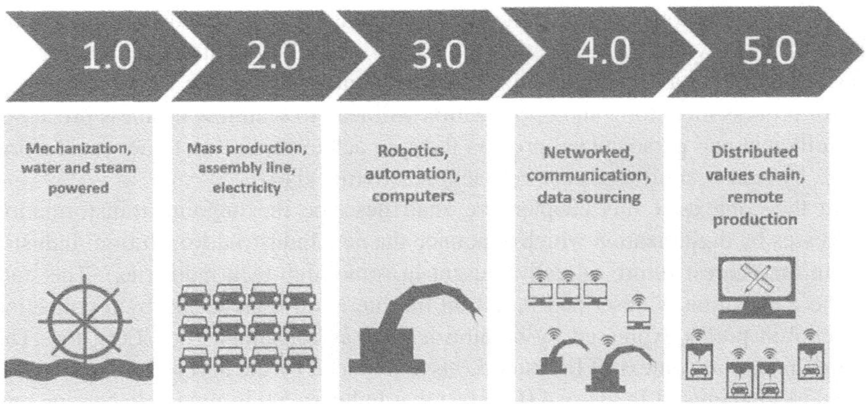


**Fig. 1** Industrial revolution steps with brief description

advent of electricity late in the nineteenth century until the early twentieth century [28, 110].

The Third Industrial Revolution (Industry 3.0), also known as the computer or digital revolution, started around 1960 and was characterized by the development and use of electronics and information technology (like semiconductors, mainframe computing, Internet, etc.) [110]. After that, since 1990, the Fourth Industrial Revolution (Industry 4.0) has been recognized and the dominant characteristics were the advancement and extended use of information and communication technology which were spread in the business and social fields. Thus, the presence of Cyber-Physical Systems (CPS), the IoT, mobile technologies, devices and applications, mobile Internet, and AI has built the digital revolution [126]. The technical integration of CPS into the business processes as well as the use of IoT solutions and the digital industrial services have and will have extensive implications for the value co-creation, change of the business models, downstream services, and work organization [56]. In addition to this debate [110], schwab has stated that:

> The changes are so profound that, from the perspective of human history, there has never been a time of greater promise or potential peril. My concern, however, is that decision makers are too of-ten caught in traditional, linear (and non-disruptive) thinking or too absorbed by immediate concerns to think strategically about the forces of disruption and innovation shaping our future.

Industry 4.0 can be described by a high level of automation by programmed controlled machines (as described by the CPS concept) as well as intensive use of Cloud Technology and Big Data [68, 127]. Furthermore, Industry 4.0 is more and more concerned with the implementation of digital solutions to take systems automation to the next level of Industry 5.0 [89, 126].

Primarily, the digitalization concept is considered as a driving force of Industry 4.0 [79]. The integration of dynamic value-creation networks with other economic sectors and industries requires standardization and security for networked systems, as well as a reference architecture as a foundation for implementing Industry 4.0 emergent technologies, infrastructure, and the processing of data provided by sensors placed on the machines, workplaces or those related to the information technology systems [126].

CPS monitors physical manufacturing processes and supply chain end-to-end using real-time communication and IoT [126]. Industry 4.0 enables the development and execution of Smart Factories with modular structures that have been based on the created "virtual copy of the physical world" which has supported data-based decentralized and autonomous decisions [29, 45].

The fundamental changes in Industry 4.0 revolution are based on technology developments like AI, IoT, Data Analytics, Cloud technology, intensive robotizarion, blockchain technology, 3D printing, Cybersecurity, etc. [89, 126]. These innovative and disruptive technologies have led to the development and extensive use of data management and big data [94]. These amounts of generated data of technology developments are revolutionizing companies how to operate and will become increasingly critical in the coming years by analysing, assessing, managing data and distinguishing trends and recommendations for customers as well optimizations of business processes [72]. Thus, there have been recognized six major benefits of Industry 4.0 for business processes as increase of efficiency, agility and scalability, intensive innovation and real-time data use, improvements related to customer experience and centricity, cost reduction, and positive impact on revenues [28, 90]. In addition, the benefits of considerable improvement of product quality are related to the high degree of automation throughout the usage of digital technologies that can be enhanced by the monitoring process and production in real time [3, 4].

In conclusion, Industry 4.0 revolution with smart production (referring to manufacturing and factories) and smart products (including the Product Service Systems and extended products) have facilitated the actual emerging state of using digital services and platforms in production systems contributing to the transformation of the whole value chains. Thus, the new Industry 5.0 is announced.

## 2.1 *Challenges and Risks of Industry 4.0*

According to the present debate in the literature, as well as in practice, many challenges and hurdles are identified and characterized during implementing and executing processes consisting of high technology according to the actual Industry 4.0 maturity level (e.g., with a high degree of digitalization, automatization, and robotization). Organizations need flexibility and agility to adequately respond to high dynamic market requirements; the actual organizational challenges such as flexibility, predictability, and robustness are determined by the unexpected external environment conditions [130]. An overview of the five main challenges is depicted in Fig. 2 and defined in the following.

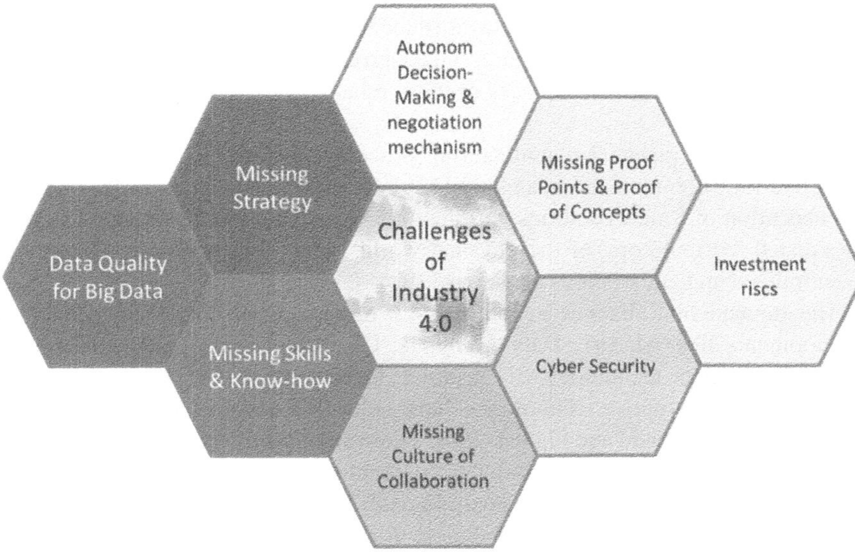


**Fig. 2** Main challenges of Industry 4.0

A successful transition and transformation to Industry 4.0 should be planned and executed in a series of steps, rather than a comprehensive and sudden change determine Andy [118], Head of International Account Management at Siemens Financial Services. It can be stated that a missing strategy for Industry 4.0 and rash transformation represent one of the biggest challenges for organizations [15]. This challenge faces the rethinking of the organization and its processes to maximize outcomes as well as the evaluating and understanding the business use case and the commercial benefits gained from each phase of investment, too [109].

As one further main challenge, missing digital skills and know-how for three key areas can be detected: digital production expertise, digital maintenance capabilities, and operating and strategic analytics. Next to the missing knowledge and experience, the establishment of a culture of collaboration by a connected, open-minded organization where multiple perspectives can be combined and viewed represents a challenge for companies [118]. This has been underlined by the research of [77]; they discovered that most studies discuss "the technical aspects but do not pay attention to managerial approaches and organizational culture that are a significant factor influencing the success" of the organization's transformation towards Industry 4.0 and following Industry 5.0 [77, 90]. Furthermore, with the increased connectivity and use of standard communications protocols, as well as an increasing number of attacks in the industry, IoT facilities the information security concerns (Cybersecurity) and the security of privacy [118, 127].

Another challenge faces the benchmark to proof points like case studies about investment in applying (adoption and integrated into the organization) emergent technologies, which clearly illustrate the level and period of return-on-investment as well as the conducting of successful pilots to gather experiences by own projects [98, 118].

Further on, intelligent decision-making and negotiation mechanism present challenges, because in an automated environment these CPS need more autonomy and sociality capabilities to react automatically to market changes [130]. In the case of using Data Analytics and Big Data technologies, it is a challenge to ensure the high quality and integrity of the data from different systems, to combine diverse databases and repositories with different semantics [3, 122]. Regarding the advancements of technology and the need to implement the new and emergent ones, there can be an investment challenge, too [128]. Without a consistent budget or access to appropriate third-party finance, an effective digital transformation is impossible to implement [118].

In addition, consulting companies in the field have underlined a collection of technical, practical (related to the specific conditions in the organization), and ecosystem-related challenges with impact on creating advanced systems of Industry 4.0 and Industry 5.0 [78, 98][1]:

- Better understanding and integration of IT and OT technologies;
- "Data compliance;
- Managing risk and lowering costs in uncertain times;
- Dealing with the complexity of the connected supply chain;
- Altering customer and industrial partner demands;
- Competition and the fact that Industry 4.0 champions gain a competitive benefit fast and new competitors can emerge;
- The eternal and human challenge (talent, future of work, employment etc.)".

In addition [74], McKinsey has identified, through a survey, the top five barriers in manufacturing organizations when they get started with Industry 4.0 implementation (Table 1). These challenges are the same as analysed before and underline the defined challenges for organizations' transformation. One new important barrier is the lack of courage to push through radical, technical, and organizational transformation changes that Industry 4.0 requires.

This underlines the necessity of entrepreneurship to take risks and make investments whose return on investment is not yet assessable. McKinsey has discovered three additional barriers which are facing the problem of data management, ownership, and data integration as well as the lack of knowledge in this field to assess if outsourcing or insourcing may be useful [74]. Especially, the concerns about data will have a big impact on companies in three core areas: improving decision-making, improving operations, and monetarization of data by becoming increasingly critical and a strategic asset for organizations [72].

[1] Retrieved from: https://www.i-scoop.eu/industry-4-0/#:~:text=Industry%204.0%20is%20the%20current,called%20a%20%E2%80%9Csmart (Access on 10–03-2021).

**Table 1** Major implementation barriers of Industry 4.0 [74]

| Manufacturers need to overcome major implementation barriers of which some are more relevant for advanced players | |
|---|---|
| Barriers for Industry 4.0 implementation (currently mentioned by manufacturers): | Additional top barriers (mentioned by advanced manufacturers): |
| 1. Gap of top management coordination—Difficulty in coordinating actions across different units in organization<br>2. Gap in change management—Lack of courage for the adoption of a radical transformation<br>3. Gap in human resources management—Lack of necessary qualified personnel and talent (e.g., data scientists) | 1. Concerns about data ownership when working with third-party services providers |
| 4. Gap of information management system on overcoming cybersecurity problems (most when operating working with third-party services providers)<br>5. Gap in BPM on clearly description of the business and the related investments and the IT architecture | 2. Operating with uncertainties about versus outsourcing and lack of information about providers<br>3. Challenges regarding data integration when operating with diverse, disparate sources |

## 2.2 *Digitization, Digitalization, and Digital Transformation*

A preliminary literature review has been developed to define and describe the main concepts connected with the phenomena of the Digital Era. As already mentioned, Industry 4.0 and digitalization challenge organizations within all economic and social sectors being affected by emerging technologies and a developing highly volatile, complex, uncertain, and ambiguous digital economy [125].

The first focus was on the analysis of the differences between Digitization, Digitalization, and Digital Transformation to create a knowledge base and clear understanding because there are still confusions [11]:

- Digitization describes the transition from analogue to digital data. Analogue business processes with, for example, "customer files, product manuals, repair handbooks were hard copies while digitization is the process of making all this information available and accessible in a digital format" [51],
- Digitalization is the process of employing digital technologies and information to transform business operations [11]. In addition, Gartner describes digitalization as the use of digital technologies to change a business model and provide new revenue and value-producing opportunities as well as the process of employing digital technologies and information to transform business operations [38],
- Digital transformation is about taking advantage of digitalization to create completely "new business concepts and processing of developing new business applications that integrate all this digitized data and digitalized applications" [51].

The second focus is on describing the characteristics of digitalization, which is used (in the literature and in organizational practice) in different interpretations linking the term digital with society, transformation, change, business processes, management, and revolution. To summarize, digitalization can be simply defined as the "transfer of analogue information into digital data effect which is triggered by it" [63, 89].

Daniel [79], CEO of Broadsuite Media Group, "describes the digitalization as the driving force for Industry 4.0 by several trends like connected consumers, empowered employees, optimized production and transformed products". This underlines the represented connection of digitalization and Industry 4.0. Further on, digitalization respects the adoption of digital technologies in business and society, as well as the related changes in the connectivity of individuals, organizational systems, and objects [89, 125].

The goal is to transform all the input information (push or pull) for an organization "into a uniform digital format, to process it electronically through the processes with the electronic documents and thus to increase the efficiency, flexibility, and the service level internally and externally to customers" [63, 89]. Simultaneously, the goal is to read the data in the documents in such a way that they can be further processed in subsequent (sub-)systems. Digitalization means to implement digital business processes in the organization's environment and over the supply chain supplemented by the customer focus [86]. The business world's rapid digitalization is breaking down the traditional barriers of the industry, so the business need arises to rethink the existing Business Models [40].

Digitalization has pushed companies to change their business processes and, finally their business models, by using new technologies of Big Data and BPM, and in the last years, they started implementing AI to nurture new customer experience, by decision and predictive analytics capabilities integrated into the application [40]. Even though AI is not a new concept (it was first launched in 1940, with the first computer program written by Ada Lovelace), it has been recently reconsidered because of the capabilities of using and storing data that have only grown recently. "Advances in data processing speeds, lower costs, big data volume and the integration of data science into technology has made practical AI a reality for many organizations" [72].

The mentioned connection between Industry 4.0 and digitalization can be seen by analysing the benefits of digitalization too. Starting with time- and location-independent availability of data, speed of access, space-saving storage, and loss-free reproduction [63], and overgrowth in revenue as the most dominant expected value among CEOs [92] increased profits and productivity, cost efficiency, enhanced information preservation, and enhanced security to stay competitive [20] "to enables an exceptional speed of change and level of connectedness over the whole supply chain" [89, 125]. Further on, seven advantages can be conducted as shown in Table 2 [25].

In comparison with the benefits of Industry 4.0, clear similarities can be seen, which leads to the conclusion that terms and currents are closely linked. Because of the continued and rapid development of digital transformation, some estimations on

**Table 2** An inventory of business digitalization advantages

| Advantages | Details |
|---|---|
| Digital presence | Presence on the Internet, by online stores, social networks, blogs, corporate pages, etc., increase an organization's visibility globally |
| Customer centricity | The new business model is placing customer, user, and client in the core of business to serve their needs |
| Improves the decision-making process | Because of the support given by data management approaches, Big Data technologies (extensive use) contribute to better exploitation of knowledge management |
| Improves efficiency and productivity | Extensive use of data management approaches, big data technologies contribute to the improvement of processes development in an organization, together with productivity increase and cost reduction |
| Encourages innovation | Business digitalization support and facilitation of innovation (open, collaborative) which transform an organization into a more flexible and agile one |
| Easier communication and teamwork | Business digitalization support and facilitation of internal communication and data exchange over the whole supply and value chain |
| Improves working conditions | Business digitalization support and facilitation of flexible working solutions such as teleworking and e-working which influence the decrease of job turnover and increase talent retention |

the future are presented in the synthesis. Even from 2018, the World Economic Forum and Accenture emerged "five key enablers to maximize the return on digital investments by the discussions with industry leaders, for future digital transformations" [129]:

1. "Agile and digital-savvy leadership—Maintaining a strategic vision, purpose, skills, intent and alignment across management levels to ensure a quick decision-making process on innovation;
2. Forward-looking skills—Infusing a digital mindset in the workforce by making innovation the focus of training;
3. Ecosystem thinking—Collaborating within the value chain (e. g., with suppliers, distributors, customers) and outside (e. g., start-ups, academia);
4. Data access and management—Driving competitiveness through strong data infrastructure and warehouse capability combined with the right analytics and communication tools;
5. Technology infrastructure readiness—Building the required technology infrastructure to ensure strong capabilities on cloud, cybersecurity and interoperability".

According to these five enablers, organizations must reconsider their transformation in strong relation with their strategy, leadership, and workforce development as well as technology selection, implementation, and execution. Further on, the digital transformations have been estimated that will dominate business strategy by exceeding tech spendings by $6 trillion over the next 3 years globally [10].

Digital analysts like "IDC's Digital Transformation Prediction Report for 2019" [48] identified several future trends which are affecting the business like the following (mentioned also by [10, 90]) and which have been confirmed during the pandemic period 2020–2021:

- AI and ML technology will continue to be a colossal force in the business world which lead to new operating models that add value to the organization;
- "By 2020, at least 55% of organizations will be digitally determined, transforming markets and reimagining the future through new business models and digitally enabled products and services;
- By 2020, 80% of enterprises will create data management and monetization capabilities, thus enhancing enterprise functions, strengthening competitiveness, and creating new sources of revenue;
- By 2023, 35% of workers will start working with bots or other forms of AI, requiring company leaders to redesign operational processes, performance metrics, and recruitment strategies;
- By 2020, 30% of G2000 companies will have allocated capital budget equal to at least 10% of revenue to fuel their digital strategies;
- By 2021, prominent in-industry value chains, enabled by blockchains, will have extended their digital platforms to their entire omni-experience ecosystems, thus reducing transaction costs by 35%".

The "IDC's digital transformation prediction report for 2020" [49] has supported the idea that:

> Even though we are experiencing a global pandemic, direct digital transformation (DX) investment is still growing at a compound annual growth rate (CAGR) of 15.5% from 2020 to 2023 and is expected to approach $6.8 trillion as companies build on existing strategies and investments, becoming digital-at-scale future enterprises. Organizations with new digital business models at their core that are successfully executing their enterprise-wide strategies on digital platforms are well positioned for continued success in the digital platform economy.

These predictions next to the stated outlooks of Accenture, McKinsey, and the World Economic Forum [129] show the importance and the actuality of the topics, and to deal with the digitalization and digital transformation for organizations. With respect to the increasing speed of innovations, the next era of Industrial Revolution can be seen on the horizon where transformation must be done.

## *2.3 A Synthesis on Digital Technologies*

Big Data, Cloud technology, IoT, and AI are the top supply chain drivers and pillars of Industry 4.0 [38, 70]. In addition, 85% of CIOs, at Fortinet's 2018 Security Implications of Digital Transformation Survey, support the idea that security issues during digital transformation had an "extremely large" business impact on organizations; by implementing new technologies and adopting business processes, security issues are on the rise. Mainly, the rapid adoption of new technologies, especially IoT and multi-cloud environments, has dramatically increased the attack surface and the number of entryways into an organizational network [80]. This can be underlined by the analyses of the European Commission about the most popular digital technologies in the last three years. They found out that most of the online discussions were focused on cybersecurity with a 53% share of averaged scores distantly followed by blockchain technology (17%) and AI (15%) [92]. The blockchain technology will not be analysed in a dedicated way as it will have no effect on the future approaches for AI and Business Process Management integration.

Next to the popularity of digital technologies, the EU Commission examined the digital awareness of the EU countries and the USA. The five leaders of digital awareness by the mentioned study were the UK with 100%, Spain with 84%, the USA with 60%, Denmark with 58%, and Luxembourg with 46% [92]. Germany reached 20% and Romania only 10%, while the EU average gained 30%. These results are possible explanations in relation to the status of digitalization as well as digital transformation within the different countries as discussed previously.

By discussing digitalization and digital transformation, one important concept is CPS which is defined by "networks of interacting elements, including sensors, machine tools, assembly systems, and parts, all connected through digital" communication networks which are controlled or monitored by computer-based algorithms [94]. The main characteristics of CPS "are decentralization and autonomous behaviour of the production process as well as continuous interchanging of data by linking cyber physical systems intelligently with the help of cloud systems in real time" [127]. Thus, CPS are parts of the IoT [94].

The IoT is an emerging technology and describes billions of physical devices that are connected to the Internet, collecting and sharing data. Furthermore, IoT is "the network of physical objects, devices, vehicles, buildings and other items which are embedded with electronics, software, sensors, and network connectivity, permitting these objects to gather and interchange data" [116].

For instance, the use of smart pallets within the logistic industry is meant to avoid loss, theft costs, and delay, which are the biggest challenges facing the supply chain business by several billion euros every year. In addition, due to missing or incorrect information, about 30% of all deliveries worldwide do not reach their destination on time. Therefore, Deutsche Telekom, Fraunhofer IML, and EPAL invented standardized smart pallets to better monitor the movements of their goods on water, rail and asphalt. "The pallet reports its status whenever there is a deviation from the plan, i.e., if it senses any unexpected shaking or temperature fluctuations, the IoT object automatically updates its data back to a dedicated portal" [132].

All these smart and IoT objects as well as CPS are generating and collecting data and are part of a value chain with several impacts. By analysing and executing these data with the support of the following technologies, the opportunity arises to gain advantages and optimize the existing value chain in different dimensions. The core characteristic of smart is the ability to connect to a network to share and interact remotely [119]: for instance, the use of smart technology in the field of sports to improve and monitor the performance, provide real-time statistics while running, playing rugby, or cycling. In addition, smartwatches and so-called wearables monitor health and well-being on the go by tracking steps, stairs, and calories to push people to more motion [71]. Smart homes or smart parenting are further examples where connected objects gather data to inform and support human beings represent in fact of small personal value chain.

Next to the Social Mobile Analytics Cloud (SMAC) technologies is a foundation for an ecosystem that enables the transition from e-business to digital business, as well as driving business innovations [23]. SMAC arises due to the exponential growth of "structured and unstructured data created by mobile devices, wearable technology, connected devices, sensors and social media is generating new business models built on customer-generated data. The integration of social, mobile, analytics and cloud together create a competitive advantage and new business opportunities". The technologies behind SMAC are [101]

- Social media which provide organizations a new way to interact with their actual and potential customers;
- Mobile technologies and devices allow working and communicating anywhere and anytime if you have a connection;
- Analytics provide the organizations' results of Big Data to understand and optimize work and predict future customer demands;
- Cloud technologies, especially Cloud Computing which provides high accessibility to technology, data, and thus organizations could be more flexible and agile.

It may be noted that the marriage of data and technology is radically changing the personal world and making it smarter while business must become smarter too [71].

### 2.3.1 Big Data and Data Analytics

Marr [71] describes Big Data as "the heart of the smart revolution" and characterizes it by the increasing leaving of a digital trace (by data), which can be used and analysed to become smarter. The logical derivation is that the ever-increasing volumes of data—by the year 2020, about 1.7 megabytes of new information will be created every second for every human being on the planet—and the increasing technological capability to mine that data are the driving forces [71]. The competitive pressure on companies to process these rapidly increasing amounts of data is accelerating and traditional technologies such as traditional databases, data warehouses, and reporting are no longer enough [21]. Therefore, it is important to collect and evaluate much more data about the market and the customers to continue to develop a competitive advantage. Due to the increasing amount of generated, tracked, and stored data, the term "Big Data" was simply derived [71].

In the literature, "Big Data describes large amounts of data that can originate from a wide range of sources and are evaluated using modern computer technology like CPS or IoT devices" [16]. The statement refers to Douglas Laney's three dimensions or the 3Vs which are described as follows [37, 111]:

- Volume refers to the huge amount of data, such as those produced daily by companies. Data volume is large, complex, and consequently they "cannot be stored or even analysed by conventional methods of data processing";
- "Velocity, which represents the speed at which data can be generated, evaluated, and further processed, required in real-time;
- Variety refers to the variety of data types and sources, which make them incompatible, inconsistent, or unstructured".

IBM Company extended the 3 Versus of Laney by a fourth V, called Veracity. In addition, Microsoft tries to maximize the business value by extending the 3 Versus of Laney by three further Versus [16]: veracity, variability, and visibility.

Briefly, "Big Data is about the entrepreneurial success to handle the acquisition of new information which must be available to a very large number of users in a very short time, with enormous volumes of data (terabytes or petabytes) from a variety of sources and different qualities, to analyse these data to take informative and smarter business decisions" [88]. To set the correct focus to the definition, the real value of Big Data is the ability to analyse vast, complex, and unstructured data with high network speed accompanied by creative techniques to improve the decision-making process [71, 111].

Data can be separated into the three following types [21]: (1) unstructured data are those not organized in a pre-defined organizational schema or do not have a pre-defined data model and can be stored in various formats, (2) semi-structured data refer to data that have an organizational schema and aspects that are arbitrary that make it easier to analyse; (3) structured data formatted in a way that makes it easy to extract insights and derivations. The data is often organized in a database ore any other formatted respository. It is already ordered and divided, which facilitates analysis.

To provide a better understanding of these differences with a clear view of the properties, the following Table 3 shows a short overview and comparison about it. It is important to understand the varieties of data regarding the collecting and processing within a value chain.

Data sources are manifold by the present technology in variety and volume but there are different primary sources for data in general like social networks, media, data warehouses, and CPS like sensors, wearables, and IoT objects [21]. Further on, in the field Big Data and Data Analytics seem to be equal and the terms are used in the same way but there are some differences to have in mind by talking about [100]

- Big Data describes a huge amount of data that cannot be stored and processed effectively by traditional techniques like mathematics or statistics as defined in detail before;
- "Big Data Analytics is the often-complex process of examining large and varied data sets, or big data, to uncover information including hidden patterns, unknown correlations, market trends and customer preferences that can help organizations make informed business decisions power by high performance data analytics technologies" [100].

**Table 3** Structured, semi-structured, and unstructured data—a comparison analysis

| Properties | Structured data | Semi-structured data | Unstructured data |
|---|---|---|---|
| Technology | It is based on Relational database table | It is based on XML/RDF | It is based on character and binary data |
| Transaction management | Matured transaction and various concurrency techniques | Transaction is adapted from DBMS not matured | No transaction management and no concurrency |
| Version management | Versioning over tuples, row, tables | Versioning over tuples or graph is possible | Versioning as a whole |
| Flexibility | It is schema-dependent and less flexible | It is more flexible than structured data but less than flexible than unstructured data | Very flexible and there is absence of schema |
| Scalability | It is very difficult to scale DB schema | Its scaling is simpler than structured data | Very scalable |
| Robustness | Very robust | New technology, not vey spread | – |
| Query performance | Structured query allow complex joining | Queries over anonymous nodes are possible | Only textual query is possible |

Big Data could be synonymous with Big Data Analytics. Big Data relies on the principle that the more it knows about anything or any situation by comparing more data sets, detecting relationships, or patterns, the further it enables us to learn and make smarter decisions. "Most commonly, this is done through an automated process that involves building models, based on the collected data, and then running simulations, optimizing the value of data sets and monitoring how it impacts the results". To settle and interpret all this messy data, Big Data projects often use cutting-edge analytics involving AI and machine learning to identify what this data represents more quickly and much reliably than human operators [71].

Applying Big Data organizations must find and characterize customers' needs, "locate customers, predict their behaviour and deliver bottom line results". Therefore, the SMART Model is a navigation system to establish and drive a smart business, as depicted in the following Fig. 3. One of the above-mentioned biggest issues to fail the digital transformation—to have a clear strategy aiming for business objectives and what the organization is specifically trying to achieve—is the first step within the model. If this is clear, the company must verify how the company could access that information to "Measure metrics and data" [71].

Once the type of data is available and accessible, the organization "Applies analytics" to extract useful insights from the data that can help to answer the raised strategic questions and "Report results". These three stages of SMART business are underpinned by technology, which helps to collect the data that are needed to measure, and will facilitate analytics, as well as convert the insights into data visualizations that can be easily and quickly understood and acted on [71].

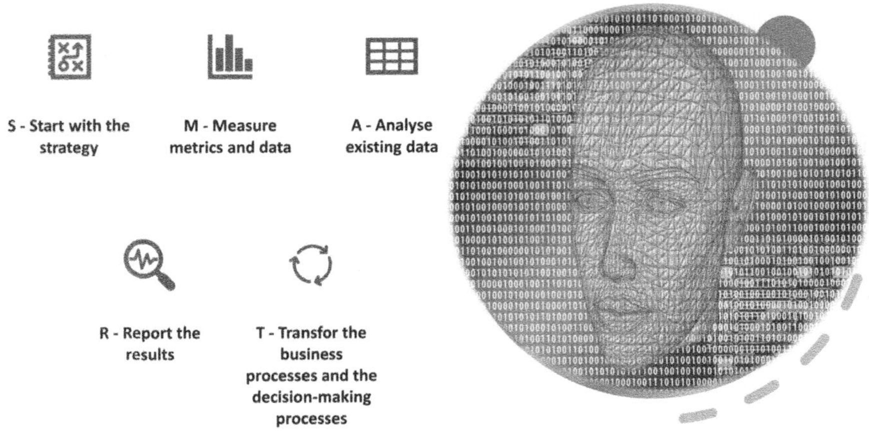


**Fig. 3** SMART big data model

Thereby, Big Data is not based on a singular technology but is rather the result of the interaction of a whole series of innovations in different fields like hardware, software, and data management [21]. While unstructured and semi-structured data types usually do not fit well in traditional data warehouses, data warehouses may not be able to handle the processing demands posed by sets of big data that need to be updated frequently or even continually [100].

Therefore, many of the organizations that collect, process, and analyse big data turn to NoSQL databases, and Hadoop. Recently, the advancement of AI and machine learning technologies has enabled vendors to produce software for big data analysis that is easier to use [100]. Using data analytics in a focused and practical perspective, the last step of the SMART model 'Transform your business' will allow us to reap the considerable rewards of improving and optimizing business processes, increasing security, reducing fraud, or driving business and people performance [71].

Furthermore, Big Data is reported to meet the business purpose of engaging with customers, increasing competitiveness, enhancing sales, and analysing information about products, services, and employees [92]. The given insights, opportunities, and advantages of Big Data execution also raise concerns and questions that must be addressed as represented in Table 4.

Organizations must face and address these challenges; they want to take advantage of data because "failure can leave businesses vulnerable, not just in terms of their reputation, but also legally and financially". Forward-looking data is changing the business world because "the ability to leverage Big Data is going to become increasingly critical in the coming years. Those companies that view data as a strategic asset are the ones that will survive and thrive and those that ignore this revolution risk being left behind" [71].

**Table 4** Big data concerns

| Concern | Data privacy | Data security | Data discrimination |
|---|---|---|---|
| Description | By posting and writing something, a lot of personal information about our lives arise, much of which we have a right to keep private | Even if we decide, we are happy for someone to have our data for a particular purpose; can we trust them to keep it safe? | When everything is known, will it become acceptable to discriminate against people based on data we have of their lives? |

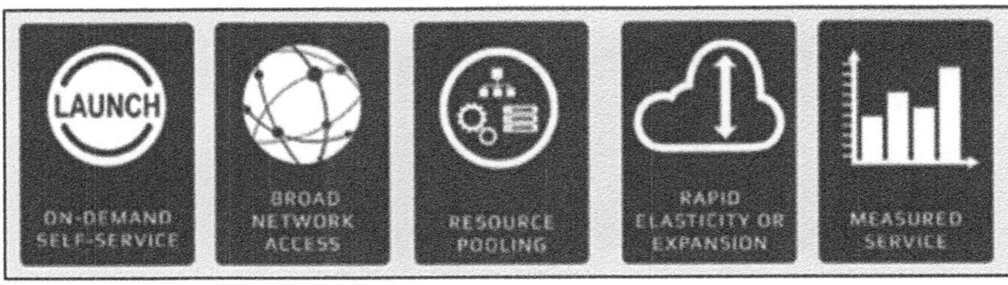


**Fig. 4** Five essential Cloud characteristic (a synthesis from [65, 66])

### 2.3.2 Cloud Computing

A central driver "of the digital transformation technology ecosystem is cloud computing which consists of several technologies, components, approaches and types of applications in an increasingly multi-cloud reality" [53]. It enables companies to achieve the desired degree of flexibility based on individually scalable service and increased data sharing, too [96, 127]. In the next three years, in the next three years, 75% of existing non-Cloud apps will be transformed into Cloud apps, while the next decade of Cloud computing promises new ways to collaborate everywhere, through mobile devices.

The term cloud is often used as a metaphor for the Internet and represents the transport of data over the Internet to an endpoint location on the other side of the Cloud. Today, the paradigm of Cloud computing can be defined as the use of various services, such as software development platforms, servers, storage, applications, and services over the Internet [96]. In other words, Cloud Computing (CC) is "a kind of outsourcing of computer programs where users can access software and applications from wherever they are while the applications are being hosted by an outside party and reside in the cloud" [103]. Simplified in a nutshell, CC provides computing, storage, services, and applications over the Internet [116]. In addition, CC can be described by five essential Cloud characteristics to define what a Cloud service is as depicted in the following Fig. 4.

On-demand self-service means "if you are not able to subscribe and configure the service without the aid of a systems administrator or outside party it is not a cloud service. If it is not possible to access the service over a network (Broad network access) using multiple platforms it is not a cloud service. Resource pooling means if there is an exclusive use of the underlying hardware or software resources. If the service not able to scale as demand increases, it is not a cloud service and described the fourth characteristic of rapid elasticity". Thus, an unmeasurable service is not a Cloud service [65]. On the other hand, the classic business applications have always been very complex and expensive because of the amount and typology of the hardware and software required to be installed, configured, tested, secure, and update them regularly. Cloud computing eliminates those disadvantages because managing hardware and software parts becomes an experienced vendor responsibility.

Next to the earlier mentioned advantages of using Cloud computing, the following "benefits continuously evolve with the addition of new services, the conducting of uncompromising research and the creation of innovations that relentlessly push the boundaries of the underlying technology and architecture and clarifies why more and more businesses are making the switch" towards CC [47, 85]:

- Cost efficiency by the elimination of investments in stand-alone software or servers as well as scalable charging models with pay-as-you-go accounting and low maintenance costs;
- Security and reliability with high availability for users by extreme physical data security, integrity, and confidentiality—all in all a more secure environment; in addition, it is hosted by a third party, businesses, and other users have greater assurance of reliability;
- Speed, scalability, flexibility, and elasticity by demand, so organizations can add storage RAM and CPU capacity according to the requirements by flexible scaling due to a changing environment. In addition, "cloud-based apps can be up and running in days or weeks—just open a browser, log in, customize the app, and start using" [47];
- Convenience and continuous availability due to easy access to information in different time zones and geographical locations which supports collaboration over the globe by updating or sharing documents in real time and 24/7 availability;
- Device diversity and location independence so the employees can access several different electronic devices from anywhere in the world.

While CC generates many benefits, it also introduces "challenges for business leaders and IT departments" related to security problems and inconsistent performance [47, 85]. The Cloud infrastructure of an organization is only as secure as they make it not only by the responsibility of the "security teams, but also with DevOps and operations teams that are charged with ensuring appropriate security controls" [47]. Within CC, the following three types can differ [67, 85, 90]:

- Infrastructure as a Service (IaaS)— it provides companies with elements of infrastructure and "computing resources including servers, networking, storage and data-center space" also providing backup, security, and maintenance by a third-party host [47].
- Platform as a Service (PaaS)— it provides organizations a Cloud-based environment to develop, run, and manage applications "with everything required to support the complete lifecycle" without investments into hardware, software, provisioning, and hosting [47].
- Software as a Service (SaaS)—it is software run on environments that are owned and operated by third parties and where a user can use an Internet browser as a usable tool.

For a better understanding, Fig. 5 illustrates the differences between the mentioned types of IaaS, PaaS, and SaaS. In addition, the classic dedicated hardware on premise as well as the shared hardware on premise are depicted for a detailed comparison.

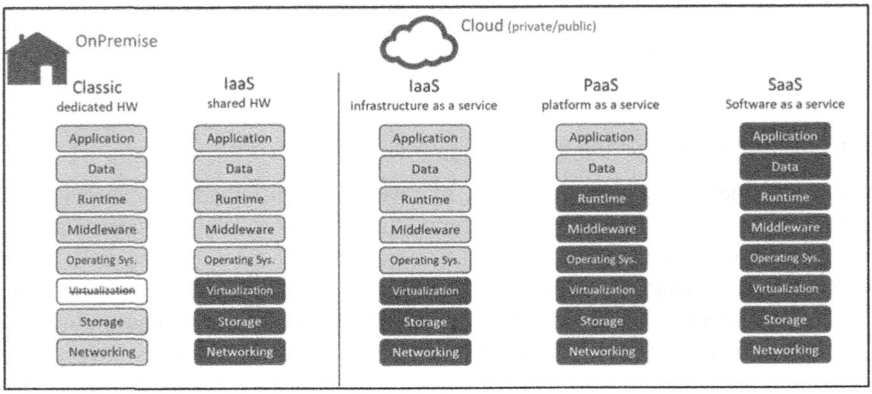


**Fig. 5** Everything as a service

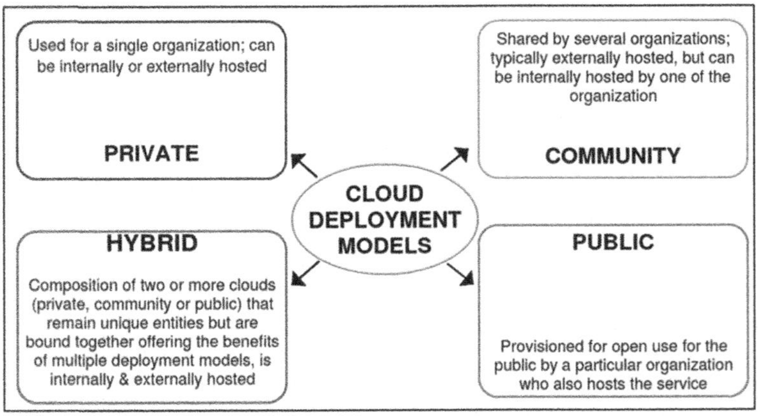


**Fig. 6** Four main Cloud computing deployment models

In practice, there are four main Cloud computing deployment models which include public, private, community, and hybrid Cloud. For a better understanding, these types are characterized in the following Fig. 6.

As a prerequisite for a successful shift to the Cloud and executing the digital transformation, all businesses and technical requirements should be considered, as well as each model's feature. "Cloud simply is essential for fast execution and informed decision making and building bridges for actionable intelligence" [53]. Therefore, the Cisco Global Cloud Index from 2016 forecasts that Cloud-based data traffic will rise to 14.1 zettabytes by 2020 and provides further details on IoT, workloads, and storage capacities in the data centres of the future [114]. In addition, the IBM five cloud prediction for 2019 supports the following ideas [97]:

1. "Hybrid multi-cloud architectures will replace the 'one-cloud-fits-all' approach;
2. Companies will increasingly embrace open cloud technology;

3. Cloud skills and culture will be the key to cloud adoption;
4. As cloud adoption rises, developers must put security first;
5. There will be an explosion of edge computing".

These five trends show the way forward. "Organizations will focus on shifting their cloud strategies from the IaaS opportunity to extracting valuable data from their business processes; integrating data across the enterprise and with external data sets; and applying new, innovative services such as AI, blockchain and analytics to that data" [97]. Especially, the combination of CC and AI technologies is introduced and "will improve several areas performance and productivity, reduce costs, improve sustainability, and increase users' satisfaction, retention, and loyalty in the long run" [32]. Therefore, "organizations will need hybrid cloud environments comprised of the right tools and infrastructure to get their jobs done more efficiently" [90, 97].

### 2.3.3 Artificial Intelligence

AI can be defined as "a section of informatics and applied computer science to pattern human proceedings of problem solving and transferring them to computers to invent efficient and new solutions as well as new course of actions. Therefore, AI is a computer program running on any possible device or data center with the skill to interact with its environment" [62, 87, 90].

Two approaches are presented in the scientific literature defining AI: "on one hand, humans are drawn as comparison and, on the other hand, AI is characterized by the list of different applications and work areas, for example in the following" [64, 90]:

> The theory and development of computer systems able to perform tasks normally requiring human intelligence. (...) The AI examines how to capture and understand the intelligent behaviour of computers, or how to solve problems by using computers that require interoperability (...) AI is a subdivision of computer science, which deals with the investigation of different problem areas like robotic, speech and flow text recognition as well as image and video processing.

### 2.3.4 Automation and Robotics

As mentioned before, the organizations must respond and ride the wave of digital transformation that comes in its wake by being faster, more adaptive, and nimbler than competitors. To do that, the phrase "business agility" is cutting-edge and means adopting a strategy that allows the organization to respond rapidly to changes in internal and external environments without losing momentum or vision [17].

Therefore, the question for every organization is how to face the challenge of? [17] constitutes that automation unlocks the door to agility and digital transformation by moving beyond simple replication of existing processes to create new sources of business value and customer engagement. This can be underlined by the IDC's

Enterprise Cloud and DevOps Management Survey. They stated that 46% of organizations agree that "increased use of automation" is among the most important drivers and requirements shaping their organization's overall IT strategy from today through 2020 [17]. In addition, consultants predict that "automation now has the potential to change the daily work activities of everyone, from miners and landscape gardeners to commercial bankers, fashion designers, welders and CEOs" [75]. Cisco [18], an American multinational technology company, complemented the following top two reasons for automating:

1. 20% of processes will not require human intervention by 2020 because it will be impossible to manage tasks manually due to scale and complexity, and
2. Rapidly changing business demands of 70% of companies in the top quartile of organizational health are agile.

Many business leaders perceived automation as a quick fix to the problem of inefficient processes [115]. Already at this point, it can be stated that automation is closely connected with processes, process optimization, and technology execution. Therefore, it is essential to form a clear definition of automation. Within the literature, there can be found many definitions of automation like the following three:

> Automation is the replacement of human activities by machine activities. (…) A device that accomplishes (partially or fully) a function that was previously carried out (partially or fully) by a human. (…) A system or method in which many of the processes of production are automatically performed or controlled by self-operating machines etc. [107].
>
> Automation refers to "a mechanical figure or contrivance constructed to act as if by its own motive power; robot" or "a person or animal that acts in a monotonous, routine manner, without active intelligence "[8].

Based on the various definitions, the term "automation" will be understood as the replacement of human activities by robots or machines to perform monotonous, routine, and standardized tasks or functions.

TAlthough process optimizations such as simplification and digitalization are done to improve routines through customization or personalization of products or services [8], "Personalization is a limited case of customization and shifting from mass customization to mass personalization is a strategic transformation which led by automation strategies" [61, 90] asset.

At this point, the relationship between process, people, and technology, known as the "the business triangle", can be assessed as depicted in the following Fig. 7. As can be seen, the three factors in combination enable business success by the integration of machines to enable processes and perform tasks as well as people who optimize, define, and follow processes and workflows by implementing and leveraging technology [62, 90].

As mentioned before, automation is about using technology for carrying out tasks to respond fast and agile to changing business conditions by adapting and optimizing organizational internal and external processes [123]. Therefore, organizations "are turning to an emerging technology practice called Robotic Process Automation (RPA) to automate simple rules-based business processes to enable business users to gain

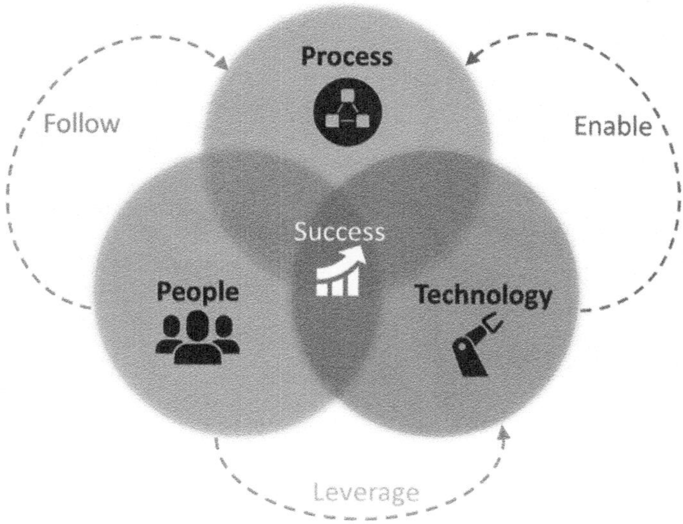


**Fig. 7** The business triangle of people, process, and technology

more time to serve customers or other higher-value work". In more detail, RPA can be defined as "an application of technology, governed by business logic and structured inputs" to carry out business processes automatically [13, 90].

Within the area of robotics, different types like soft robotics, swarm robotics, serpentine robots, and humanoid robots exist. In the context of this work, robotics will be understood as a class of software "robots" "that replicates the actions of a human being interacting with the user interfaces of other software systems" [75]. If the term robot is used in a different context, this is explicitly pointed out.

RPA is of great interest for an organization's management because it can eliminate manual and repetitive tasks, improve the delivery services, and reduce operational costs [104]. This opportunity is underlined by a study of [75], which stated that "16% of available working hours across industries in the United States are spent on data processing and 17% on data collection, which are activities that are mainly performed by humans with an automation potential of 69% in data processing and 64% in data collection". That generates an opportunity of approximately 30% of the available working hours and more than 60% automation potential. Thereby, [75] researched the activities with the highest automation potential and identified:

1. Predictable physical activities by 81%;
2. Processing data by 69%;
3. Collecting data by 64%.

The potential benefits of implementing RPA are depicted in Fig. 8 (a synthesis from [90, 104]).

Organizations "implementing process automation have a competitive advantage over their competitors in operability and reliability by reducing operative costs and

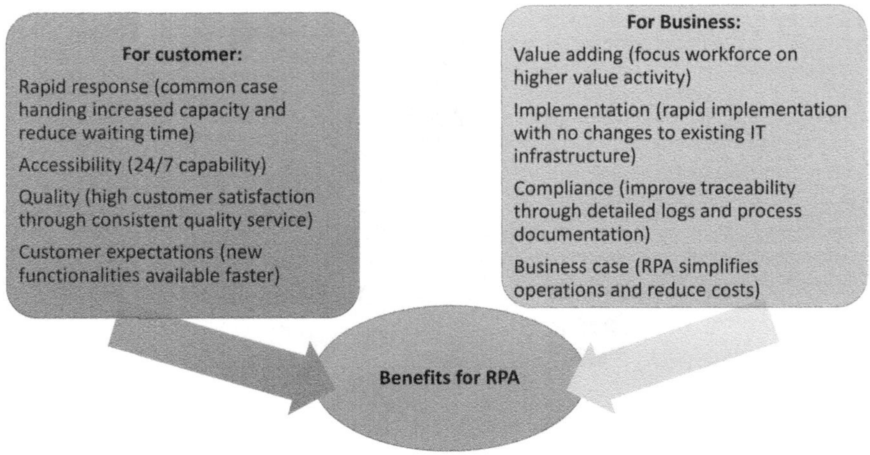


**Fig. 8** Benefits of RPA

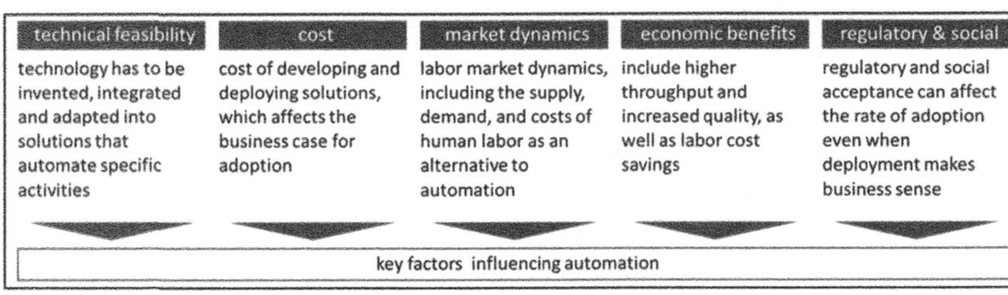


**Fig. 9** A synthesis of the five key factors influencing automation

increases speed and reliability of task implementation, development, and support". Further on, process automation improves, and optimizes strategic analysis by the simplification of application development processes and by the automation of job execution [13, 81]. Further advantages are the visibility and control of workflows and tasks, providing continuous monitoring and controlling of the processes in progress and incoming tasks as well as enhancing work continuity and satisfying demands for IT departments and computer systems in virtual and hybrid Cloud environments [29, 81]. It shall be considered that RPA "best use after simplifying and standardizing the business processes to magnify the efficiency, whilst leading to improved effectiveness, control, and business value, so Bill Gates is still fundamentally right concerning his quote" [115].

The pace and extent of automation will be influenced by the following five key factors illustrated in Fig. 9 [75].

Rutaganda et al. [104] predict "RPA and AI together will penetrate a wide segment of daily life in the next decade, with huge implications across various industries. Others see RPA as a stopgap route to intelligent automation (IA) via machine learning

(ML) and AI tools, which can be trained to make judgments about future outputs" [13, 90].

"Cognitive process automation (CPA) widens the application of RPA to more knowledge-based work by extracting information from unstructured sources and enhancing decision-making with a kind of self-learning capability to act and learn from experience, from humans, and even on their own, thereby developing the ability to interact with their own environment". The combination of RPA and CPA will provide greater customer satisfaction and will contribute to the increase of the revenues. The cognitive component—AI "can make predictions about process outcomes by identifying patterns and prioritizing actions depending on predicted outcomes while the use of Natural Language Processing (NLP), speech, and image recognition can facilitate understanding of free flow sentences and convert speech audio, text, or images into structured information" [104].

Centralized management and monitoring of automated processes will be integral for end-to-end visibility and control of IT and business processes [17]. Automation will change the organization of companies, the structure, and base of competition of industries and business models [75].

### 2.3.5 Cybersecurity

The ongoing DX by "integration of business systems, information- and operational technology that enables data-driven decision-making poses new security challenges because these newly connected systems can also accelerate the speed and damage of attacks across enterprise networks". The 2018 Security Implications of Digital Transformation Report of Fortinet which outlined 85% of CISOs said security issues during digital transformation had a "somewhat" to "extremely large" business impact [80]. Especially, the adaption of "IoT and multi-cloud environments have increased the attack surface and the number of entryways into a network" [76, 80]. In addition, the European Commission supports the idea that "cybersecurity has never been more essential, as companies have more digital valuable assets than ever before". The increasingly used hybrid Cloud architecture and use of mobile devices by employees requires a new approach to cybersecurity [92].

To better understand the characteristics of cybersecurity, the following definition will be fundamental:

> Cybersecurity is the strategic (mission-focused and risk optimized = management of information technology and systems, which maximizes confidentiality, integrity, and availability using a balanced mix of technology, policy, and people while perennially improving over time [44].

Cyberattacks have become more sophisticated, therefore "organizations must be mindful of compliance requirements and turn to best-in-class, certified products, processes, and people to ensure a reasonable level of risk management" [80]. While hackers are perfecting their skills and refining their techniques, three business trends make companies more vulnerable to cyberattacks, McKinsey and Company stated [76]:

- "The digitization of the entire value chain means that assets from the design of new products and services to distribution networks and customer data are now at risk";
- Increasing complexity of corporate digital value chains with thousands of people, countless applications servers, workstations, and other devices;
- "Outsourcing and offshoring generates the weakest link of a company's value chain due to a supplier or other third party".

In the past, cyberrisk has primarily affected IT by, for example, trivial entry points such as a WiFi-enabled camera. Nowadays, companies hook their production systems and products onto the Internet whereby the number of vulnerable devices is increasing dramatically. McKinsey predicts 50 billion IoT devices by 2020, with millions or tens of millions of endpoints within a large corporate network where many of them will at least partly be outside corporate control. In effect, the number of potential entry points for hackers and the value at risk will be multiplying across all industries and supply chains [76]. According to Kaspersky Labs, 323.000 new malware types are discovered daily [33].

To prevent cybersecurity issues, often IT departments of organizations receive the responsibility but defending a business requires a sense of the value at risk, based on an understanding of business priorities, a company's business model and value chain, the corporate risk culture, roles, responsibilities, and governance [44, 76]. Furthermore, the mindset of organizations must switch from thinking about cybersecurity as a cost issue to doing business to understanding it as a growth enabler.

Towards the appropriate security level for organizations, the four key aspects of a complete security architecture—prevent, detect, respond, and predict—should be considered as depicted in Fig. 10 [33]. Furthermore, organizations must accept Cybersecurity as a business discipline with a starting point at the highest executive level where leaders focus on risk opportunities and threats [44].

Digital transformation will continue to impact how organizations approach business and technology. As their networks become more complex, organizations must adjust their approach to security to ensure there are no gaps in protection [80]. "With the growing capabilities in the field of AI/Machine Learning, cybersecurity experts are expecting that AI and Automation would help deter and manage cybersecurity threats" [121].

"Automation involves a range of competencies such as Process Automation or Security Automation whereby the last one is designed to reduce risks, operational errors to address the Cyber Security Threat". "Security tasks are often prone to error when it comes to processing large volumes of data therefore Security automation helps security analysts to be more proactive and innovative to focus on complex types of attacks and be prepared for them. Behavioural analytics and machine learning are the advanced forms of automation for Security Automation leverage the alert, monitoring, and prioritization tasks to the next levels by learning from gradual training and failures which can easily and immediately catches any abnormal behaviour.

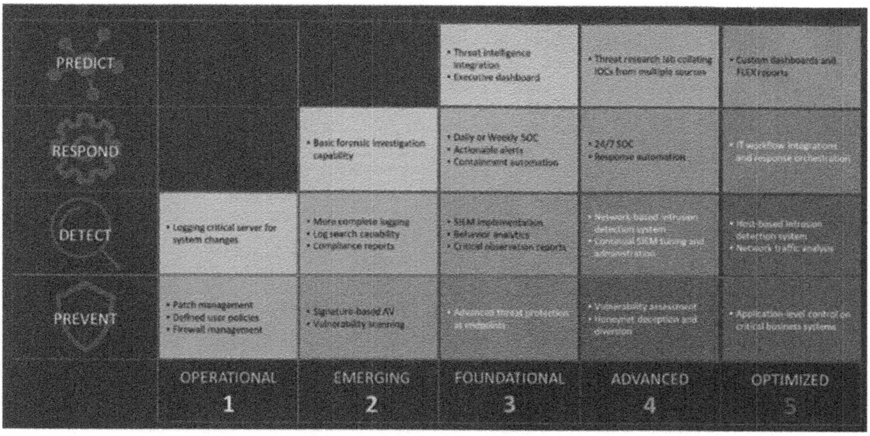


**Fig. 10** A synthesis of the cybersecurity maturity model

Improving the quality of threat intelligence is the next step to pass more control to AI but there needs to be a balance between operational control and critical exercise that can escalate up to humans. This will ensure that AI applications for cybersecurity defence are truly effective for a better decision making, quick resolution, consistent and stable root cause analysis and predictive analysis, diagnosis, and recommendations" [121].

In the nearest future, it is estimated that a mature AI system and Big Data algorithms could better support complex decision-making processes which presently require human intelligence [71, 90, 121].

## 3 Going Beyond Industry 5.0 and Society 5.0

Starting from the "Industry 4.0 advances in robotics, AI and Machine Learning are launching in a new age of automation, as machines match or outperform human performance in a range of work activities, including ones requiring cognitive capabilities" [75]. Therefore, it is not new that production conditions and economic framework conditions change over time.

Joseph Schumpeter, the famous Austrian economist and professor at Harvard University, described this phenomenon as "creative destruction" or sometimes known as Schumpeter's gale, about 100 years ago. In a nutshell, "creative destruction" describes the industrial change that constantly revolutionizes the existing economic structure by destroying the old one and creating a new one; thereby, the "creative destruction" is triggered by innovation [90].

Regarding the duration of the individual industrial phases, the creative destruction by technological innovations is progressing rapidly; the "first three Industrial Revolutions took decades, actual Industrial Revolutions only last until industrial

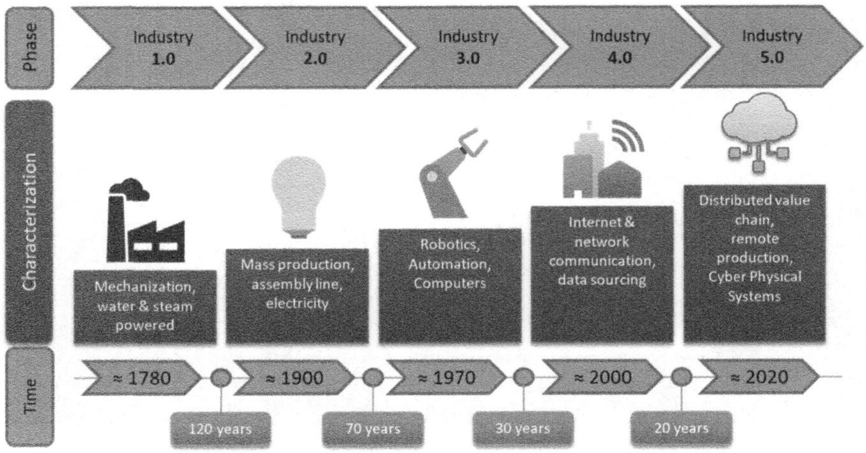


**Fig. 11** Phases and duration of industrialization

implementation is complete" (Fig. 11) [134]. Less than 8 years "has passed the literature started talking about Industry 4.0, which is al-ready moving towards the next revolution: Industry 5.0" [83, 99, 106].

Industry 5.0 is considered (even in the scientific literature) as the next industrial revolution with an impact on industry and economy, civil society, governance, and structures as well as human identity [127]. The consumers' high demand for individualization will be one mandatory reason for the next step of the Industrial Revolution [82]. Further on, "products with distinctive mark of human care and craftsmanship are these where the customers will pay most for like designer items of every kind, fine watches, craft beers or black hand dyed salt from Iceland with local coal. This demand of human touch will be raising in the future much more because consumers seek to express their individuality through the products they buy, which outlines a new kind of personalization, feeling of luxury society where the business must deal with" [82].

In the examination of the term Industry 5.0, it became clear that the first states of the term were already announced several years ago in 2016 not only by different authors and researchers but also by different consulting companies' studies. Within the last years, the term Industry 5.0 became more often used in the scientific and political discourse related to the Fifth Industrial Revolution.

Industry 5.0 is considered to be in its early stages, but for competitive organizations this new Industrial Revolution will affect the way they operate. During the acceleration of the digital transformation which has been accompanying the pandemic period, practitioners and researchers have recognized the need for innovation, creative ideas, a new way of thinking, and approaches that can support the unknown future (called even the New Normal), together with agile, flexible, and lean business [59, 108]. Furthermore, Stuart Scanlon, managing director for epic ERP, explained [89]:

> People have different opinions when it comes to predicting the start of Industry 5.0 but, if you consider the speed of transformation in technology, I believe it's going to be here for sooner than most people think. The future is happening now, and we need to rise to the challenges if we are to thrive in the next revolution [108].

Therefore, it is important to understand the definition and characteristics of Industry 5.0 as well as the differences in comparison with Industry 4.0. These characterizations are described in the following.

### *3.1 Detailed Definition and Characterization of Industry 5.0*

The business transformation into a digital smart chain will undoubtedly determine the future world and is the keystone of the industry of the new millennium [106]. The research on Industry 5.0 concept and theory shows many uncertainties about how it will disrupt business processes. But on the contrary, it is going to break down barriers between the real and virtual world through new technologies which are "going to be faster, more scalable and adopted globally" [29, 108].

Industry 5.0 will increase the collaboration between humans, "which take creative side to take on more responsibility and increased supervision of systems to elevate the quality of production across the board and smart systems" like robots which are working on monotonous, repetitive tasks. This outlook can be underlined by a survey with more than 500 manufacturing executives worldwide from Accenture. They figured out that "85% of the participants foresee a collaborative pro-duction line between humans and robots in their plants by 2020" [2, 89].

To be more concrete, the European Economic and Social Committee (EESC) describes Industry 5.0 as "… focused on combining human beings' creativity and craftsmanship with the speed, productivity, and consistency of robots" [30].

Another definition focused on the implementation of AI as the core element, to enhance the common human lives and their collaboration with the aim of enhancing the human capacity and returning the humans to the "Centre of the Universe" by naming "Society 5.0" as a more exact term instead of Industry 5.0 [113].

Another concept of Industry 5.0 describes it "as faster, more scalable and more people concerned than the former ones through the kind of technology" [102]. "This will happen by the drive towards more advanced human–machine interfaces by enhanced technological integration, improved automation of robots paired with the power and creativity of human brains, which will lead to improved productivity" [89, 112].

The mentioned collaboration "between humans and technology will affect the economy, ecology, and the social world" as well [112]. "These influences are accompanying by a waste prevention perspective applied in industrial upcycling" [89, 93]:

- "Process waste—overproduction, empty trucks on the roads, overstock;

- Social waste—People willing to work but having no opportunity and people who are not willing.
- Physical waste—general trash and trash of production lines and logistics;
- Urban waste—not necessarily needed Greenfields, empty spaces, inadequate infrastructure".

Another vision describes "factory workers and robots may end up working together on the design and sharing of workload across a variety of manufacturing processes. While robots are excellent for producing standard products in standardized processes in a high volume of production, customizing each individual product can be a challenge where robots need to be guided" by humans within production processes [24]. To exploit automation "to its full potential there is a need of human creativity that influences the processes" in collaboration of collaborative robots or "Cobots" working with humans [84].

Phill Cartwright, executive chairman of the Centre for Modelling and Simulation, underpinned the previous definitions by focusing on manufacturing. He characterizes the Fifth Industrial Revolution by the important cooperation of human intelligent workers in harmony with cognitive computing, so the cooperation of humans and machines and added "Industry 5.0 starts when you begin to allow customers to customize what they want" [99]. Workforces will be upskilled to provide value-added tasks in production by putting humans back into industrial production with collaborative robots, called cobots [99]. Industry 5.0 takes that concept of personalization and customization to the next level [99].

In addition, "Artificial gene synthesis (artificial DNA or DNA printing from synthetic biology filed), new raw materials or sustainable uses of re-sources are further aspect of Industry 5.0 characterization with potential impacts on business. These fields put the human factor to the center, too in relationship with nature and physical integrity" [89, 105].

According to SAP, we are already talking of Industry 5.0, which shifts the focus on the customer as a key player and "user contributor" in the process and product value chain by new technologies [42]. In more detail, clients and customers (B2B or B2C) "purchase and use intelligent, connected, and self-aware products capable of sharing information and data about their health, location; wear and tear; usage level, type, and timing; storage conditions" and so on to the company which will be able to process these data and gain a big advantage by putting themselves in their customers' shoes to understand the behaviour, demands, and so on. Thereby, the "customer intimacy increases with maturity, and each customer becomes a vital contributor of continuous insights. Assuming looking ahead, SAP expects Industry 5.0 will create new and more agile offerings centred around customers' demands, likes, needs, and preferences" [42].

In conclusion, Industry 5.0 focuses on a new way of involving humans into the industrial production chain and environment, working together, and be supported by AI and further innovative technologies to enhance existing working procedures and processes by executing tasks by an increasing level of speed and accuracy with sustainable uses of resources [83, 84, 99, 105, 106, 113]. Next to the return of humans

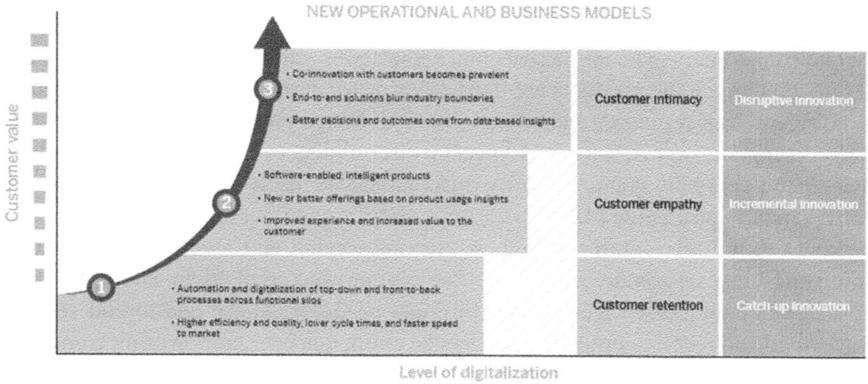


**Fig. 12** A model for going beyond Industry 4.0

into the production chain, customers will be an essential part of the product value chain by its needs, demands, and service expectations [42]. The improving interaction of human operators with computers and robots supported by AI will be significant to the technological field to take the concept of personalization to the next level to satisfy customer needs best [24, 99].

## 3.2 *Business Value and Benefits of Industry 5.0*

Looking to Industry 5.0, SAP "anticipate the further restructuring of product development and production in ways that will redefine not only processes but also what a product is and the value it provides to customers". Connected, interactive products support companies in gathering data and information on how customers use products, and their satisfaction. These types of data used by a stronger involvement of the customers create a unique window of opportunities into the customer experience for decision makers.

As part of the product value chain and by sharing their data, customers have been offered personalized products and value added, together with data-based services portfolio. The new market leaders will continuously "listen" to customers' opinions "through streaming product usage data and respond swiftly with customer-driven products" [42]. This so-called co-innovation with customers as part of the end-to-end value chain can be defined as the highest level of customer value-driven business models as can be seen within Fig. 12.

Level 1 is characterized by connected machines, processes, and people by automating, integrating, and "digitalizing top-down and front-to-back processes across functional silos to improve product quality, efficiency, and speed to market" by digitalizing horizontally and vertically [42].

The second level is distinguishing by "software-enabled, intelligent products" that help customers to be an "active contributor to the value chain" by providing data about "how much the products are used, rates of wear and tear, when products need to be repaired or replaced, and so on" [42, 99]. Based on these insights on product usage, companies can develop and "invest with confidence in features and capabilities that customers really want, and truly differentiate themselves from competitors" to deliver meaningful value [42, 99].

Level 3, the level with the highest customer value and level of digitalization, companies use customer usage data "to co-innovate with them and even build digital, value-added services around physical products. These end-to-end services can extend and redefine product value for the entire markets, empowering companies to differentiate even the most commoditized products in new ways" by [42]

- Providing recommendations;
- Improving operations;
- Reducing costs;
- "Share best practices with the customer on the optimal use of the product;
- Given the design parameters;
- Advise on how to achieve better outcomes and maximize return on investment or implement predictive maintenance to maximize longevity and minimize downtime".

These cases exemplify that manufacturers can support customers to make better data-driven decisions and thus, improve their business [42].

Besides, the collaboration between advanced technologies and humans as well co-innovation between manufacturer and customer, for instance, can lead to robots completing the automated manufacturing of goods, IoT, and smart devices submitting important production and usage data while Business Process Management (BPM) software will monitor and control the data. If an error occurs, it triggers a process automatically that alerts the employee to act and decide what to do within the process chain. Therefore, Industry 5.0 will promise companies "improved adaptability, change readiness and a responsive working environment" by digitalizing and automating their critical business processes while keeping humans at the centre of decision-making [26, 59, 89]. Through this collaboration, the companies can realize digital transformation and attain greater productivity, agility, and profitability by the integration of customers into the value chain [59, 99].

A further benefit will be the fact that "machines do not replace human workers and create job losses to improve productivity". Furthermore, [82] underlined this by his experience at Universal Robots, where they have analysed that companies who deploy cobots end up employing more people. In addition, Mitsuru Kawai, Executive Vice President of Toyota, says (mentioned by ([12])

> Robots do not improve processes. Only humans can improve processes. That is why they should always be the focus of attention.

Equally, Tesla's manager Elon Musk says in the meantime that he had progressed too fast in terms of automation. "People are underestimated", he wrote in a tweet sold in April after his company received bad press due to manufacturing defects [12].

"Rather, the lack of qualified human workers with missing skillsets has driven the conversation to Industry 5.0 and the implementation of robotic systems, which result in greater productivity" [95]. The Fifth Industrial Revolution will create "higher-value jobs because the freedom of design responsibility gets back to the human" [84, 99].

The employee "has more responsibility and ends up with a bigger, lighter, and safer environment and becoming more involved in the design process rather than the manufacturing process, which is automated". That means, if companies "have true, seamless data between the field, the manufacturing process and the design, humans can take out of the manufacturing route, but they will be more involved in how the product is being used and how it can be designed and optimized because they have more usage information from the customer" [99].

"Industry 5.0 anticipate the further restructuring of product development and production in ways that will redefine the manufacturing processes and what a product is and the value it provides to customers" [42]. Through this change, it is important to understand that "as more products incorporate digital technologies, they will move from being a bundle of functionality to become platforms for value creation for the user and the maker of products" [42]. Success in this area will require integrated ecosystems, or "industry mashups", of companies "in different industries working together to pro-vide distinctive, significant value to customers through the creation and delivery of more complete and agile offerings centered around customers' demands, likes, needs, and preferences" [42, 90, 106].

## 3.3 Technology of Industry 5.0

One core technology that will drive Industry 5.0 is robotic process automation (RPA), which is robotic software that companies can use to capture and interpret the user interface and actions of other applications used in various business processes [26, 59]. In general Robot dexterity will rise by teaching themselves to handle the physical world [58].

RPA software in combination with AI will allow understanding specific processes and then automatically process transactions, input and analyse data as well as communicate with other existing transactional and non-transactional systems. It will mirror the way humans work with applications but faster and better. If something goes wrong in one of the RPA's automated processes or an important change needs to be made in a process, BPM software will kickstart a workflow that notifies an employee to look at the issue and provide all gathered information needed to make an educated decision that will produce the best business outcomes [59].

Next to RPA, Gartner predicts that 25% of digital workers will use virtual employee assistants daily by 2021 "for Business helping employees delegate tasks such as scheduling meetings and logistics operations, helping engineers find answers as they perform complex tasks or diagnose problems" to increase employee productivity and foster constructive engagement through AI. In addition, "Gartner predicts that, by 2023, 25% of employee interactions with applications will be via voice so AI-enabled speech-to-text and text-to-speech hosted services are improving rapidly". Van Baker, vice president at Gartner [37], underlined this by saying:

> Enabling voice interaction with applications will ultimately enhance the digital dexterity of workers who have access to them.

AI, robotics, RPA, and BPM are setting the stage for Industry 5.0 [59]. In addition, Cybersecurity and cyberdefenders are more important than before to "proactively hunt over companies anonymized security data for the most important threats, such as human adversary intrusions, hands-on-keyboard attacks and advanced attacks like cyberespionage", Ann Johnson, Microsoft CVP Cybersecurity Solutions Group, mentioned [55]. By horizontal and vertical integration as well as adding mobile objects and IoT devices of customers or manufacturer to the value chain to gain information, you will never know if there arises a security issue and how big it will be. To cover this uncertainty, companies should use Cybersecurity defenders to protect their business and data.

By "pervasiveness and exponential growth of data and software as well as increased standardization of application interfaces, integration points, and automation technologies, which reduce friction", the expectation of the rapid evolution of platform economy is reality [42, 106]. "As more products become platforms for innovation and value creation, opportunities will open for companies to be platform adopters and innovate on the platforms of the players" [42]. Therefore, it can be assumed that the integrated platform development will be one important part of Industry 5.0.

The debate in the literature about Industry 5.0 is already extensive. Therefore, the following main findings regarding Industry 5.0 are key things to remember by talking about Industry 5.0:

1. Industry 5.0 put humans in the centre and uses Robots to support them and make the lives of humans better and easier;
2. Consumers will increasingly demand craftsmanship, personalized products and services, something special and unique. This will be created by humans through automation and cobots and a streamlined manufacturing process;
3. "Industry 5.0 is meant to optimize human efficiency and productivity through to meet the manufacturing complexity of the future in dealing with increasing customization through an optimized robotized manufacturing process" [34];
4. Industry 5.0 is inevitable or like the European Economic Social Commitee (EESC) said:

> The proliferation of robotic automation is inevitable [31].

## 3.4 Society 5.0—The Super Smart Society

In anticipation of global trends, Japan raised the core concept "Society 5.0" by the 5th Science and Technology Basic Plan [106]. "The humanity is now in a new era, one in which innovation driven by enabling technologies like described before, are bringing significant changes to the economy and society" [43].

Like the explained industrial revolution steps from Industry 1.0 to Industry 4.0, a comparable evolution in society took place [90]. The timelines of all the phases do not exactly map with that of the industrial revolution. In the future, the industrial and social revolutions are coming closer and the next jump can be a natural merge of these two. According to the Japanese government, "Society 5.0 represents the 5th form of society in human history" [9, 46, 106]:

- Society 1.0 (Hunting society)—Till 13,000 BC, groups of people hunting and gathering in harmonious coexistence with nature;
- Society 2.0 (Agrarian society)—From 13,000 BC to the eighteenth century, people formed "groups based on agricultural cultivation, in-creasing organization and nation-building";
- Society 3.0 (Industrial society)—From the end of the eighteenth century to the end of the twentieth century, people formed "a society that promoted industrialization through the Industrial Revolution by implementing mass production";
- Society 4.0 (Information society)—From the end of the twentieth century to the end of the twenty-first century, people saw the beginning of an information society with the invention of the computer and information sharing;
- Society 5.0, related to the twenty-first century, a super-smart society built upon Society 4.0, which aims the creation of "a society where people can resolve various social challenges by combining innovations such as AI, robots and big data into society".

Society 5.0 is [46]:

> A social state where material and information are highly integrated and enables society to provide certain goods and services to people who need them at an appointed time, and free people from inconveniences caused by age, gender, region and language differences.

In addition, Japan's Prime Minister Shinzo Abe said at the International Conference of the Future of Asia in 2017 [54]:

> The essence of Society 5.0 is that it will become possible to elicit quickly the most suitable solution that meets the needs of each individual. We will become able to solve challenges that have defied resolution until now.

In combination with the statements of [82] regarding Industry 5.0:

> People will pay a premium if they believe a product will help them meet their deeper more personal goals.

Østergaard [82] appreciates that there is a direct linkage between Society 5.0 and Industry 5.0 that can be underlined and can be determined in the field of the fulfilment of individual needs and personal goals. From the derivative from the previous quotations, Society 5.0 can be defined as a "society of intelligence", "in which physical space and cyber-space are strongly integrated and innovation in science and technology occupies a prominent place, with the aim of balancing social and societal issues that need to be solved, while ensuring economic development" [106].

Furthermore, Society 5.0 predicts a "sustainable, inclusive socio-economic system, powered by digital technologies and cyber physical system, in which cyberspace and the physical space are tightly integrated, becomes a universal technological mode" [124]. At this point, the connection to Industry 5.0 can also be drawn based on the closer linkage of cyber-physical systems and people. Three major characteristics of Society 5.0 were identified [35, 46].

- Society 5.0 is emphasizing on how to "optimize the man-hour responsibility to get the job done";
- "Society 5.0 focuses on the effectiveness of optimizing knowledge workers assisted by intelligent machines";
- Society 5.0 is "meant for the harmonization of work with the help of intelligent machines for the benefit of the workers to put humans at the centre of innovation".

The people's lives of a Society 5.0 are characterized by a higher level of convenience and sustainability as before through optimally delivered products or services tailored to their individual needs as well as the "help to overcome chronic social challenges such as an ageing population, social polarization, depopulation, and constraints related to energy and the environment" [124]. Furthermore, Society 5.0 is characterized by [124]

- "Autonomous vehicles and drones will bring goods and services to people in depopulated areas;
- Customers will be able to choose the size, colour, and fabric of their clothing online directly from the garment factory before having it delivered by drone;
- A doctor will be able to consult her patients in the comfort of their own home via a special tablet;
- While she examines them from a distance, a robot may be vacuuming the carpet;
- At the nursing home down the road, another robot may be helping to care for the aged;
- In the nursing home's kitchen, the refrigerator will be monitoring the condition of stocked foods to cut down on waste;
- The town will be powered by energy supplied in flexible and decentralized ways to meet the inhabitants' specific needs;
- On the outskirts, autonomous tractors will be toiling in the fields while, downtown, advanced cyber physical systems will be maintaining vital infrastructure and standing by to replace retiring technicians and craftspeople, should there not be enough young people to fill the gap".

In addition, [106] named "adaptability, agility, mobility, and reactivity as key words in the life of society 5.0, which entails the fact that mutations, changes and evolution are an observable daily constant, which is also reflected by infrastructure, knowledge and skills". Society 5.0 is a fresh "New World" in which the exchange between business, universities, society, and further stakeholder is one key requirement.

In summary, Society 5.0 is about the digitalization and advanced IT technologies across all levels of the society to address key problems and resolve social issues like labour shortage or ageing workforce rather than making simple productivity improvements [9, 113].

The close interlocking with the properties of Industry 5.0 shows that the two evolutions (Industry 5.0 and Society 5.0) have adapted over time and have nearly similar objectives. Therefore, Industry 5.0 and Society 5.0 are used as synonyms in the following.

### 3.5 *Leadership 5.0*

Holger Arians, co-founder of PLDx.org and the CEO of Dominet Digital Corporation, said (as mentioned by [5]):

> There is no question about it: technology is the driving force behind the new style of leadership.

Furthermore, successful leaders understood these paradigms how much connected devices and AI can support to manage and develop people and empower companies. Therefore, the most meaningful ways are [5]

- Giving your employees more flexibility by a positive work/life balance through collaborative working by a decentralized business environment to force a positive behaviour and a comfortable feeling for being creative and engaged;
- Be an influencer and inspire employees, motivate the sales teams, build customer loyalty, and connect with peers through various social media platforms and blogging.

In addition, Klaus Schwab calls for leaders [73]:

> … putting people first, empowering them and constantly reminding ourselves that all of these new technologies are first and foremost tools made by people for people.

Leaders must understand and adopt the spirit of change by digitalization and must take out the fear of employees regarding the predicted job loss in the age of automation and IA where 800 million people might be out of a job by 2030 because of automation [60, 69, 73, 131]. Therefore, leaders are needed with the skills to manage organizations through these shifts with the foresight to better prepare people for the flexibility and critical thinking skills they will need in the future workplace [73].

While Leadership 4.0 was "doing the right thing, but for the wrong reason" like empowerment, high motivation of employees but totally profit-driven, at the expense of the customers, Leadership 5.0 is doing the right things well for the right reason

[19]. This simple approach can be supported by the three principles for augmenting leadership [7]:

1. “Human centricity: using technology to boost human qualities of ingenuity, judgment, contextualization, creativity, and social interaction by putting human beings in the center of the approach;
2. Full circle: Strategy without action is just fantasy. To fulfil strategy, people must be able to effectively collaborate, and their experiences must be captured to create a source of shared learning that improves future efforts;
3. Followership: Leadership is ultimately defined by its ability to create followership by setting a goal, be authentic, be part of the team as well as coach in one person to re-quires people to be truly engaged with the tasks at hand, and to have a greater collective role in defining the organization’s future”.

By the faster changing frameworks of business and economy as well as social implications, a new style of leadership must be executed to generate a positive and focused team who are able to adapt to the changes and to shape them in a positive way for the organization. Therefore, the leader should be able to show the sense of the tasks and have a vision in which the team and the organization can participate. Industry 5.0 “recognizes that man and machine must be interconnected which will lead to a human and robot workforce” [82].

This “will call for a new executive role, the chief robotics officer (CRO), who will be responsible for planning and managing” all actions related to robotics and intelligent systems. This assumption can be underlined by a report from “Myria Research which forecasted that by 2025, more than 60 percent of manufacturing, logistics and supply chain, agri-farming, and the oil, gas and mining sectors will have CROs” [34].

## 4 Conclusions

Industry 4.0 revolution is characterized by progressive automation, digitization, and increasing convenience of customers. Despite Industry 4.0, the need for personalized products that can be made available to customers in a high quality and at the same time cost-effective manner without waiting times presents companies with challenges. Summarized in a metaphor, it could be expressed as “Me as a customer wants All, Immediately Everywhere” [90].

Nevertheless, thanks to the technological developments, the Fourth Industrial Revolution generates a lot of added value for companies and customers in various stages of the value chain. Productivity increases through intelligent systems, reduction of downtimes and predictive maintenance, reduction of scrap, and better forecasting possibilities as well as a faster time to market are just some of the value drivers. A lack of strategy is one of the biggest risks and challenges for the implementation of Industry 4.0. In addition, poor data quality, a lack of skills, and a lack of cooperation mentality are the main reasons why the digital transformation of companies fails.

The definition and implementation of a digital business strategy are essential for the realization and achievement of the desired added value. It should be noted that the data generated by the supply chains is an important component when this data is analysed and processed with the help of data analytics. The underlying data strategy should be linked to the business strategy to establish a targeted strategy for the company. Within this strategy, the technology to be used should also be considered. In the context of the fourth interdisciplinary revolution, Cloud computing and platform technology as well as automation, robotics, and cybersecurity are the focal points. Big Data and AI are becoming increasingly essential and are based on the above-mentioned technologies, creating new opportunities in digitization. To handle and implement all these topics, skilled and motivated employees and leaders are needed.

Furthermore, "automation is needed as a part of digital transformation and many companies understood importance of all these driving factors in company's development but there still is lot of areas to improve" [8].

Already today, Industry 5.0 is starting to offer our society and business a people-based economy with the focus on collaboration supported by machines and AI within different fields of action but is expected to do so even more in the future. By putting people back at the centre of production, Industry 5.0 will provide consumers the products they want today and offers jobs that are more meaningful than factory jobs have been [82].

Like the Fourth Industrial Revolution, this progress is driven by technological developments, society impacts as well as the demands of human beings to solve existing humanity problems. Therefore, we can speak in parallel with the development of Society 5.0, which places people in the foreground and meets the needs of everyone through targeted services and product solutions. The degree of satisfaction and comfort that can be achieved is also based on technological development as well as the use of AI.

The leaders of the future must therefore be able to generate visions and ideas for the further development and use of technologies, on the one hand, and on the other hand, motivate employees for new goals and approaches despite increasing automation. Therefore, the leaders of the future represent an essential component for achieving business goals as well as the state of Society 5.0 and Industry 5.0.

In conclusion, "all visionaries or future observers see the main ambition of Industry 5.0 in the human touch back by the collaboration between humans and machines" [89]. Thus, Industry 5.0 will lead to

- "The evolution of a global society;
- Waste reduction and waste prevention activities" [93, 112];
- "Strong bioengineering and biotechnology influence business and society" [105].

It is impossible to ignore the industrial evolution to Industry 5.0 [34], as well as Society 5.0. Dealing with this topic, there are important questions that the business, leaders and policymakers need to address soon to be competitive on a global market. Therefore, it can only be recommended to deal with these issues at an early stage in order not to miss the next steps of the Industrial Revolution as well as the changes in society.

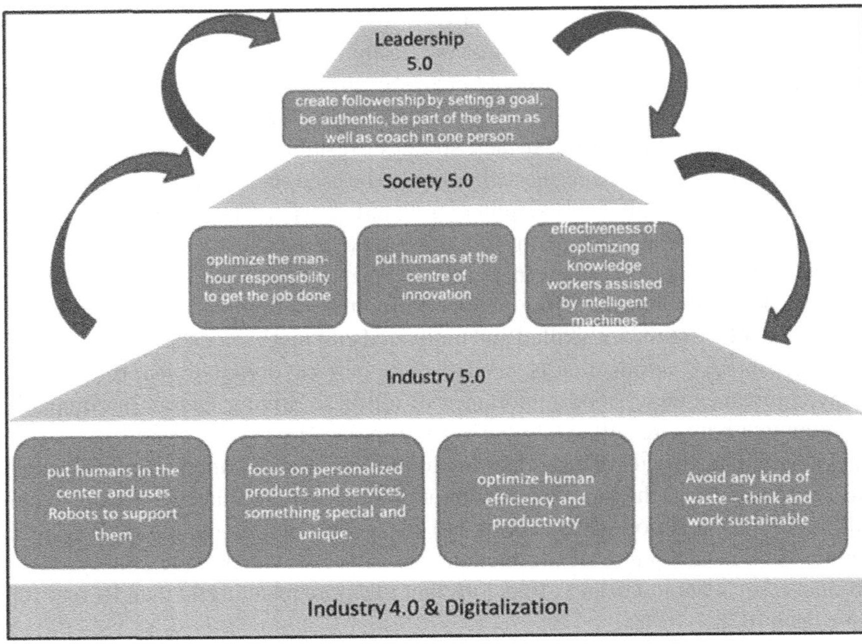


**Fig. 13** The pyramid of industrial revolution

In the combination of the different thematic fields, not only the relationship between them but also their dependence can be seen. If we visualize this network of relationships, it becomes clear that Industry 5.0 is only possible based on digitalization and Industry 4.0. Industrial development, on the other hand, makes it possible to optimize the use of human labour, thereby establishing better resource usability and sustainability. The right leadership behaviour can then be abstracted through the corresponding orientation as can be seen in Fig. 13.

Inspired by the representation and the model of Maslow's pyramid of needs, the different levels of the pyramid build on each other and should be fulfilled before the next level is reached. However, there is also an interaction in this pyramid. For example, in the fact that society shapes industry, or an innovative thought leader with great reach shapes an entire society. Examples of this can be found in the daily media channels, for example, Elon Musk, Donald Trump, and even Bill Gates or Jeff Bezos have the power to change society and thereby influence industrial development [90].

In the end, the only question that remains is whether the Industrial Revolution (technology), Society, and the Leaders of companies and nations have begun the path to the next industrial revolution. This movement has been accelerated starting from 2020 and it will be strongly visible in the next years.

## References


1. Acatech (2015) Umsetzungsstrategie Industrie 4.0 Ergebnisbericht der Plattform Industrie 4.0, BITKOM e.V., VDMA e.V. und ZVEI e.V
2. Accenture (2016) Machine dreams making the most of the connected industrial workforce. https://www.accenture.com/us-en/_acnmedia/PDF-13/Accenture-Connected-Industrial-Workforce-Research. Accessed 02 Mar 2021
3. Albulescu CT, Draghici A, Fistis GM, Trusculescu A (2016) Does ISO 9001 quality certification influence labor productivity in EU-27? Procedia Soc Behav Sci 221:278–286
4. Almada-Labo F (2017) Six benefits of industrie 4.0 for businesses, Retrieved from: https://www.controleng.com/articles/six-benefits-of-industrie-4-0-for-businesses/. Accessed 03 Apr 2021
5. Arians H (2017) The impact of technology on leadership, published on: 10.09.2017. https://peopledevelopmentmagazine.com/2017/09/10/technology-leadership/. Accessed 02 Apr 2019
6. Atwell C (2017) Yes, industry 5.0 is already on the horizon. https://www.machinedesign.com/industrial-automation/yes-industry-50-already-horizon. Accessed 08 Mar 2021
7. Aurik J, Anscombe J, Jonk G (2018) How technology can transform leadership—For the good of employees. https://www.weforum.org/agenda/2018/03/how-technology-can-transform-business-performance-for-human-good/. Accessed 08 Mar 2021
8. Balic T, Ebrahimi H (2017) Automation and digital transformation the ways that automation solutions can support digital transformation within ICT companies, Master of science thesis work in IT Management Report nr. 053, Swede
9. Basu K (2018) The journey from industry 4.0 to society 5.0 with IOT, big data and AI, Published on: 14.02.2018. https://www.linkedin.com/pulse/journey-from-industry-40-society-50-iot-ai-big-data-kallol-basu/. Accessed 13 July 2019
10. Bayern M (2018) 10 digital transformation predictions that will shape the future of IT, posted: 08.11.2018. https://www.techrepublic.com/article/10-digital-transformation-predictions-that-will-shape-the-future-of-it/. Accessed 28 Mar 2021
11. Bloomberg J (2018) Digitization, digitalization, and digital transformation: confuse them at your Peril. https://www.forbes.com/sites/jasonbloomberg/2018/04/29/digitization-digitalization-and-digital-transformation-confuse-them-at-your-peril/#6d4e70452f2c. Accessed 24 Mar 2021
12. Bork H (2019) Toyota feuert die Roboter, Published on: 04.01.2019. https://www.tagesspiegel.de/themen/reportage/kuenstliche-intelligenz-toyota-feuert-die-roboter/23821418.html. Accessed 20 Mar 2021
13. Boulton C (2018) What is RPA? A revolution in business process automation, published on: 03.09.2018. https://www.cio.com/article/3236451/business-process-management/what-is-rpa-robotic-process-automation-explained.html. Accessed 17 Feb 2021
14. Bpmonline (2018) 5 BPM trends essential for your digital transformation in 2018, https://www.bpmonline.com/5-BPM-trends-that-will-be-critical-for-your-digital-transformation. Accessed 20 Jan 2021
15. Brunet-Thornton R, Martinez F (eds) (2018) Analyzing the impacts of industry 4.0 in modern business environments. IGI Global
16. Buyya R, Calheiros RN, Dastjerdi AV (2016) Big data: principles and paradigms, 1st edn. Morgan Kaufmann, USA. ISBN 9780128053942
17. CA Technologies (2018) Automation the imperative for digital transformation. https://www.ca.com/content/dam/ca/us/files/ebook/automation-the-imperative-for-digital-transformation.pdf. Accessed 16 Feb 2021
18. Cisco (2017) Why automation for digital transformation?, Published on: 2017 https://www.cisco.com/c/dam/en/us/solutions/collateral/digital-transformation/why-cisco-for-automation.pdf. Accessed 16 Feb 2021
19. Cope A, Martin M, Peach J (2018) Leadership: the multiplier effect. John Murray Press Learning, Published in Great Britain, UK

20. Corp A (2018) 10 advantages of digitization and data capture you must know, Published on 14th Jul 2018. https://www.aptaracorp.com/blog/10-advantages-digitization-and-data-capture-you-must-know. Accessed 21 Mar 2021
21. Dasgupta N (2018) Practical big data analytics: hands-on techniques to implement enterprise analytics and machine learning using Hadoop. NoSQL and R, Packt Publishing, Birmingham-Mumbai, Spark
22. De Nul L, Breque M, Petridis A (2021) Industry 5.0—Towards a sustainable, human-centric and resilient European industry, Directorate-General for Research and Innovation (European Commission). https://doi.org/10.2777/308407. https://op.europa.eu/en/publication-detail/-/publication/468a892a-5097-11eb-b59f-01aa75ed71a1/. Accessed 03 Apr 2021
23. Debashis D (2016) Mobile cloud computing: architectures. CRC Press, Algorithms and Applications
24. Demir KA, Cicibas H (2017) Industry 5.0 and a critique of industry 4.0. In: Proceedings of the 4th international management information systems conference, Istanbul, Turkey
25. Dominguez A (2018) 8 advantages of digitalization of business, published on Jun 12, 2018. https://ehorus.com/digitalization-of-business/. Accessed 07 Mar 2021
26. Draghici A, Olariu C, Rozman T (2014) Certified business process manager: a transfer of innovation experience. Int J Innov Learning 16(1):12–33
27. Draghici A, Mocan M, Draghici G (2011) On-line training and certification solution for business process managers. In: International conference on enterprise information systems. Springer, Berlin, Heidelberg, pp 380–389
28. Draghici A, Baban CF, Ivascu LV, Sarca I (2015) Key success factors for university-industry collaboration in open innovation. In: 8th annual international conference of education, research and innovation, Spain
29. Dragoi G, Draghici A, Rosu SM, Radovici A, Cotet CE (2010) Professional risk assessment using virtual enterprise network support for knowledge bases development. In: International conference on enterprise information systems. Springer, Berlin, Heidelberg, pp 168–177
30. EESC (2018a) Industry 5.0, Published on: 22.11.2018. https://www.eesc.europa.eu/en/agenda/our-events/events/industry-50/. Accessed 22 Feb 2021
31. EESC (2018b). Artificial intelligence and robotics: inevitable and full of opportunities, published on: 16.03.2018. https://www.eesc.europa.eu/en/news-media/news/artificial-intelligence-and-robotics-ineviTable-and-full-opportunities/. Accessed 22 Feb 2021
32. El Khatib MM, Al-Nakeeb A, Ahmed G (2019) Integration of cloud computing with artificial intelligence and its impact on telecom sector—A case study. iBusiness 11:1–10
33. EventTracker (2018) Cybersecurity is an investment, not a cost center, publishes on: 14.05.2018. https://www.eventtracker.com/blog/2018/may/cybersecurity-is-an-investment-not-a-cost-center/. Accessed 17 Feb 2021
34. Fazzi C (2018) 3 things you need to know about industry 5.0, published on: 28.12.2018 https://www.mastercontrol.com/gxp-lifeline/3-things-you-need-to-know-about-industry-5.0/. Accessed 13 Mar 2021
35. Ferreira CM, Serpa S (2018) Society 5.0 and social development. Manag Organ Stud 5:26–31. https://doi.org/10.20944/PREPRINTS201811.0108.V1
36. Finlay S (2017) Artificial intelligence and machine learning for business, 2nd edn. Relativistic Books, Lancashire
37. Gartner (2017a) IT glossary. knowledge management (KM). http://www.gartner.com/it-glossary/km-knowledge-management/. Accessed 14 Aug 2019
38. Gartner (2017b) IT glossary; digitalization. https://www.gartner.com/it-glossary/digitalization/. Accessed 14 Mar 2021
39. Geng H (2016) Internet of things and data analytics handbook. Wiley, Palo Alto, USA
40. Gerlitz L (2016) Design management as a domain of smart and sustainable enterprise: business modelling for innovation and smart growth in industry 4.0. Entrep Sustain Issues. 3(3):244–268
41. Glichrist A (2016) Industry 4.0: the industrial internet of things. Springer, New York

42. Guttman U, Papbst J, Merlo R, Kane D, Bieser G, Grob O (2017) Industry 4.0: what's next an SAP Point of view, SAP
43. Harayama Y (2017) Society 5.0: aiming for a new human-centred society Japan's science and technology policies for addressing global social challenges. Hitachi Rev 66(6):558–559
44. Hasib M (2014) Cybersecurity leadership: powering the modern organization. published in the United States of America by tomorrow's Strategy Today
45. Hermann M, Pentek T, Otto B (2015) Design principles for industrie 4.0 scenarios: a literature review. Technische Universität Dortmund Fakultät Maschinenbau & Audi Stiftungslehrstuhl Supply Net Order Management
46. Hong'e M (2019) What is 'Society 5.0' at the G20 summit?, published on: 29.06.2019 http://www.ecns.cn/news/society/2019-06-29/detail-ifzkrnzp2015002.shtml/. Accessed 14 Feb 2021
47. IBM (2018) Cloud computing: a complete guide. https://www.ibm.com/cloud/learn/cloud-computing. Accessed 14 Feb 2021
48. IDC (2019) Worldwide digital transformation 2019 predictions. https://www.idcdxawards.com/worldwide-digital-transformation-2019-predictions/. Accessed 03 Mar 2021
49. IDC (2020) IDC FutureScape: worldwide digital transformation 2021 predictions. https://www.idc.com/getdoc.jsp?containerId=US46880818/. Accessed 01 Apr 2021
50. Ibarra D, Ganzarain J, Igartua JI (2017) Business model innovation through industry 4.0: a review. In: 11th international conference interdisciplinarity in engineering, INTER-ENG 2017
51. Irniger A (2017) Difference between digitization, digitalization and digital transformation. http://di.univ-blida.dz:8080/jspui/bitstream/123456789/799/1/008-2019.pdf. Accessed 03 Mar 2021
52. i-sscoop (2019a). https://www.i-scoop.eu/industry-4-0/#Benefits_of_Industry_40. Accessed 28 Feb 2021
53. i-scoop (2019b). https://www.i-scoop.eu/cloud-computing/digital-transformation-present-future-of-the-cloud-in-digital-business/. Accessed 27 Feb 2021
54. Jao N (2017) Japan's 'Society 5.0' will integrate cutting-edge tech at all levels, published on: 29.06.2017. https://news.itu.int/japans-society-5-0-will-integrate-cutting-edge-tech-at-all-levels/. Accessed 27 Feb 2021
55. Johnson A (2019) Announcing new cloud-based technology to empower cyber defenders, published on: 28.02.2019. https://blogs.microsoft.com/blog/2019/02/28/announcing-new-cloud-based-technology-to-empower-cyber-defenders/. Accessed 02 Mar 2019
56. Kagermann H, Wahlster W, Helbing J (2013) Securing the future of German manufacturing industry recommendations for implementing the strategic initiative INDUSTRIE 4.0 final report of the industrie 4.0 working group, acatech—National Academy of Science and Engineering
57. Kagermann H (2015) Change through digitization-value creation in the age of the industry 4.0. management of permanent change, Wiesbaden, Springer; Katte A (2020) artificial intelligence and schumpeter's creative destruction. https://analyticsindiamag.com/artificial-intelligence-and-schumpeters-creative-destruction/. Accessed 04 Apr 2021
58. Knight W (2019) 10 breakthrough technologies 2019 how we'll invent the future, by Bill Gates, published on: 27.02.2019. https://www.technologyreview.com/lists/technologies/2019/. Accessed 02 Mar 2021
59. Kospanos V (2017) Industry 5.0—When man meets machine, published on: 11.01.2017 http://www.pnmsoft.com/industry-5-0-man-meets-machine/. Accessed 14 Mar 2021
60. Kroker M (2019) Ohne digitales Know-how gibt's keine Digitalisierung!, published on: 08.02.2019. https://www.xing.com/news/insiders/articles/ohne-digitales-know-how-gibt-s-keine-digitalisierung-2046276. Accessed 14 Mar 2021
61. Kumar A (2008) From mass customization to mass personalization: a strategic transformation, Published online: 24 April 2008 Springer Science+Business Media, LLC 200 A. Kumar (&) Seidman School of Business, Grand Valley State University

62. Kumar Illa P (2018) Integrations: key technology enablers in digital transformation, published on: 01.10.2018. https://www.dataversity.net/integrations-key-technology-enablers-digital-transformation. Accessed 14 Mar 2021
63. Köhler-Schute C (2016) Digitalisierung und transformation in unternehmen: strategien und konzepte, methoden und technologien. Praxisbeispiele, KS-Energy-Verlag Berlin, Germany. ISBN 9783945622025
64. Lämmel U, Cleve J (2012) Künstliche Intelligenz. Carl Hanser Verlag GmbH & Co. KG München, Germany. ISBN 9783446427587
65. Lazari C (2017) What is cloud computing?, published on:15.08.2017. https://chrislazari.com/what-is-cloud-computing/. Accessed 14 Feb 2021
66. Lee D (2013) Cloud computing. https://slideplayer.com/slide/7992289/. Accessed 10 Mar 2021
67. Lee R (2019) Big data, cloud computing, data science and engineering, Springer, Switzerland
68. Lewis A (2017) Guide to industry 4.0 & 5.0. https://blog.gesrepair.com/category/manufacturing-blog/. Accessed 11 Mar 2021
69. Malito A (2017) 800 million people might be out of a job by 2030 because of automation, published on: 29.11.2017. https://www.marketwatch.com/story/800-million-people-might-be-out-of-a-job-by-2030-because-of-automation-2017-11-29. Accessed 24 Jan 2021
70. Markoff R, Seifert R (2019) The real industry 4.0 challenge, April 9, 2018, technology, operation, business process, industry 4.0, internet of things, supply chain. http://www.europeanbusinessreview.com/the-real-industry-4-0-challenge/. Accessed 14 Jan 2021
71. Marr B (2015) Using SMART big data, analytics and metrics to make better decisions and improve performance. Wiley, UK
72. Marr B (2017) Data strategy: how to profit from a world of big data. In: Analytics and the internet of things. USA, Kogan Page Ltd.
73. Marr B (2018) The 4th industrial revolution is here—Are you ready? https://www.forbes.com/sites/bernardmarr/2018/08/13/the-4th-industrial-revolution-is-here-are-you-ready/#2b6b3e2628b2. Accessed 16 Mar 2021
74. McKinsey and Company (2016) Industry 4.0 after the initial hype where manufacturers are finding value and how they can best capture it, McKinsey Digital, McKinsey Global Institute
75. McKinsey and Company (2017) A future that works: automation, employment, and productivity, McKinsey Global Institute
76. McKinsey and Company (2018) Digital and risk a new posture for cyber risk in a networked world, published on: 03.2018. https://www.mckinsey.com/de/~/media/mckinsey/locations/europe%20and%20middle%20east/deutschland/publikationen/2018%20compendium/a%20new%20posture%20for%20cybersecurity%20in%20a%20networked%20world/kompendium_03_cyberrisk-2.ashx. Accessed 17 Feb 2021
77. Mohelska H, Sokolova M (2018) Management approaches for industry 4.0—The organizational culture perspective. VGTU Press
78. Müller JM, Däschle S (2018) Business model innovation of industry 4.0 solution providers towards customer process innovation. Switzerland, MDPI. https://www.i-scoop.eu/industry-4-0/#:~:text=Industry%204.0%20is%20the%20current,called%20a%20%E2%80%9Csmart. Accessed 10 Mar 2021
79. Newman D (2018) Four digital transformation trends driving industry 4.0. https://www.forbes.com/sites/danielnewman/2018/06/12/four-digital-transformation-trends-driving-industry-4-0/#156d285e604a. Accessed 03 Apr 2021
80. Nguyen-Duy J (2018) How digital transformation has impacted security and how to minimize risk, published on: 26.07.2018. https://www.csoonline.com/article/3292927/security/how-digital-transformation-has-impacted-security-and-how-to-minimize-risk.html. Accessed 12 Mar 2021
81. Ojeda M (2017) Process automation: advantages and disadvantages for IT companies, published on: 28.11.2017. https://www.gb-advisors.com/process-automation/. Accessed 08 Mar 2021

82. Østergaard EH (2018) Welcome to industry 5.0. https://info.universal-robots.com/welcome-to-industry-5.0. Accessed 16 Mar 2021
83. ÖOzdemir V, Hekim N (2018) Birth of industry 5.0: making sense of big data with artificial intelligence, "the internet of things" and next-generation technology policy. Omics 22:65–76
84. Ozkeser B (2018) Lean innovation approach in industry 5.0. In: The Eurasia proceedings of science, engineering and mathematics (EPSTEM), vol 2, pp 422–428
85. Pandey US, Choudhary K (2014) Cloud computing. S. Chand & Company PVT.LTD, India
86. Paschek D, Ivascu L, Draghici A (2017) Knowledge management—The foundation for a successful business process management. In: Proceedings of 14th international symposium in management 2017
87. Paschek D, Luminosu CT, Draghici A (2017) Automated business process management in times of digital transformation using machine learning or artificial intelligence. In: Proceedings of 8th international conference on manufacturing science and education—MSE 2017 "Trends in New Industrial Revolution", Sibiu, Romania, In: MATEC web of conferences, vol 121, p 04007. EDP Sciences. https://www.matec-conferences.org/articles/matecconf/pdf/2017/35/matecconf_mse2017_04007.pdf. Accessed 26 Feb 2021
88. Paschek D, Mocan A, Dufour CM, Draghici A (2017) Organizational knowledge management with big data. The foundation of using artificial intelligence. In: Balkan region conference on engineering and business education, vol 2, no 1, pp 301–308. Sciendo
89. Paschek D, Mocan A, Draghici A (2019) Industry 5.0—The expected impact of next industrial revolution, MakeLearn 2019: thriving on future education, industry, business and society. In: Proceedings of the make learn and TIIM international conference, Piran, Slovenia. ToKnowPress, pp 125–132. https://pdfs.semanticscholar.org/3007/06e2ee34fe3b6348a05a8d4608a7a3a59044.pdf. Accessed 19 Feb 2021
90. Paschek D (2020) Artificial intelligence business process management—an important requirement and success factor for automation of business processes for industry 5.0, Ph.D. thesis, Politehnica University of Timisoara, Romania
91. Prause G, Atari S (2017) On sustainable production networks for industry 4.0. Entrep Sustain Issues 4(4):421–431. https://doi.org/10.9770/jesi.2017.4.4(2)
92. Probst L, Lefebvre V, Martinez-Diaz C, Bohn NU, PwC, Klitou D, Conrads J (2018) CARSA. Digital transformation scoreboard 2018. EU businesses go digital: opportunities, outcomes and uptake. Publications office of the European Union. Luxembourg, p 7. https://ec.europa.eu/growth/tools-databases/dem/monitor/sites/default/files/Digital%20Transformation%20Scoreboard%202018_0.pdf/. Accesseded 28 Jan 2021
93. Rada M (2018) Industry 5.0 definition, Published on: 21.01.2018. https://medium.com/@michael.rada/industry-5-0-definition-6a2f9922dc48. Accessed 02 Feb 2021
94. Rasche C (2014) Industry 4.0 a discussion of qualifications and skills in the factory of the future: a German and American perspective. VDI the association of German engineers. Rendall
95. Rendall M (2017) The new terminology: CRO and industry 5.0. published on: 15.06.2017, Retrieved from: https://www.automation.com/automation-news/article/the-new-terminology-cro-and-industry-50. Accessed 21 Mar 2020
96. Rittinghouse JW, Ransome JF (2017) Cloud computing: implementation, management, and security. CRC Press, US
97. Robinson S (2018) 5 cloud predictions for 2019, Published on: 31.12.2018. https://www.ibm.com/blogs/cloud-computing/2018/12/31/2019-cloud-trends-predictions. Accessed 15 Feb 2021
98. Rose J, Lukic V, Milon T, Cappuzzo A (2016) Sprinting to value in industry 4.0. https://www.bcg.com/publications/2016/lean-manufacturing-technology-digital-sprinting-to-value-industry-40.aspx. Accessed 22 Feb 2021
99. Rossi B (2018) Manufacturing gets personal in Industry 5.0, published on: 07.03.2018 https://www.raconteur.net/technology/manufacturing-gets-personal-industry-5-0. Accessed 23 Feb 2021

100. Rouse M (2018) Using big data platforms for data management, access and analytics, published on: September 2018. https://searchbusinessanalytics.techtarget.com/definition/big-data-analytics. Accessed 19 Feb 2021
101. Rouse M (2017) SMAC (social, mobile, analytics and cloud), published on: December 2017. https://searchcio.techtarget.com/definition/SMAC-social-mobile-analytics-and-cloud. Accessed 23 Feb 2021
102. Rundle E (2017) The 5th industrial revolution, when it will happen and how, Published on: 27.12.2017. https://devops.com/5th-industrial-revolution-will-happen/. Accessed 13 Feb 2019
103. Rupra SS, Omamo A (2020) A cloud computing security assessment framework for small and medium enterprises. J Inf Secur 11(4):201–224
104. Rutaganda L, Bergstrom R, Jayashekhar A, Jayasinghe D, Ahmed J (2017) Avoiding pitfalls and unlocking real business value with RPA. In: Automation henley business school—Capco institute paper series in financial services, The capital markets Company BVBA
105. Sachsenmeier P (2016) Industry 5.0—The relevance and implications of bionics and synthetic biology. Elsevier Engineering, Engineering 2:225–229
106. Salgues B (2018) Society 5.0: industry of the future, technologies, methods and tools. Wiley, Hoboken, NJ, USA
107. Satchell P (2018) Innovation and automation. Routledge, USA
108. Scanlon S (2018) Now prepare for the 5th industrial revolution, published on: 04.06.2018. https://gadget.co.za/now-prepare-for-the-5th-industrial-revolution/. Accessed 10 Mar 2021
109. Schröder C (2017) The challenges of industry 4.0 for small and medium-sized enterprises. Bonn, Friedrich-Ebert-Stiftung
110. Schwab K (2017) The fourth industrial revolution. World Economic Forum, Switzerland
111. Sedkaoui S (2018) Data analytics and big data, Willey, United Kingdom
112. Shelzer R (2017) What is industry 5.0—And how will it affect manufacturers?. https://blog.gesrepair.com/industry-5-0-will-affect-manufacturers/. Accessed 15 Feb 2021
113. Skobelev PO, Borovik SY (2017) On the way from industry 4.0 to industry 5.0: from digital manufacturing to digital society. Industry 4.0 2(6):307–311
114. Srocke D, Karlstetter F (2016) Cisco global cloud index 2015–2020 Public Cloud auf Vormarsch. Published on 22.11.2016. https://cloudcomputinginsider.com. Accessed 25 Mar 2021
115. Stancombe C (2015) Tempted to rewrite bill gates' rules on automation?, Published on: 02.01.2015. https://www.capgemini.com/2015/01/tempted-to-rewrite-bill-gates-rules-on-automation/. Accessed 10 Mar 2021
116. Stergiou C, Psannis KE, Kim B-G, Gupta B (2016) Secure integration of IoT and cloud computing, vol 78. Elsevier, Part 3, pp 964–975. https://ruomo.lib.uom.gr/bitstream/7000/228/1/10.1016%40j.suscom.2018.06.003.pdf. Accessed 12 Feb 2021
117. Stock T, Seliger G (2016) Opportunities of sustainable manufacturing in industry 4.0. Procedia CIRP 40:536–541
118. Swadlo A (2018) Creating a practical pathway to industry 4.0, München, Siemens AG, Siemens Financial Services
119. Techopedia (2018) Smart device. https://www.techopedia.com/definition/31463/smart-device. Accessed 13 Jan 2021
120. Thanh Dat N (2017) Six sigma-based knowledge management and its application in IT systems management. Ph.D. thesis. University "Lucian Blaga" of Sibiu, Romania. http://doctorate.ulbsibiu.ro/wp-content/uploads/2017FinalThesisSummary-ThanhDatNguyen-v2017.07.11.pdf. Accessed 13 Mar 2021
121. Thippanna A (2018) AI and automation—A combo to manage cyber security threats, Published on: 18.01.2018. https://blogs.informatica.com/2018/01/18/ai-and-automation-a-combo-to-manage-cyber-security-threats/#fbid=XFVC03RGxdq. Accessed 13 Mar 2021
122. Thoben KD, Wiesner S, Wuest T (2017) Industrie 4.0 and smart manufacturing—A review of research issues and application examples. Int J Autom Technol 11(1):4–16

123. Tripathi AM (2018) Learning robotic process automation: create software robots and automate business processes with the leading RPA tool—UiPath, Packt, UK
124. UNESCO (United Nations Educational, Scientific and Cultural Organization) (2019) Japan pushing ahead with Society 5.0 to overcome chronic social challenges; published on: 21.02.2019. https://en.unesco.org/news/japan-pushing-ahead-society-50-overcome-chronic-social-challenges. Accessed 13 Mar 2021
125. Urbach N (2018) Digitalization cases, how organizations rethink their business for the digital age, Cham, Switzerland Springer Nature Switzerland AG
126. Ustundag A, Cevican E (2018) Industry 4.0: managing the digital transformation, Cham, Switzerland Springer Nature Switzerland AG
127. Vaidya S, Ambad P, Bhosle S (2018) Industry 4.0—A glimpse. Procedia Manuf 20:233–238
128. Valdeza AC, Braunera P, Schaar AK, Holzinger A, Ziefle M (2015) Reducing complexity with simplicity—Usability methods for industry 4.0. In: Proceeding of the 19th triennial congress of the IEA, Melbourne
129. WEF and Accenture (2018) Digital transformation initiative maximizing the return on digital investments. World Economic Forum, Switzerland. http://www3.weforum.org/docs/DTI_Maximizing_Return_Digital_WP.pdf. Accessed 08 Mar 2021
130. Wang K, Wang Y, Strandhagen JO, Yu T (eds) (2018) Advanced manufacturing and automation VII, vol 451. Springer
131. Wazed S (2019) The future of recruiting in the age of automation and artificial intelligence, published on: 04.01.2019. https://www.forbes.com/sites/forbeshumanresourcescouncil/2019/01/04/the-future-of-recruiting-in-the-age-of-automation-and-artificial-intelligence/#6f1d8ad51830. Accessed 22 Mar 2021
132. Weidmann M (2018) The gold nuggets in the supply chain: a tracker that makes shipping pallets smart, posted on: 15.10.2018. https://www.telekom.com/en/media/media-information/archive/telekoms-tracker-makes-shipping-pallets-smart-547120. Accessed 18 Mar 2021
133. Wellers D, Elliot T, Noga M (2017) 8 ways machine learning is improving companies work process. https://hbr.org/2017/05/8-ways-machine-learning-is-improving-companies-work-processes. Accessed 23 Feb 2021
134. Zambon I, Cecchini M, Egidi G, Saporito MG, Colantoni A (2019) Revolution 4.0: industry versus. Agriculture in a Future Development for SMEs. Processes, 7, 36

# Chapter 3
# Smart Manufacturing Systems Management

**George Carutasu and Nicoleta Luminita Carutasu**

**Abstract** The influence of IT in the manufacturing systems became not only visible but also a necessity of operational management. The manufacturing systems evolution is presented using a comparison between the industrial revolutions and Maslow' hierarchy of needs, up-to-date. However, the presented evolution used life quality indicators, to explain the need for continuous adaptation of the manufacturing systems to people, the end customer. The virtualization of the enterprise IT system process and the reference cloud architecture is also presented, in order to clearly estate the prerequisites for Industry 4.0, with a detailed map of the internal cloud processes. The study also presented the major software providers prediction, before and during the COVID pandemic time, which greatly impacted consumer behaviour. Furthermore, the changes of the customer behaviour, cumulated with the globalization limitation, and workforce polarization, affected businesses in terms of rethinking the supply chain, with the target on flexibility, prediction and resilience and the attitude regarding the employees. The conclusions section summarizes the key findings of the study and exposes the authors' prediction for the short-time evolution of the manufacturing systems.

**Keywords** Manufacturing systems · Industrial revolutions · Organizational management · Cloud computing · Industry 4.0 · Industry 5.0 · Smart manufacturing

---

G. Carutasu (✉)
Romanian-American University, Faculty of Computer Science for Business Management, Bucharest, Romania
e-mail: carutasu.george@profesor.rau.ro

N. L. Carutasu
University Politehnica of Bucharest, Faculty of Industrial Engineering and Robotics, Bucharest, Romania
e-mail: nicoleta.carutasu@upb.ro

 
A. Draghici and L. Ivascu (eds.), *Sustainability and Innovation in Manufacturing Enterprises*, Advances in Sustainability Science and Technology,
https://doi.org/10.1007/978-981-16-7365-8_3

## 1 Manufacturing Systems Evolution and Previous Forms of Organization

The evolution of the manufacturing systems and industry, in general, came to a closer comparison with Maslow' human hierarchy of needs [39]. The following figure reflects the reason for comparison, enhancing each need covered by the industrial revolutions.

However, having as a starting point, the First Industrial Revolution, which came in the late eighteenth century, and an ending with a projected state with no human engagement in the production of goods or services, it must be further exploited, using several indicators, like life expectation, income, and gross domestic product (GDP) per capita, whether those evolutions are reflected into people's everyday life (Fig. 1).

Using the data available for the range of late 1700 to today from various sources [48, 54, 66], the industrial revolution impact on the major life quality indicators is presented in Table 1.

With a 40.75 years world average life expectancy, for the period 1778–1889, covering the First Industrial Revolution, the beginning of the era, and from the available data in the UK and Sweden, a starting life expectancy is 35.5 years in Sweden and 37.7 years in the UK. The same countries finished the First Industrial Revolution

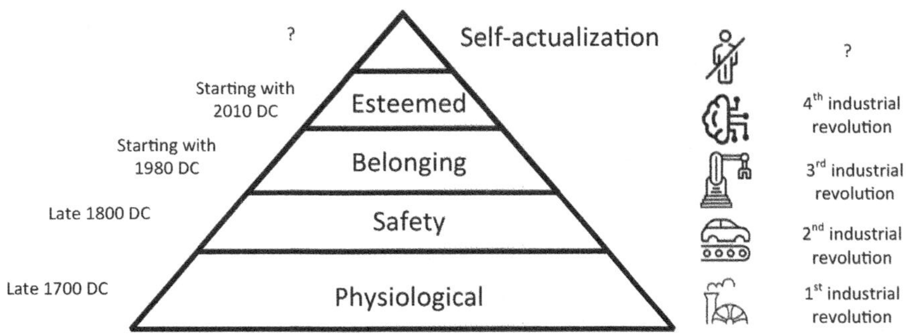


**Fig. 1** Close comparison between human need and industrial revolutions

**Table 1** Life expectancy versus industrial revolutions

| Industrial revolution/time span | Average world life expectancy (year) |
|---|---|
| First Industrial Revolution (1784–1869) | 40.75 |
| Second Industrial Revolution (1870–1969) | 53.07 |
| Third Industrial Revolution (1970–2010) | 63.42 |
| Fourth Industrial Revolution (2011–present) | 71.45 |

with 45.9 years, in the case of the UK and 52.26 years for Sweden. The impact of the First Industrial Revolution on the health, wealth and population was closely analysed by Buer in [5], appreciating the rising of the population in the London city, based on multiple historical sources for England and Wales, currently in interest because of the large-scale industrial development in this geographic area. The birth rate was estimated roughly to 34 births/1000 inhabitants, almost steady in the mentioned period. However, the population growth was given by the infancy death rate reduction, from 85% in the case of age 5 (at the beginning of the period) to 35% in London city (at the end of the period). The reduction of the infancy date rate was influenced by the better access of the population to healthcare facilities, vaccination and improving hygiene conditions in town, including sewers. The natural growth rate in towns (without immigration) came positive after the beginning of the nineteenth century when the number of births (evaluated by baptism records) overcome the number of deaths. The general population almost triple it in the mentioned period. The estimated world population rose from an average estimated 750 billion people (in 1750) to 1,200 billion people (in 1850). The time span was altered to fit the existing estimation.

The First Industrial Revolution attracted the population from villages to towns, assuring the basic goods and healthcare services, possible by concentrating the population in towns. According to the source [53], the GDP per capita rose from 2,092.98 £ (for 1784) to 3,683.75 (for 1869). The values were adjusted with the inflation rate and prices for 2013. The new social class, the workers, replaced the old form of slavery in most countries. The wage is the payment for the disposed work. The work schedule and work conditions are overstressed, limiting the life expectancy of the workers. As a general conclusion, the First Industrial Revolution, in terms of management structure, established the company as the form of organization, introduced a new social class as workers, and introduces the wage as the form of payment. The need that has been accomplished by the First Industrial Revolution using the Maslow hierarchy can be appreciated as physiological, with access to basic goods and services, necessary for living.

The Second Industrial Revolution came with more disruptive technologies and management structures, including the assembly line, labour division and business functions. Mass production lowered the price of most commodities and gave access to the workers. The sanitation and healthcare services were largely improved, including the life quality in cities. The large-scale sales of automobiles and trains as mobility support revolutionized the living, by adding mobility to workers, and the positioning of factories. For the sake of comparison, the same sources of data are used. The average world life expectancy, started from 41.22 years in 1869, calculated for eight European countries, for the existing recorded data, to 55.63 years in 1969 (with uneven distribution around the continents, Asia 53.93, Americas 64.63, Europe 70.15 and Africa 45.15). The United Kingdom started in 1870 with a life expectancy of 40.65 years and ended in 1969 with an astonishing 71.84 years. In the same manner, for the US, the first record appreciated the life expectancy in 1880 from 39.41 to 70.58 years in 1969. In terms of population growth during the Second Industrial Revolution, the average estimated number of people was 1,650 billion in 1900 and more precise 3,625.68 billion, doubling the population in almost 70 years. Regarding the GDP per capita, the evolution indicated a rise from 3,927 £ for the UK in 1870 to 12,224.58 £ [53] actualized with the inflation rate and price for 2013. The same ratio (over 400%) is registered also for the US [53, 67] in 1870; the GDP per capita was estimated to be 4,803 USD and 24,165 USD in 1969 (data adjusted for 2013). The agricultural production was for the first time overrun by industrial production in the US in 1900, according to [28]. The mass production of goods and services rose the issue of mineral resources shortage, such as iron, oil, coal etc. The shortage of resources caused two World Wars, with a great impact on the population and GDP growth. Also, the mass production generated several economic crises when the offer overwhelmed the request. A new social class appeared, the middle class, with sufficient purchasing power to access the most category of goods.

Starting from the first scientific production management architecture, introduced by Frederick Winslow Taylor by "The principles of the scientific management" and simultaneously were applied into the automotive factory by Henry Ford, to the smart manufacturing gathers over a century of the struggle of the companies to best fit to market demand. The second step to formalize the enterprise was done by Henri Fayol in [27].

The business function division proposed by Fayol covered the technical function, with reference to the technical capacity to produce a product, the commercial function, not only stressing the importance of supplying but also disposing of the products made, the financial function, with the role of secure the capital flow necessary for the operationalization of the other functions, the security function, covering ensuring the integrity of the assets held and ensuring the smooth management of the activity, under conflicting conditions, with the role of prevention, the accounting function as a tool for assessing the operationalization conditions and providing accurate and immediate information that can be used to assist the decision, and administrative function, taking into account decision-making, at an operational level and the implementation of management principles.

The hierarchical approach of the manufacturing systems continued in the Second Industrial Revolution, with new influences for the time and movement study (Frank and Lillian Gilbreth), to increase the productivity and limit the fatigue of workers, quality assessment and cost control (Edward Deming, Joseph Juran). The cooperation of the workers in a company was early studied by Cherster Barnard and Mary Parker Follet and deeply explained by Max Weber.

The second way of expressing the functions of the enterprise, in this case in the number of four, is presented by Boris Evgrafoff in 1970, in the paper [20], which reorganizes the internal business processes in the form of [1, 15, 49, 50]:

- Management function, by collecting information assisting decision-making, decision-making, information and awareness of the members of the organization regarding the decision taken
- Distribution function, with reference to the conduct of market research, sales, sales management and after-sales services
- Production function, aiming at production preparation, production management, quality control etc.

- Logistical function, which ensures financial, human and material resources for the production process, including research and development.

The Second Industrial Revolution, as conclusion might be seen as corresponding to accomplish the safety level of the Maslow hierarchy of needs, where large access to commodities, goods and services rose the life expectancy, is also positively influenced by large access to healthcare services and large use of antibiotics. The Second Industrial Revolution add the middle class as a new social class, being the backbone of a developed society. The industrial production exceeded agricultural production for the first time in history, and the offer overwhelmed the request, generating an economic crisis. However, in this period the manufacturing process is largely dependent on the human factor, the working condition and crises, leading to unions organization and strikes as a form of reposting [49, 50].

The Third Industrial Revolution is assessed by using large-scale automation and later by computers in the manufacturing process. The enhancing of the coordination of the manufacturing facilities make possible the birth of a new paradigm of manufacturing, largely based on collaboration and independence.

In the paper [51], M. Porter proposes the concept of the value chain, organizing internal processes according to the flow of transformation of a system's inputs into outputs, thus proposing the following functions, which will be used, by addition, in the current statistical system:

- Core functions directly refer to the physical creation, sale, maintenance and support of a product or service. They consist of the following:
  - Inbound logistics, representing the processes of reception, handling and storage, with a view to being transformed into operationalization into finished products
  - Operations, including processes for the transformation of raw materials and materials into finished products that are sold to customers: production planning, processing, assembly, technical quality control etc.
  - Outbound logistics, all processes of collection, handling, storage and distribution of the product or service produced to the customer
  - Marketing and sales, customer awareness processes of the value, products or services offered, bidding, sales and sales management
  - Service, the totality of product value maintenance processes throughout the lifecycle
- Support functions, which support the main functions, in the process of transforming system inputs into outputs:
  - Procurement, the process of obtaining the necessary resources for operation, the selection of suppliers and the purchase of raw materials, equipment, materials necessary for the transformation of system inputs into outputs
  - Technological development, development processes and knowledge base protection, including research and development and improvement of existing products and processes

- Human resources management includes recruitment, hiring, training and employee motivation processes
- Organization infrastructure by providing support systems for day-to-day operations: management, accounting, finance, strategic planning.

The paper [45] presents a comparative study of the models of the company's functions, used for statistical purposes, from 2007 to 2012, by the 2007 International Survey (Eurostat), 2012 International Sourcing/Global Value Chains Survey (Eurostat), 2010 National Organization Survey (USA: [4]), 2009/2012 Survey of Innovation and Business Strategy (Statistics Canada), which distinguishes the following observations, on their identification and grouping [16, 24]:

- Only one main function (core business function) is retained, which the authors, as well as 2009/2012 Survey of Innovation and Business Strategy (Statistics Canada), propose to be divided into the production of goods and the production of services
- Support functions vary between the models presented, keeping a common trunk represented by the distribution and logistics function, the Information Technology and Communication) service function, the management and administration function
- Differences are given by highlighting marketing, sales and post-sales services, including helpdesk and call centre services, which is divided into the 2010 National Organization Survey (USA: Brown and Sturgeon), 2009/2012 Survey of Innovation and Business Strategy (Statistics Canada), and the research development function, in the 2012 International Sourcing/Global Value Chains Survey (Eurostat) and the 2010 National Organization Survey (USA: Brown and Sturgeon) are combined with engineering and related technical services.

The model used by EUROSTAT [19] for reporting maintains the R&D function, as a distinct support function from engineering and technical services, in a model with a single main function and six support functions (distribution and logistics, marketing, sales and after-sales services, information and communication technology services, administration and management, engineering and associated technical services and research and development).

The collaboration, as a mark of "Belonging" need of the Maslow hierarchy of needs, came into the manufacturing paradigms, with a series of approaches cumulatively presented in Table 2.

The challenges regarding the new possibility of coordinating multiple manufacturing facilities, using the widespread of computer networks and computer software in all aspects of operational management and production management, correlated with the globalization of enterprises, were addressed using biological models (bionic manufacturing systems) or abstract, using mathematical models (fractal company). The main concept was to enhance the business units with a similar function that may play on the local market. Also, the business concept of a single enterprise was replaced with temporary alliances (virtual enterprise). The switch from mass production to a customer-centred approach, because of overproduction, was generating the development of new needs and sophistication of the existing ones.

**Table 2** Collaboration architectures of the manufacturing systems

| Manufacturing architecture | References |
|---|---|
| Bionic manufacturing systems | Tharumarajah et al. [62], Ueda [64] |
| Flexible manufacturing systems | Berman and Maimon [3], Kassicieh and Schultz [33], Slomp and Zijm [59] |
| Autonomous and distributed manufacturing systems | Duffie and Prabhu [18], Kádár et al. [35], Ryu and Jung [55], Srai et al. [59] |
| Agile manufacturing | Shewchuk [57] |
| Multi-agent autonomous systems | Franklin and Graesser [22] |
| Holonic enterprise | Mella [41], Tharumarajah et al. [62], Tharumarajah et al. [63], Ulieru et al. [65] |
| Fractal company | Deng et al. [14], Warnecke [70], Warnecke and Warnecke [71] |
| Convergent enterprise | Nau et al. [44], Shah and Rogers [56] |
| Virtual enterprise | Camarinha-Matos and Af- sarmanesh [8], Carutasu and Aurite [9], Guran et al. [17], Dragoi [26] |
| Next-generation manufacturing systems | Bunce et al. [6], Kurihara et al. [34], Okabe et al. [46] |

In conclusion, the Third Industrial Revolution brought as a novelty to the customer-centric approach, the sliding from the hierarchical manner of an organization to multi-polar and coordination organization, imposes the automation of processes, using IT deeper in the operational management and processes management. The average world life expectancy rose by almost 10 years in only 40 years' time. The GDP per capita continues to rise and the work condition largely improved, the human resource becoming the most valuable resource of a company. However, the difference between countries regarding technological advancement and access to innovation has been deepened. Thus, it explains the polarization of wealth among the world, with a high differentiation from primary resources (food, water, shelter) to more elevated ones (arts, education, medical care). The Third Industrial Revolution correspond to "Belonging" level of Maslow hierarchy, with a great accent to "collaboration" between business units, replacing the hierarchical way of thinking.

The claimed Fourth Industrial Revolution, purely announced by the German consortium as a desiderate, without having proper implementation or proofs [32] gathered various emergent technologies and foreseen an integration in the manufacturing process.

The model of the new was enthusiastically adopted by a large community, being a great marketing driver for the software industry. Nevertheless, the concept itself has a great value, by introducing the machine-to-machine collaboration, as a strong possibility, without human intervention, human-to-machine collaboration, as pairs or avatars, and the old human-to-human collaboration, but renewed with new augmented tools. The real impact, after using the sensing ability to decision-making artificial

**Table 3** Referential for new Fourth Industrial Revolution terms

| Terms/concepts | References |
|---|---|
| Industry 4.0 | Cotet et al. [12], Gorecky et al. [25], Marr [38], Melanson [40], Shrouf et al. [58] |
| Industry 5.0 | Cotta et al. [13], Özdemir and Hekim [47], Paschek et al. [50], Sołtysik-Piorunkiewicz and Zdonek [60], Vogt [69], Zengin et al. [73] |
| Smart manufacturing | MacDougall [37], Qi and Tao [52], Vater et al. [68] |

entity, is the passing the decision-making process from human to robots (in the form of physical or software form of artificial intelligence). The next step was to prevent the human totally replacement by robots in the enterprise architecture, by the Industry 5.0 concept (released by EU Commission [13]), with a human-centric view, with a sustainable and resilient approach to manufacturing. The term smart manufacturing was a generic term where "smart" symbolizes the use of sensors and artificial intelligence, which was also debated in the period. A selective referential for the presented terms is enhanced in Table 3.

The main barriers to the adoption of Industry 4.0 are presented in the paper [36]:

- Interaction between operators and equipment by limiting the ability to make decisions according to the state of operation of the equipment in real time
- Machine centres, by grouping identical machines, forecasting methods that do not consider the particularities of operation and the operating environment
- Quality of products and processes, by the absence of a control reaction in the manufacturing process
- State and speed of operation of cloud technologies and the analysis of large volumes of data, and these technologies being at the beginning of the lifecycle.
- Network of sensors and controllers, by their degradation or malfunction, which may lead to inappropriate decisions.

It is noted that starting from the requirements of the evolution of integrated information systems for the transition to cloud architectures, presented above, the concept of Industry 4.0 is in fact a customization of it, by using these in the industrial environment.

The new concept of Industry 4.0 has no significant impact yet and it is too early to be considered as an Industrial Revolution. However, significant steps forward were made in the conceptualization of the human-centric approach, the human-to-machine and machine-to-machine collaboration. The Fifth Industrial Revolution, with no human intervention, is more like an ending point right now, with an expected concentration of human attention to more altruistic and creative goals, having no pressure of fulfilling basic or more elevated needs.

## 2 Virtualization on the Architecture of Information Systems at Organizational Level

In cloud infrastructure, computing power is provided by data centres, which include servers and data storage systems. It can accommodate most types of integrated information systems [11]. The customers pay flexibly, depending on the resources used, based on a monthly fee. Cloud service users also reduce TCO by eliminating usage license spending, hardware architectures needed to store data, unnecessary space to place such equipment and increased data security [10].

From a business model point of view, the components of the cloud services are:

- Customers, who access the service via mobile devices or computers, with the benefits of reducing the costs of purchasing and maintaining hardware infrastructure, reducing security costs, low energy consumption etc.
- Data centres, which consist of server collections. They can be arranged in the same space, in the same building or in a space outside the organization and may contain virtualized servers with installed applications
- Distributed servers, which are in the same organization, but in different geographical locations, to ensure disaster redundancy (power outage, etc.).

*Data centres* can have several functions, such as transaction processing centres, multimedia content delivery centres; data centres to perform complex simulations and data processing operations for integrated information systems.

General models of virtualization contain the following types of cloud services [31]:

- IaaS (Infrastructure as a Service) is a cloud service model in which a provider leases a technology infrastructure, i.e., virtual servers, that can replace existing systems completely or partially within the company. IaaS allocates the entire suite of infrastructure resources and facilities (electricity, cooling solutions, etc.) for hosted hardware platforms
- PaaS (Platform as a Service) is a type of cloud service in which a provider offers software development and hosting solutions. It is used by companies to develop and market host solutions on demand or provide services to other companies. PaaS is created using an IaaS infrastructure that adds to an additional level of integration with different application development environments
- SaaS (Software as a Service) is a model in which a provider provides web services for applications that it makes available to end-users. Such services are generally intended to replace applications installed by users on their local systems. SaaS (Software as a Service) is based on a PaaS infrastructure, providing an autonomous operating environment used to provide the ultimate user experience, including ERP systems, multimedia applications, accounting programmes etc. (Figs. 2 and 3).

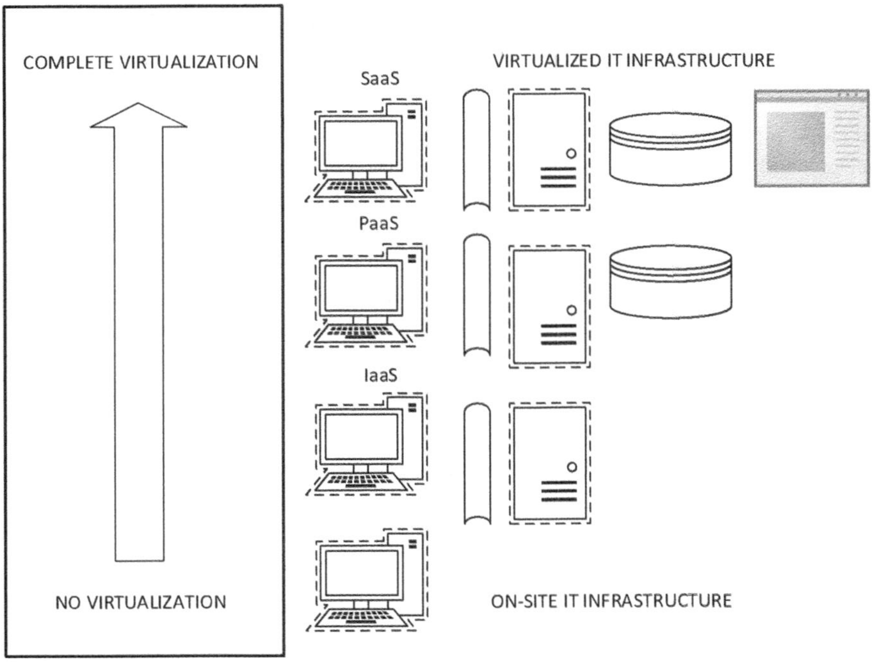


**Fig. 2** IT infrastructure transition to cloud architecture

Another study concerning cloud adoption and related technologies states, in 2020, a massive option to the companies for hybrid architectures where the private cloud tools are combined with public ones, rising to 78% of total users [21].

The emergence and development of IoT (Internet of Things) devices lead to the need for organizational integration of such devices. The Garner report [23] shows that even during the pandemic period, the IT investments that dropped during 2020 is expected to rise during the next period, using cloud technologies and IoT as a method to reduce costs, with examples. The following figure shows a reference architecture for a hybrid cloud platform, using Microsoft Azure, in which one can identify the following:

- Company's infrastructure, organized in the form of a private cloud, managed by an internal data centre, which manages various applications as well as the integrated ERP, CRM or SCM computer system, which ensures increased data privacy
- Public cloud platform, providing specific collaboration tools through SharePoint and Office 365
- Azure IoT platform, necessary for integrating IoT devices into the company's workflow.

The detailed map of the IoT Azure services is presented graphically below, having as distinctive levels data connectivity, data processing analytics and management, and presentation and business connectivity (Fig. 4).

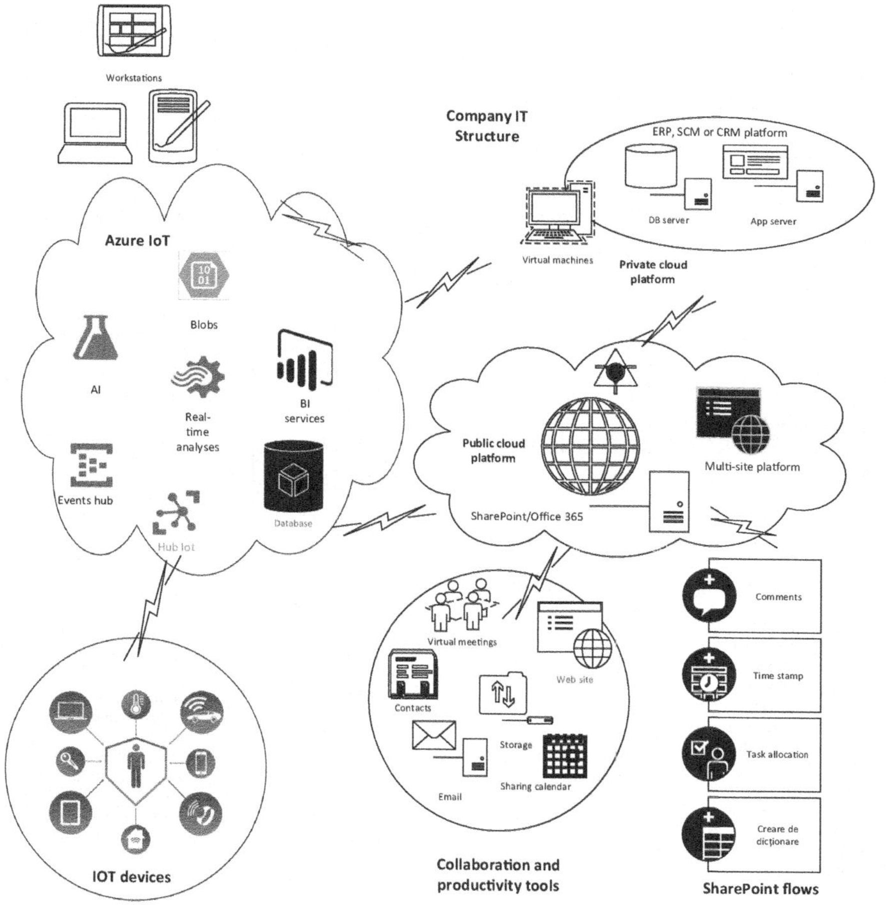


**Fig. 3** Hybrid reference architecture for cloud services

# 3 Production Systems—Trends and Perspectives

From the point of view of current production systems, the benchmark architecture expected to be implemented in the future is Industry 4.0, with the updates of Industry 5.0. The Industry 4.0 concept enables the integration of the activity of cyber-physical systems by developing the communication and real-time cooperation capabilities of human operators and robots, based on the following technologies:

- Implementation of the sensor capacity of the IIoT (Industrial Internet of Things) production systems, aimed at monitoring, controlling and organizing production and predictive maintenance
- Analysis of the large volumes of data provided by IIoT sensors and their use in optimization and prediction algorithms

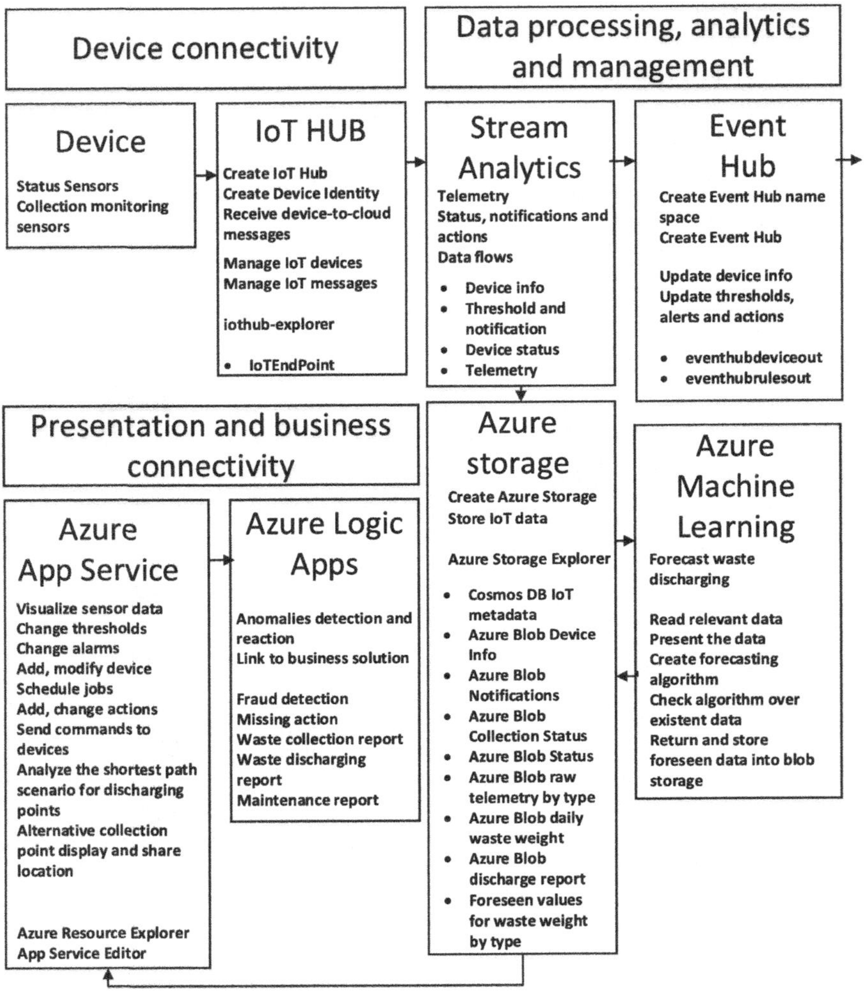


**Fig. 4** Detailed map of IoT Azure services

- Systems integration used at the organization level, in the form of extensive platforms, integrating suppliers and customers, with enterprise resource management modules and product data management, throughout the lifecycle, simulation of the production process and optimization of production flow
- 3D printing, by additive manufacturing processes, aimed at reducing production time for unique products and prototypes and eliminating transport times, for certain categories of parts, in the service activity (Fig. 5)

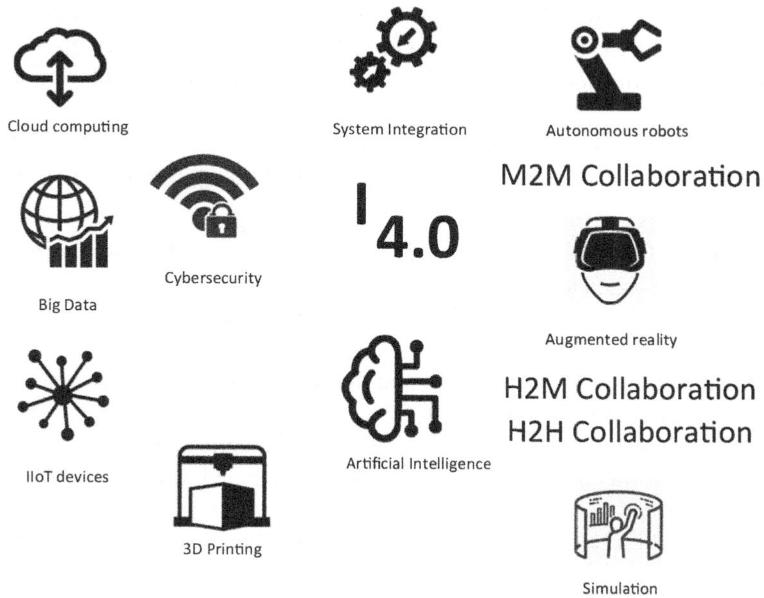


**Fig. 5** Industry 4.0 technologies

- Augmented reality, by accessing contextual information for operations carried out by human operators as decision-making support in the production and service activity
- Cloud computing, as a support structure for the integration of information systems and sensor data
- Simulation and artificial intelligence, concerning the existence of a digital clone of physical systems with a degree of advanced behaviour fidelity, with which simulations of real-time production streams can be carried out, their optimization and production control
- Autonomous robots, by connecting them to the computer structure of the organization, having the capacity to communicate and work together, both with human operators and other machines, equipment, and production facilities in real time
- Cybersecurity, in view of the increasing volume of data being circulated, the implementation of mandatory security policies and the potential as a source of bottlenecks or production shutdowns following cyber-attacks.

It is noted that starting from the requirements of the evolution of integrated information systems for the transition to cloud architectures, presented above, the concept of Industry 4.0 is in fact a customization of it, by using them in the industrial environment.

The forecasts presented in [29], by IQMS, manufacturer of ERP systems, integrated into the 3DExperience platform owned by Dassault Systèmes, identify the main trends in production systems, how manufacturers intend to achieve their revenue growth targets, operational improvements, digital transformations and the launch of new products and services in 2019, and in [30]. The studies identified the following opportunities for the development of production systems (presented by comparison in Table 4).

Autodesk Company presented in 2019 five major trends for production systems on the integration or increasing share of various technologies at the organizational level:

- Adoption of Industry 4.0 production architecture as a standard for future production systems, characteristically having a large volume of data used and the possibility of the informational interconnection of the components of the production system
- Increase purchase costs for robots and drones, tasks such as 3D exploration or visual inspections to be automated
- Improving interconnectivity by adopting the 5G standard, allowing the widespread adoption of artificial intelligence and data collection and equipment control using IoT in real time
- Inclusion of blockchain technology in the information architecture of production systems, prerequisites for increasing the accuracy of available data through the existence of shared data block, data can be viewed and used in decision support at all levels of the production system
- 3D printing expansion as a manufacturing process by reducing manufacturing time and costs for various applications.

The same company presented in [2] the prediction for 2021 in the case of the manufacturing systems evolution:

- Orienting to more personalized production, the customers valuing the real individual needs, pressing to the manufacturers to provide more individualized products
- Refining the smart products' design, with more understandable data and aggregating into more comprehensive form, strongly related with the individual behaviour, sensing capability being not enough without a proper interpretation of the data obtained, using phone apps or web-based portals
- Imposed quarantine periods, and the temporary incapacity of workers in 2020, due to the COVID pandemic, impose the automation of most repetitive tasks, extended to all industries
- Shortening and flexibilization of the supply chain, from the same reasons explained above, during the COVID pandemic, with closed borders and production stopped because of workers temporary incapacity to work
- Integrating on-site data centres, in the construction industry and extensive use of prefabrication, shortening the construction time and saving space needed to store the materials,

Another assessment of production system trend is presented by Microsoft, identifying six trends in [42] the development of production systems:

**Table 4** Prediction for the manufacturing systems by IQMS in 2019 and 2021

| 2019 predictions | 2021 predictions |
|---|---|
| Increasing the share of the use of analysis and business intelligence tools, chain-wide suppliers, stores and management teams, through custom-integrated platforms, which will use data provided by existing systems and which will include predictive analysis or machine learning tools | Rethinking of the workers responsibilities, by automation of annoying tasks, the companies struggling to attract trained employees with enhanced digital skills |
| Obligation to use an enterprise resource planning (ERP) system to increase customer-centred competitiveness | ERP remain a "must have" for the industry, with real-time capabilities, and more advanced financial and sales reports, being considered the backbone of the IT system |
| Integrating supply chains, from the manufacturing industry, into smart grids with the real-time integration of tracking systems, aiming at increasing delivery accuracy, and improving the quality of raw materials | Use of the automated configure, price, quote tools (CPQ), to avoid the errors from human interaction, enhanced with 3D images of the products, able to implement more accurately the market strategies for a limited time |
| Use of Wi-Fi and IoT sensors, to the detriment of PLC or other previous systems, leading to the creation of a real-time data stream that provides information on product quality, performance and performance of the use of means of production, also used as decision support in solving operational problems by manufacturers | Adoption of real-time reporting for the supply chain and manufacturing, given by the instability of COVID pandemic, shortening the usual need of the reporting period to almost instant |
| Widespread adoption of robots, due to the chronic lack of labour in the manufacturing industry, accompanied by increased demand for large-volume parts and assemblies among mid-level producers, with a view to increasing and maintaining market share | Due to the COVID pandemic instability, the customer preferred the suppliers who can deliver faster and predictable, the manufacturers integrating the ERP with the CAD systems |
| Imposing as a standard of manufacture intelligent machinery and equipment at all levels of production | The IoT ability inclusion in the machine tools, enhancing the existing predictive maintenance function with dimensional control, integrated with the quality control module |
| Widespread adoption of IIoT (Industrial Internet of Things) devices, particularly not only in real-time monitoring of operations but also in predictive maintenance or quality assurance | Expanding the traceability for the products, especially for food and beverages, increasing the production conformity with various faster changing regulation, companies being interested in obtaining certifications or to make external audit to prove the quality and traceability of the products |
| Implementation of blockchain data structures at the production activity level, through relationships with similar structures at supplier level, operational planning, and the emergence of new business models | Increasing the sustainability for the manufacturing, in terms of reducing the generated waste, or to acquire more efficient facilities for energy production |
| Migrating production to smart products with Product as a Service business models by ensuring a steady flow of sales and revenue, respectively, by increasing the duration of use and the possibility of updating the performance of the product | More extensive the use of digital twins and simulation in the smart factory, at workplace level, to improve the safety and the efficiency of the workers |

(continued)

**Table 4** (continued)

| 2019 predictions | 2021 predictions |
|---|---|
| Increase cybersecurity spending and improve the organization's internal data protection | |

- Integrating the IT system into operational technology by adding communication and data collection capabilities to old equipment, using cloud-computing infrastructures, integrating existing command and control software systems, implementing IoT at all levels of the production system, increasing human–robot interaction and adopting efficient and environmentally friendly technologies
- Adoption of the Product as a Service model, by sharing products or shortening the supply chain of the product, due to the decrease in purchasing power and the tendency to federalize and depersonalize the means of production. A trend with potential is Manufacturing as a Service, in which the production order is transmitted through control platforms, the actual production facility being unknown to the beneficiary, the decision of location belonging to the proprietary group of production facilities. Similarly, emerging services are supported by industry profile, such as Design as a Service, Experimentation as a Service, Equipment as a Service, Simulation as a Service, Management as a Service, Maintenance as a Service and Integration as a Service
- Widespread implementation of the smart production concept, through the integration of artificial intelligence and machine learning at the operational level, leading to lower production costs, shortening production times, avoiding bottlenecks and failures through productive maintenance, adopting blockchain technology at the organizational level and including supply in the smart value chain
- Replication of physical production systems in the virtual environment and use of these models for design, operation, simulation and diagnosis, with the aim of reducing manufacturing defects at the project level. Also, it is foreseen to extend additive and subtractive production processes in the production of goods, as well as the use of new materials in these processes. Another trend is the use of autonomous equipment and machinery capable of operating without human intervention or by limited intervention in automated workflows. The expansion of human–machine interaction capacity, as well as the capitalization of the operating ability, is estimated to be achieved by using augmented and virtual reality as a component part of operations
- Adapting business processes to the available human resource by increasing the share of the active population after 55 years, correlated with the competition of digital natives, leading to changes in the way of inter-personal interaction, by adapting IT tools as a way of conversation and information, at the expense of direct conversation. Another direct consequence will be the change in the way of working in companies, requiring a focus on employee availability, and supporting the entrepreneurial spirit, by participating in the decision-making process. It is to

be expected, by changing the age structure and adopting information technology, to have a gap, through the prism of people's abilities,
- Increasing the uncertainty of the business environment by introducing new rules (GDPR) or excessive politicization by delineating new markets through new trade agreements and lack of predictability on industrial development. In this respect, the additional costs of ensuring the protection of personal data at the company level, requirements for concrete data protection measures at the company level and the obligation to report security incidents are underway. In terms of the delimitation of new markets and trade agreements, the most eloquent examples of 2019 are the trade agreements and customs policy of the United States of America, in particular with China, but also the imminent exit of the United Kingdom from the European Union.

The 2020 point of view of Microsoft, regarding the future of the manufacturing systems is presented in [43], concerning the next tendencies:

- Data unification and availability extension is the most important trend for the manufacturing systems, the same data being acquired, shared and used among the company, by different worker categories from various devices
- Shifting from new emergent technology adoption to an increased experience of the customer, regarding the goods and services, the customer being more careful to spend, expecting to cheaper, better and faster-delivered products, forcing the manufacturers to develop more innovative processes to reduce cost, to include more additional services and to find a faster-delivering solution, using the unified data collected in the process
- Spreading the smart manufacturing as standard, with IoT and operation technology connected, enhanced with AI capacities
- Enhancement of sustainable manufacturing and increase the safety of workers at the workplace, because of the regulatory bodies, communities and NGO's pressure
- Adapting to a more polarized workforce, with an aged population and the rise of Z-generation, necessitating more efforts of training and changing the workstyle, the workforce should adapt to very fast-changing technologies, including the organizational way of conducting business
- Increase the R&D spending, with a strong accent on rapid prototyping, digital twin, augmented reality, nanotechnologies, and simulation, to fasten the time-to-market
- Rethinking of the supply chain, as an important differentiation from the competition, adopting on large scale the traceability and prediction of delivery time, including autonomous vehicles or drones, for delivery and autonomous robots to sort and store the products inside the warehouse
- Continuous adaptation of businesses to increased uncertainty, the COVID pandemic impacted the global economy, by exposing the supply chain weakness, changing the consumer behaviour to a smarter purchase and the suppliers' temporary indisposition as a result of the imposed quarantine.

## 4 Conclusions

The current trends of the manufacturing systems are the result of a long evolution of technology and management style. The comparison of the industrial revolution with Maslow' hierarchy of needs, offer a clearer image of the challenges and limitations of each time span. The life quality indicators, such as average world life expectancy, GDP per capita and general population revealed a continuous improvement of the general life because of the industrial revolutions. The first section of the chapter enforced the time placement and the opportunity of the manufacturing systems evolution. Using extensively IT and other related technologies, in the manufacturing process, and in all aspects of everyday life, impacted massively the quality of life. However, Industry 4.0 become possible after the manufacturing collaborative paradigms presented (e.g., bionic, fractal, multi-agent etc.) which established the principle of operation. The impact of Industry 4.0 at the society level is too early, in historical terms to be evaluated. Furthermore, the initial concept was updated with sustainability and resilience dimensions, protecting the resources and the human-centred paradigm. Even if autonomous robots became a necessity in today industry, the role of the human workers remain decisive in the manufacturing systems, exposed also by the pandemic crisis. Globalization also showed limitations, with delivery outages or unpredictability, and more and more countries orienting to relocate the critical industries to their homeland.

The COVID pandemic impacted the manufacturing systems, by shifting the focus from the product to the customer needs. The consumer, because of cutting incomes, became more aware of the product value, carefully spending on smarter products, more oriented to their needs. The pandemic situation exposed the globalization weakness, in terms of concentration various industries in a limited geographical area, depending on international transportation to deliver goods on the local market. The supply chain flexibility and prediction were the most valuable asset in the past year. However, the already implemented IoT and AI in the manufacturing information systems must be more integrated and available to all working positions with real-time data, and more important, interpreted data using AI or other decision support systems techniques. The faster than ever-changing technologies used in the manufacturing process and in the operational chain, cumulated with the ageing of the workforce, conducted a more careful management of the human resources, the worker being exposed to a large volume of data, to support the working process. The ageing of the workforce and changing of the generation to "native digitals" polarize the teams and impose continuous training programmes. The pandemic also changes the work-style to remote work, changing the office with the employees' home, for a period of time, or alternating the office style with home. Thus, reveals an immediate necessity to offer the same data and tools by internet-based solutions. The pandemic practice might become a standard in the future, with a great impact on the workforce mobility, shortening the operation costs with offices operation cost and increasing the workers general quality of life, by eliminating the travel time to the office.

Concluding the above issues and relevant for further development, we mention the main forecast directions for the development of manufacturing systems:

- Organization of an integrated management system at the organization level, including mandatory and optional management subsystems in various fields (quality assurance, environmental protection, social responsibility, GDPR etc.) and the implementation of common procedures and measurable outcome indicators, which can be then implemented in the organization's IT system by providing decision support
- Large-scale implementation, including large companies and medium-sized and small enterprises, of an integrated enterprise resource management system capable of exchanging data with similar systems and client platforms, as well as updating real-time data on production activity
- Increasing sensory possibilities (IoT) at the manufacturing workshop level and using the data obtained in the planning, simulation and organization of ad hoc production, at the expense of centralised modelling, simulation and planning systems
- Adoption of new manufacturing strategies and processes to ensure a decrease in production time, meeting customer requirements as well as delivery forecasting, through advanced production technologies and the widespread use of artificial intelligence to reduce bottlenecks in production
- Increasing the gap in available workforce skills by segmenting it by age and digital skills. The progressive replacement of the human factor in the production process and the structural change in the way of working are also expected.

## References


1. Albulescu CT, Draghici A, Fistis GM, Trusculescu A (2016) Does ISO 9001 quality certification influence labor productivity in EU-27? Procedia Soc Behav Sci 221:278–286
2. Autodesk (2021) What will manufacturing look like in 2021? The 5 top trends to watch. https://redshift.autodesk.com/manufacturing-trends-2021/
3. Berman O, Maimon O (1986) Cooperation among flexible manufacturing systems. IEEE J Robot Autom 2(1):24–30. https://doi.org/10.1109/JRA.1986.1087033
4. Brown SP (2008) Business processes and business functions: a new way of looking at employment. Monthly Labor Rev
5. Buer MC (1926) Health, Wealth and population in the early days of the industrial revolution (Routledge & Kegan Paul Ltd, Ed.; 1st ed.). https://www.routledge.com/Health-Wealth-and-Population-in-the-Early-Days-of-the-Industrial-Revolution/Buer/p/book/9781138879836
6. Bunce P, Limoges R, Okabe T (1997) NGMS—Next generation manufacturing systems (IMS Project). In: Enterprise engineering and integration. Springer, Berlin, Heidelberg, pp 274–283. https://doi.org/10.1007/978-3-642-60889-6_30
7. Camarinha-Matos LM, Afsarmanesh H (1999) The virtual enterprise concept. https://doi.org/10.1007/978-0-387-35577-1_37
8. Camarinha-Matos LM, Afsarmanesh H (2001) Virtual Enterprise modeling and support infrastructures: applying multi-agent system approaches. Springer, Berlin, Heidelberg, pp 335–364. https://doi.org/10.1007/3-540-47745-4_16

9. Carutasu G, Aurite T (2005) Partners evaluation in virtual enterprise. Univ Politeh Buchar Sci Bull Ser D: Mech Eng 67(4):25–36
10. Carutasu NL, Carutasu G (2016) Collaborative virtual environments using ERP and cloud tools. In: 10th annual international technology, education and development conference valencia, Spain, pp 5934–5944
11. Carutasu G, Botezatu MA, Botezatu C, Pirnau M (2016) Cloud Computing and Windows Azure. In: 2016 8th international conference on electronics, computers and artificial intelligence (ECAI), pp 1–6. https://doi.org/10.1109/ECAI.2016.7861168
12. Cotet GB, Carutasu NL, Chiscop F (2020) Industry 4.0 diagnosis from an iMillennial educational perspective. Educ Sci 10(1):21. https://doi.org/10.3390/educsci10010021
13. Cotta J, Breque M, de Nul L, Petridis A (2021) Industry 5.0 towards a sustainable, human-centric and resilient European industry. https://op.europa.eu/en/publication-detail/-/publication/468a892a-5097-11eb-b59f-01aa75ed71a1/
14. Deng X, Peng J, Huang H (2009) Research on the fractal company modeling based on competence. In: IE and EM 2009—Proceedings 2009 IEEE 16th international conference on industrial engineering and engineering management, pp 2136–2140. https://doi.org/10.1109/ICIEEM.2009.5344226
15. Draghici A, Mocan M, Draghici G (2011) On-line training and certification solution for business process managers. In: International conference on enterprise information systems, pp 380–389. Springer, Berlin, Heidelberg
16. Draghici A, Baban CF, Ivascu LV, Sarca I (2015) Key success factors for university-industry collaboration in open innovation. In: 8th annual international conference of education, research and innovation, Spain
17. Dragoi G, Draghici A, Rosu SM, Radovici A, Cotet CE (2010) Professional risk assessment using virtual enterprise network support for knowledge bases development. In: International conference on enterprise information systems, pp 168–177. Springer, Berlin, Heidelberg
18. Duffie NA, Prabhu VV (1996) Heterarchical control of highly distributed manufacturing systems. Int J Comput Integr Manuf 9(4):270–281. https://doi.org/10.1080/095119296131562
19. EUROSTAT (2019) Glossary: business functions—Statistics explained
20. Evgrafoff B (1973) Le gouvernement de l'entreprise. A. Colin
21. Flexera (2021) State of the cloud report
22. Franklin S, Graesser A Is it an agent, or just a program?: A taxonomy for autonomous agents. https://doi.org/10.1007/bfb0013570
23. Gartner (2021) Internet of things (IoT)—Key business insights. https://www.gartner.com/en/information-technology/insights/internet-of-things
24. Gogan LM, Rennung F, Fistis G, Draghici A (2014) A proposed tool for managing intellectual capital in small and medium size enterprises. Procedia Technol 16:728–736
25. Gorecky D, Schmitt M, Loskyll M, Zuhlke D (2014) Human-machine-interaction in the industry 4.0 era. In: 2014 12th IEEE international conference on industrial informatics (INDIN), pp 289–294. https://doi.org/10.1109/INDIN.2014.6945523
26. Guran M, Dobrescu T, Patrascu G, Carutasu G (2005) Adapting mechanical engineering curricula to virtual enterprise environment. In: 8th international conference on management on innovative technologies MIT 2005, pp 241–245
27. Henry F (1917) Administration industrielle et générale; prévoyance, organisation, commandement, coordination, controle. H. Dunod et E. Pinat
28. Hostetler LM (1936) 75 Years of American finance : a graphic presentation, 1861–1935. https://fraser.stlouisfed.org/title/75-yrs-american-finance-162/chart-5373?start_page=15
29. IQMS (2019) Top 10 manufacturing trends for 2019—IQMS. https://erpblog.iqms.com/top-manufacturing-trends-2019/
30. IQMS (2021) 2021 manufacturing trends: connecting workers will be key. https://erpblog.iqms.com/2021-manufacturing-trends/
31. International Telecommunication Union-Telecommunication Sector, S. of I (2013) Y.2060 : overview of the internet of things. https://www.itu.int/rec/T-REC-Y.2060-201206-I

32. Kagermann H, Lukas W-D, Wahlster W-D (2011) Industrie 4.0: Mit dem Internet der Dinge auf dem Weg zur 4. industriellen Revolution—vdi-nachrichten.com. https://web.archive.org/web/20130304101009. http://www.vdi-nachrichten.com/artikel/Industrie-4-0-Mit-dem-Internet-der-Dinge-auf-dem-Weg-zur-4-industriellen-Revolution/52570/1
33. Kassicieh SK, Schultz CR (1987) Decision support flexible manufacturing systems. Omega 15(6):495–502. https://doi.org/10.1016/0305-0483(87)90007-7
34. Kurihara T, Jordan J, Bunce P (1997) Next generation manufacturing systems in the IMS program. In: Information infrastructure systems for manufacturing. Springer US, pp 418–430. https://doi.org/10.1007/978-0-387-35063-9_36
35. Kádár B, Monostori L, Szelke E (1998) An object-oriented framework for developing distributed manufacturing architectures. J Intell Manuf 9(2):173–179. https://doi.org/10.1023/A:1008876114686
36. Lee J, Kao H-A, Yang S (2014) Service innovation and smart analytics for industry 4.0 and big data environment. Procedia CIRP 16:3–8. https://doi.org/10.1016/j.procir.2014.02.001
37. MacDougall W (2014) Industrie 4.0: smart manufacturing for the future. https://www.gtai.de/GTAI/Content/EN/Invest/_SharedDocs/Downloads/GTAI/Brochures/Industries/industrie4.0-smart-manufacturing-for-the-future-en.pdf
38. Marr B (2016) What everyone must know about industry 4.0. Forbes. https://www.forbes.com/sites/bernardmarr/2016/06/20/what-everyone-must-know-about-industry-4-0/#4eccfdfd795f
39. Maslow AH (1943) A theory of human motivation. Psychol Rev 50(4):370–396. https://doi.org/10.1037/h0054346
40. Melanson A (2015) Robotics and industry 4.0—Aethon Blog. http://www.aethon.com/industry-4-0-means-manufacturers/
41. Mella P (2009) The Holonic perspective in management and manufacturing. Int Manag Rev 5(2):19
42. Microsoft (2019) 2019 manufacturing trends report
43. Microsoft Dynamics (2020) Top eight trends every COO should know about the future of manufacturing
44. Nau DS, Zhang G, Gupta SK, Karinthi RR (1993) Evaluating product machinability for concurrent engineering. In: Concurrent engineering. Springer US, pp 264–279. https://doi.org/10.1007/978-1-4615-3062-6_14
45. Nielsen PB, Sturgeon TJ (2014) Using business functions to measure international trade and economic globalization. In: International conference on trade and economic globalization
46. Okabe T, Bunce P, Limoges R (1998) Next generation manufacturing systems (NGMS) in the IMS program. In: Advances in production management systems. Springer US, pp 43–54. https://doi.org/10.1007/978-0-387-35304-3_5
47. Özdemir V, Hekim N (2018) Birth of industry 5.0: making sense of big data with artificial intelligence, "The Internet of Things" and next-generation technology policy. OMICS: J Integr Biol 22(1):65–76. https://doi.org/10.1089/omi.2017.0194
48. Parente SL, Prescott EC (2003) Successes and failures in real convergence. National Bank of Poland
49. Paschek D, Luminosu CT, Draghici A (2017) Automated business process management–in times of digital transformation using machine learning or artificial intelligence. In: MATEC web of conferences, vol 121. EDP Sciences, p 04007
50. Paschek D, Mocan A, Draghici A (2019) Industry 5.0—The expected impact of next industrial revolution. In: hriving on future education, industry, business and society; proceedings of the makelearn and TIIM international conference 2019, pp 125–132. https://ideas.repec.org/h/tkp/mklp19/125-132.html
51. Porter ME (1985) The competitive advantage: creating and sustaining superior performance. Free Press
52. Qi Q, Tao F (2019) A smart manufacturing service system based on edge computing, fog computing, and cloud computing. IEEE Access 7:86769–86777. https://doi.org/10.1109/ACCESS.2019.2923610

53. Roser M (2013) Economic growth—Our world in data. Our Wolrd in Data. https://ourworldindata.org/economic-growth
54. Roser M, Ortiz-Ospina E, Ritchie H (2019) Life expectancy—Our world in data. https://ourworldindata.org/life-expectancy
55. Ryu K, Jung M (2003) Agent-based fractal architecture and modelling for developing distributed manufacturing systems. Int J Prod Res 41(17):4233–4255. https://doi.org/10.1080/0020754031000149275
56. Shah JJ, Rogers MT (1988) Functional requirements and conceptual design of the feature-based modelling system. Comput-Aided Eng J 5(1):9–15. https://doi.org/10.1049/cae.1988.0004
57. Shewchuk, JP (1998) Agile manufacturing: One size does not fit all, in strategic management of the manufacturing value chain. In: Proceedings of the International Conference of the Manufacturing Value-Chain August '98, Troon, Scotland, UK, pp 143–150. https://doi.org/10.1007/978-0-387-35321-0_16
58. Shrouf F, Ordieres J, Miragliotta G (2014) Smart factories in industry 4.0: a review of the concept and of energy management approached in production based on the Internet of Things paradigm. In: 2014 IEEE international conference on industrial engineering and engineering management, pp 697–701. https://doi.org/10.1109/IEEM.2014.7058728
59. Slomp J, Zijm WHM (1993) A manufacturing planning and control system for a flexible manufacturing system. Robot Comput Integr Manuf 10(1–2):109–114. https://doi.org/10.1016/0736-5845(93)90032-F
60. Sołtysik-Piorunkiewicz A, Zdonek I (2021) How society 5.0 and industry 4.0 ideas shape the open data performance expectancy. Sustainability (Switzerland) 13(2):1–24. https://doi.org/10.3390/su13020917
61. Srai JS, Kumar M, Graham G, Phillips W, Tooze J, Ford S, Beecher P, Raj B, Gregory M, Tiwari MK, Ravi B, Neely A, Shankar R, Charnley F, Tiwari A (2016) Distributed manufacturing: scope, challenges and opportunities. Int J Prod Res 54(23):6917–6935. https://doi.org/10.1080/00207543.2016.1192302
62. Tharumarajah A, Wells AJ, Nemes L (1996) Comparison of the bionic, fractal and holonic manufacturing system concepts. Int J Comput Integr Manuf 9(3):217–226. https://doi.org/10.1080/095119296131670
63. Toh KTK, Harding JA, Thompson D (1999) An holonic approach for the modelling of enterprise functionality and behaviour. Int J Comput Integr Manuf 12(6):541–558. https://doi.org/10.1080/095119299130119
64. Ueda K (1992) A concept for bionic manufacturing systems based on DNA-type information. In: Human aspects in computer integrated manufacturing. Elsevier, pp 853–863. https://doi.org/10.1016/b978-0-444-89465-6.50078-8
65. Ulieru M, Brennan RW, Walker SS (2002) The holonic enterprise: a model for Internet-enabled global manufacturing supply chain and workflow management. Integr Manuf Syst 13(8):538–550. https://doi.org/10.1108/09576060210448125
66. University of Missouri (n.d.) Introduction—Prices and wages by decade—Library guides at University of Missouri Libraries. https://libraryguides.missouri.edu/pricesandwages. Accessed 20 Apr 2021
67. University of Missouri (2021) Prices and wages by decade—Library guides at University of Missouri libraries. https://libraryguides.missouri.edu/pricesandwages
68. Vater J, Harscheidt L, Knoll A (2019) A reference architecture based on edge and cloud computing for smart manufacturing. In: Proceedings—International conference on computer communications and networks, ICCCN. https://doi.org/10.1109/ICCCN.2019.8846934
69. Vogt J (2021) Where is the human got to go? Artificial intelligence, machine learning, big data, digitalisation, and human–robot interaction in industry 4.0 and 5.0. AI & SOCIETY. https://doi.org/10.1007/s00146-020-01123-7
70. Warnecke H-J (1993) The fractal company. In: The fractal company. Springer, Berlin, Heidelberg. https://doi.org/10.1007/978-3-642-78124-7
71. Warnecke H-J, Warnecke H-J (1993) The fractal factory—An integrating approach. In: The fractal company. Springer, Berlin, Heidelberg pp 137–217. https://doi.org/10.1007/978-3-642-78124-7_4

72. Wortmann JC, Browne J, Sackett PJ (1993) The system of manufacturing: a prospective study
73. Zengin Y, Naktiyok S, Kaygın E, Kavak O, Topçuoğlu E (2021) An investigation upon industry 4.0 and society 5.0 within the context of sustainable development goals. Sustainability (Switzerland) 13(5):1–16. https://doi.org/10.3390/su13052682

# Chapter 4
# Manufacturing Processes Automation and Their Intelligent Monitoring

**Dorian Stef and George Draghici**

**Abstract** Today, industrial values have led to a new step in evolution. industrial, namely the implementation of Industry 5.0. The Artificial Intelligent systems are internal optimization systems, they lead the processing processes based on the information entered by the process disruptors. Machining processes on machine tools are considered complex processes, due to the great diversity of controlled or random factors that influence their evolution over time. Thus, the Artificial Intelligent, control systems of machine tools can have independent functions or can be coupled with data management systems.

**Keywords** Industry 4.0 · Industry 5.0 · Machine tools · Artificial intelligence · Manufacturing · CNC · Flow optimization · Flow layout

## 1 Introduction

A century ago, in "The Principles of Scientific Management" (1911), Frederick Winslow Taylor published his ideas about the modern manufacturing model, which remain rational and valid to this day. Analysing the ideas proposed by Taylor, everything is starts from the elementary processes "Analysing the production activity on the elementary processes with these scientifically-based methodologies, creates a benefit in the economic efficiency of the enterprises and its workers" [4]. Currently, industrial values have led to a new step in the evolution of the industry, namely by the implementation of Industry 4.0. In the present, several steps are being taken into the implementation of this stage, which has a substantial influence on the industry and is based on the development of intelligent factories, products, and services, which have the capacity to integrate into the Internet of Things [1]. At the same time, the development of the next stage in the evolution of the industrial field began by developing

D. Stef · G. Draghici (✉)
Politehnica University Timisoara, Faculty of Mechanical Engineering, Timisoara, Romania
e-mail: george.draghici@upt.ro

D. Stef
e-mail: dorian.stef@upt.ro

 
A. Draghici and L. Ivascu (eds.), *Sustainability and Innovation in Manufacturing Enterprises*, Advances in Sustainability Science and Technology,
https://doi.org/10.1007/978-981-16-7365-8_4

the idea of Industry 5.0, which is based on the development of the workforce and the protection of the environment, based on the most efficient management [1, 19].

In an increasingly competitive economic market, oriented on the product and production, it has led to constraints in all departments of companies and requires working in a system of continuous improvement, high productivity, and efficient management, thus, "it is necessary to apply new techniques that allow the generation of values" and ensure the sustainability and stability of the market [15]. Technological advancement in new production systems requires the implementation of agile technological processes and the efficient management of information, "to create organizational synergy and offer competitive advantages within the production system and the value chain" [5].

In the academic environment, but also in the industrial one, different theories have been formulated, some even divergent, in the field of company management, with impact on the activity and work processes, of an enterprise that uses a mentality based on the industry 4.0 system; generating a gap between them and those who do not meet these criteria, leading to a rapid transformation of production and achievements by companies that have adopted the processes in the new system [15]. This evolution towards an Industry 4.0 offers enormous opportunities for sustainable production, "using the ubiquitous information and communication technology (ICT) infrastructure" [19].

In this context, the present chapter will present "a review of Industry 4.0 based on recent research and practices", but also of Industry 5.0 based on the ideas of sustainable production. In this case, the "macro and micro perspectives of Industry 4.0 will be viewed and analysed" [9]. Subsequently, sustainable production approaches are combined with Industry 4.0 requirements. Starting from the idea enunciated by Taylor more than 100 years ago, namely every-thing must proceed from the elementary processes, we propose the integration of the elementary systems of production, and the first step is to develop machine tools that have a control system based on artificial intelligence [14].

The integration of basic systems is the first step towards implementing the Industry 4.0 vision and achieving the Industry 5.0 objectives. The systems are analysed, considering the production flow [3]. "Structural changes are proposed in the organization and management of physical objects", especially machine tools, establishing connections with computers systems [5, 9].

This study consists of the implementation of an artificial intelligence control system on one of the most elementary elements of a production system, namely on machine tools. Starting from the basic processes creates a benefit in the economic efficiency of enterprises and their workers, this will allow the identification of gaps and opportunities to develop other blocks considered in the industry.

## 2 The Main Ideas of Industry 4.0

In 2011, Kagermann is the first that introduced Industry 4.0 and is the creator of the foundation of the "Industry 4.0 manifesto published by the German National Academy of Science and Engineering" in 2013 [6, 9]. Industry 4.0 concept started in October 2012, to the recommendation of the German Federal Government was presented a set of Working Groups for Industry to implementing the Industry 4.0. Those who are was proposed for these workgroups are now recognized as the founding fathers of the Industry 4.0. [7]. Industry 4.0 workgroups are:

- The Smart Factory;
- The Real Environment;
- The Economic Environment;
- Human Beings and Work;
- The Technology Factor.

"The current information and communication technology (ICT) infrastructure enables the industry to adopt quickly and efficiently the Industry 4.0 that had increasing with high expectations. Finding the human with the capacity to develop analytical algorithms to develop self-learning intelligence by taking advantage the current infrastructure is the biggest challenge. The technology future is conditioned to the scientific area creations, being an obstacle to the emerging technology adoption" [3, 15].

As Industry 4.0 is implemented the technological advances that are currently used in manufacturing will be modify, in addition the whole production process will be transformed considered the new technological model [12]. For example, "independent manufacturing cells will be unified as a fully integrated production flow, "Intelligent" machines and products will have the possibility to communicate with each other and some decisions will be made autonomously". The relation between man and machine will be change and this will conduct to a new era of relations with the suppliers, customers, and other companies [12, 13].

"Technological advance in production systems requires corporations with agile operational processes and efficient information management, to create an organizational synergy and provides competitive advantages within production system and value chain" [15].

The Industry 4.0 paradigm is highlighted by three dimensions [1, 5, 19]: "(1) horizontal integration across the entire value creation network, (2) end-to-end engineering throughout the product life cycle, as well as and (3) vertical integration and network manufacturing systems".

Horizontal integration represents "a complex relationship between the strategic and operational objectives at the different levels of the manufacturing systems that requires the modeling of an intelligent manufacturing system" [16].

Vertical integration of manufacturing systems in the network describes the cross-linking and intelligent digitization in the different hierarchy levels of a value creation

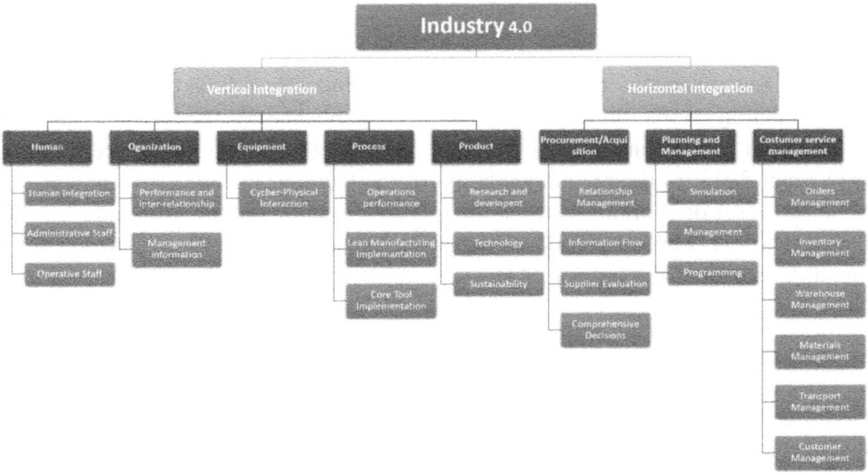


**Fig. 1** The paradigm of Industry 4.0 after [8]

module. Vertical integration or internal integration has the role of evaluating the system, to identify the crucial areas [6].

Industry 4.0 paradigm (Fig. 1) is considered a step towards "creating a more sustainable" industrial value. In the scientific literature, "this stage mainly characterized as a contribution to the ecological dimension of sustainability". The allocation of resources, products, materials, and energy can be done more efficiently, based on intelligent modules for creating cross-linked values [8].

In addition, the Industry 4.0 has a great opportunity to achieve sustainable industrial growth on all three dimensions of sustainability: economic, social, and environmental [15]. The macro perspective of Industry 4.0 is summarized as the sustainable production opportunities, and they are: "Business Models and the Value Creation Network". For the micro perspective, to provide a synthesis of the opportunities, the values underlying this perspective are:

- **Product**—In industry 4.0, the approach for products sustainable design is focuses on achieving "closed-loop life cycles for products", allowing the re-use and re-manufacture of the specific product or by applying the principles of product life cycle is often a capital asset whit a long depreciation period (a long-life cycle), up to 20 years or in some cases more. Retrofitting allows an easy and efficient way to modernize existing production equipment with sensor and drive systems, as well as the related control logic to overcome the heterogeneity of factory equipment management. Different approaches to product design, in this approach, can include the customer-oriented design. These approaches can be supported by the application of identification and recognizing customer needs systems [8].
- **Process**—"These are the activities that give life to the product, so they considered to be a sustainable process and to have a key role in the value chain of creation, as the means to achieve the objective". The holistic approach to resource efficiency

in industry 4.0 is designing appropriate manufacturing process chains by using new technologies and through the implementation of communications systems between different equipment [15, 16].
- **Equipment**—The equipment in the factory floor [3]. Retrofitting can thus use as an approach to achieve a CPS within a value creation module, such as a factory with existing production equipment [13].
- **Human**—In Industry 4.0 the people will continue to be considerate the creators in the value chain of creation [19]. Three different sustainable approaches can used to meet the social challenge in Industry 4.0 [15]:
  - "Increasing the efficiency of vocational training of workers by combining new IT technologies;
  - Increasing intrinsic motivation and encouraging creativity by establishing new approaches based on CPS of work organization and design;
  - Increased extrinsic motivation by implementing individual stimulation systems for the worker".
- **Organization**—A decentralized organization with a sustainable orientation in a smart factory is focused on the efficient allocation of products, materials, energy and considering "the dynamic constraints of the CPS". This concept leads to a "global resource efficiency" and is described as one of the essential benefits of industry 4.0 [6, 9].

## 3 The Main Ideas of Industry 5.0

In these days, the discussions that everybody has in industry and academic environment is related to the Industry 4.0, that "is being considered as a next industrial revolution, but is more a systemic transformation that includes impact on civil society, governance structures, and human identity in addition to solely economic/manufacturing ramifications, the Industry 5.0 founders prefer to speak about next step in evolution" [7, 12].

The first time the term Industry 5.0 was presented was in 2016, by the Michael Rada who presented results achieved by implementation of the principles of Industry 5.0.

- **Design principles**—The principles that stay at the base of Industry 5.0 are based "on simple tools, which are 6R Methodology and Logistics Efficiency Design" (LED). In order the 6R are recognize, reconsider, realize, reduce, reuse, recycle. The principle of 6R is in general applicable in the entire life cycle of the product not in one segment [7].
- **Meaning**—In general, the Industry 5.0 meaning is to create a sustainable instrument that can conduce to a continues process of development equal with evolution of nature, and whit the same principles regarding the recycling [1].

- **Effects**—The Industry 5.0 initiative have the same systematic instrument that have as result "waste prevention tool that is applicable to industrial environment. The effects of application are to be understand on three levels":
  - "First effect is on Economy and is the result of waste and wasting prevention projected to the value of material and products, including the logistics costs connected. The achievement of zero waste environment, mean cut off all the cost related to waste management, but that is not the total number";
  - "Second effect is on Environment is the result of same activity. The systematic utilization of the On-The-Ground-Mines results in lack of need of further natural resources exploitation" [7, 10],
  - "Third effect is on Social environment concentrate on link, which has been completely forgotten in Industry 4.0, on Human. The efficient cooperation between Human and Machine, utilizing its unique skills and abilities establish highly efficient working environment for production of high-quality products. Industry 5.0 does not concentrate just on "workforce in right age group" but define the possibilities open to Junior, so as Senior workers, so as groups, which are being for some reason not considered as useful" [10].
- **Challenges**—"Challenges which have been identified and must be solved for Industry 5.0 development: legal issues caused by discrepancy between technology development, social evolution and the changes reflected within the society and business environment", the new demography of human factor with a high problem in the technical training field, "overproduction, lack of transparency" implemented in many processes and industries, application of inadequate tools, technologies, or machine tools for the product manufacture, increasingly dependent on IT and electricity infrastructure, general reluctance to change and implementing new ideas by stakeholders from industry [7, 12].
- **The Waste Prevention**—In an era that the production numbers are higher than ever, and most of the modern industry is orientated on gain easy profit, the waste generating in the production process, the supply chain is at a level unthinkable. In industrial upcycling are four types of waste identified in the industry starting from the simple waste, up to complex production waste [10].
  - Physical waste—trash, product waste
  - Social waste—number of people that want to work but do not have the knowledge to be employee.
  - Urban Waste—the greenfield and the infrastructure between production spaces are too small or inadequate.
  - Process Waste—overproduction, supply chain unsteadies due to the bad logistical management
- **Impact of Industry 5.0**—The most notable areas that Industry 5.0 impacting in the industry with a proposition by affected areas are: "services and business models; profitability of business; reliability and continuous increase of efficient productivity; IT security; machine and human safety; product lifecycles increase;

industry value chain environmental impact decrease; workers education, skills, and involvement increase; socio-economic factors improvement" [10].

## 3.1 The 2022 Jobs Landscape

"Recent projections of the extent of structural change in the global labour market depend significantly on the time horizon taken into consideration" [17]. In addition to the rate of technological advancement, a range of other considerations must be taken into account "such as ease of commercialization, public adoption of new technologies and existing labour laws-influence the rate at which these developments accelerate workforce transformation" [2, 11].

From the companies' point of view, they estimate that in five years is need for implementing in the entire labour market, worldwide, this transformation [20].

Emergent roles groups "will gain significantly in importance over the coming years, while other groups of job profiles are set to become increasingly redundant" (Fig. 2). "Across all industries, by 2022, the emerging professions groups is set to increase its share of employment from 16 to 27% of the total employee base", "whereas the employment share of declining roles is set to decrease from currently" 31–21% (Fig. 3). Therefore, in quantitative terms, the expectation emerging from the estimates of employers surveyed is that, by 2022, "the structural decline of certain types of jobs (10% decline) will be fully counterbalanced by job creation and the emergence of new professions (11% growth). About half of today's core jobs-making up the bulk of employment across industries-will remain somewhat stable in the period up to 2022" [4, 20].

By implementing Industry 4.0 the rate of automation machinery will be increasing with approximate 20% by 2025, this aspect will lead in the industrial field to have the

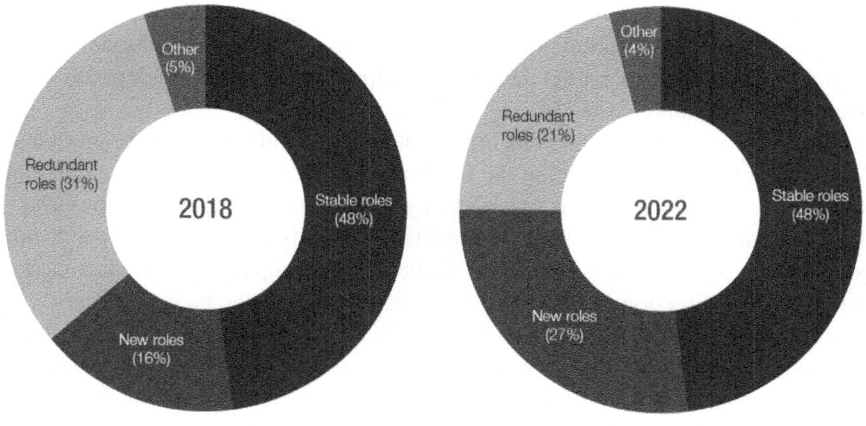


**Fig. 2** New redundant role 2018 versus 2020 [20]

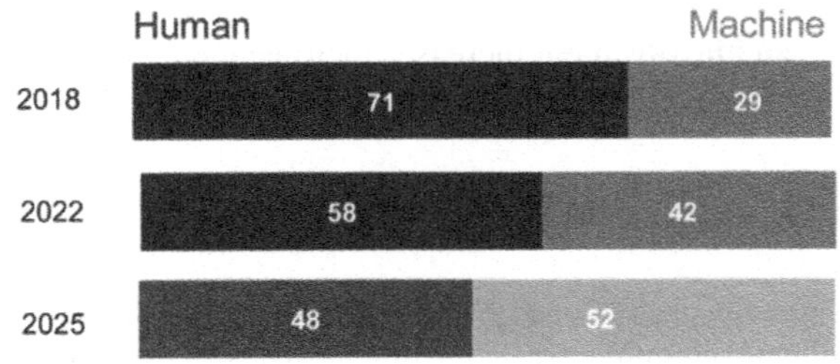


**Fig. 3** Automation machinery rate in industry [20]

ability to acquire new skills like "analytical thinking and innovation", emotional intelligence, "complex problem-solving", system analysis and evaluation. For acquiring those skills, the World Economic Forum estimates that everyone in the industry needs an extra 101 days of learning.

Thus, it can be concluded that an important step in the implementation of Industry 4.0 systems, is to develop a technological system starting from an elementary element, which can have management-oriented towards strategic planning, towards human force, with sustainable production [20]. To meet these needs, it is necessary to implement a command system with artificial intelligence in terms of the development of the manufacturing systems (machine tools) to answer the requirements of Industry 4.0 and 5.0.

## 4 The Proposed Idea of Artificial Intelligence System for Elementary Components of Production

The increase of the performances of any production system is a sine-qua-non condition, these being indispensably related to the technical progress, to the ways and modalities through which it can be introduced in the sphere of production.

A high-performance production system in terms of vertical and horizontal integration must respond to market demand, to adapt quickly to it through productivity, quality, and minimum costs [16].

One of the important factors that bring dynamism and efficiency of the production processes is the technical level of the machine tools that are included in a manufacturing structure, a level that is expressed by the degree of automation, the level of flexibility, precision, and quality of products, and finally by product and production costs. In this sense, the permanent and assiduous preoccupation of the machine tool builders in the direction of increasing their technical level can be interpreted.

Regarding the manufacturing on to machine tools, the general trend is related to their inclusion in an optimal derivative of technical objectives (production capacity, precision, and quality) or economic (minimizing processing costs, increasing productivity etc.). Achieving such goals is possible by intelligent technological systems, defined as artificial intelligence (AI), which can adapt in real-time to certain parameters that define the work process, to concrete conditions, to optimize certain criteria by using pre-established algorithms.

Artificial intelligence systems are internal optimization systems, they guide the production processes based on the information entered by the disruptive factors of the process. Thus, the machine tools whit artificial intelligence control systems can have independent functions or can be coupled with data management systems.

Machining processes on machine tools are considered complex processes, due to the great diversity of controlled or random factors that influence their evolution over time. Due to the increasing requirements of productivity, precision, quality, and costs as low as possible, in the field of production systems and especially of mechanical processing, there is an interest and a permanent concern to find the ways and means necessary to satisfy them.

The concerns are equally addressed, on the aspects of the managerial invoice of organization, control, coordination, and correlation of the industrial activities as well as on those of technical–economic invoice, seen through the performances that can be obtained in the exploitation of technological machinery and equipment.

In a unitary vision, all these aspects, as well as others, that characterize the production systems, seen through the prism of automation, presuppose a unitary structuring and hierarchy, which make them compatible, connected to an efficient and flexible information system, in other words, a permanent integration of the flow of materials with the informational one.

The hierarchical aspect in a manufacturing system implements the very flexibility of the system, and this can be understood as the ability to adapt the manufacturing system to the variation of the manufacturing load in economic conditions, conditions ensured by automation and robotization.

The degree of automation directly influences the degree of flexibility of a manufacturing system, which must respond to variations in production load. The ability to adapt the manufacturing system to the challenge introduced by the variation of the production load in its diversity involves an adaptation to each of the elements and modules in the system structure, machine tool, transport system, handling, storage etc.

Modern concepts on production processes have an integrative character, they address equally the aspects of management, manufacturing, economic and psychosocial, a complete and complex analysis from this point of view can only be done in a unitary vision, with considering all the connections and interdependencies between the factors that characterize a production process.

In the case of machining processes on machine tools, their efficiency is unconditionally related to their optimization for reasons of productivity, quality, or processing costs. Depending on the established option, algorithms are built based on which this process is carried out. In these conditions, the machine tool, through its construction, through the structure and functions of the control system that equips it, must meet the imposed requirements.

The satisfaction of these requirements can be seen from at least two points of view: the degree of loading of the machine tool and the utilization coefficient of the

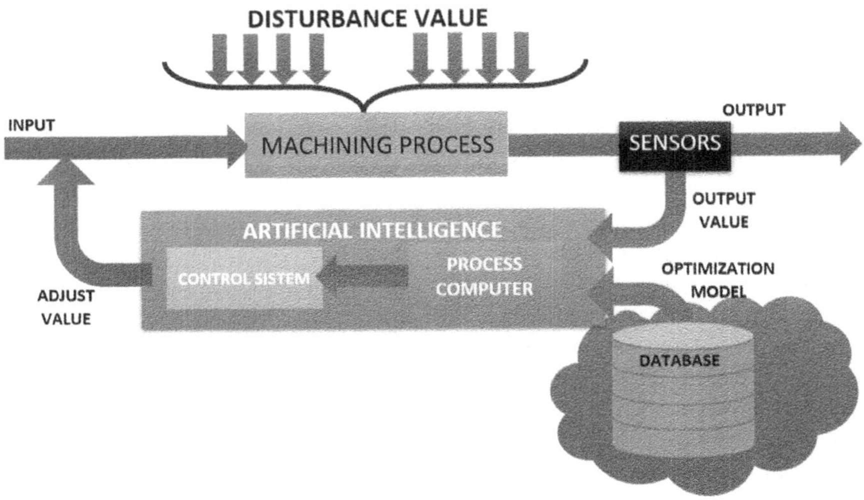


**Fig. 4** The proposed architecture of artificial intelligence system

machine tool. Within these essential landmarks, any work process must take place, and mathematical models that describe optimization algorithms based on economic or quality criteria, inevitably contain at least one term that refers to the requirements stated above.

Details are given in Fig. 4.

Optimal development of machining processes on machine tools is a complex condition that considers many influencing factors, external and internal, generated by the elastic structure of the technological system (machine tool, tool, device etc.) in their interdependence with the processing.

The large number of phenomena that accompany the machining processes on machine tools make them be in a quasi-permanent transient state, which creates difficulties in generating parts surfaces in cases where they are made in a variety of conditions and precision, quality, and productivity.

It can be appreciated that the machining processes on machine tools in their interdependence represent continuously disruptive dynamic systems, in which the automatic control through artificial intelligence systems constantly (sometimes discreetly) follows the regulation of process parameters based on algorithms (mathematical models) leading to the elimination of imbalances and therefore to the stabilization of processes.

The concept underlying the artificial intelligence systems of machine tool processing is based on the principles of automation and can be represented by the block diagram represented in the Fig. 4. It is noted that in the processing process acts many factors considered input values that can be controlled or disruptive, under the influence of which the output values are generated.

In general terms, the management of the processing process imposes the input values that aim to achieve a goal, which the processing process needs to satisfy,

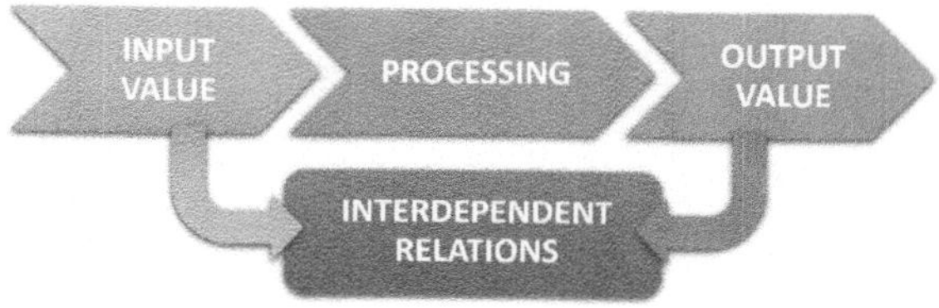


**Fig. 5** The interdependence relations

this goal can be expressed depending on the output values and the interdependence between them and the inputs values.

When the two categories of values (input and output) are in a relation through which the satisfaction of a condition pursued expressed by a physical quantity (e.g., productivity, precision, economy etc.), that focused so that the process satisfies in time the declared objective function, without the intervention of the human operator; this process is associated with the automatic control of the processing process.

With the help of the interdependence relations according to the Fig. 5, it is built on the algorithm through which the satisfaction of the purpose function is pursued. The schematic representation of the processing structure for the manufacturing processes obtained (showed in the Fig. 6) highlights the factors influencing them, the input and output parameters are specified, through which the manufacturing processes are defined.

It is noted the existence of the primary components, which ensure the working regime, it includes in their structure, the machine tool, the device, the processing tool, the processed material, as well as the technological environment. Under the action of the primary parameters, the manufacturing process takes place with the mechanical, chemical, and elastic phenomena that accompany it, and the relationship between them is achieved through the main function and as the secondary function expresses the connection between the manufacturing process and secondary parameters, evidence by the machining accuracy, surface accuracy, tool stability, tool life, productivity, and even economic aspects.

The primary parameters are specified in the concept and technological design stage, and they refer to the machine tool and its characteristics, the necessary devices and tools, the material to be processed, and the manufacturing process to be used. The relation between the two categories of parameters is expressed as a physical phenomenon through the manufacturing process, with all the phenomena that accompany these processes.

In the manufacturing processes on the machine tools, the parameters that define the efficiency or performance of processing can be evaluated analytically, quantitative, or qualitative, by measurable expressions given by output parameters such as productivity, quality, or processing costs.

Maximizing the efficiency of manufacturing on the machine tools is the result of the optimal conditions imposed on the process by considering the input parameters, mainly those that define the cutting regime, as well as those the influencing factors derived from the interdependencies and influences that the machining process induces

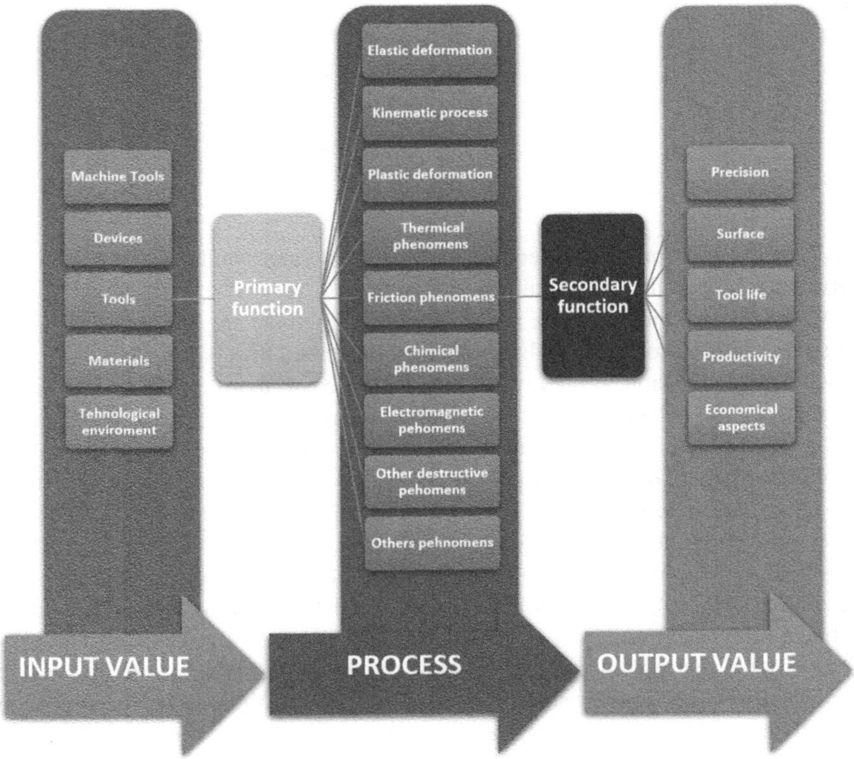


**Fig. 6** The schematic representation of the processing structure

on the other factors, expressed by the elastic and plastic deformations, thermal effects etc.

The analytical modeling of this independence by considering a number as large as possible of influencing factors, mainly expressed by constant values, experimentally determined, lead to satisfying the goal function, and in the most general term defines the algorithm that underlies the manufacturing process.

The automatic control of the machining processes on machine tools having as a basic process the real-time imposed algorithm transmits to the machine tools and/or technological equipment that compose the system through which the work process is directed to the desired goal, without subjective interventions of the human operator that he could introduce.

Therefore, we propose two forms of artificial intelligence systems for controlling the manufacturing processes and they are classified as such:

1. Artificial intelligence systems to increase processing accuracy;
2. Artificial intelligence systems designed to reduce processing costs (increasing productivity).

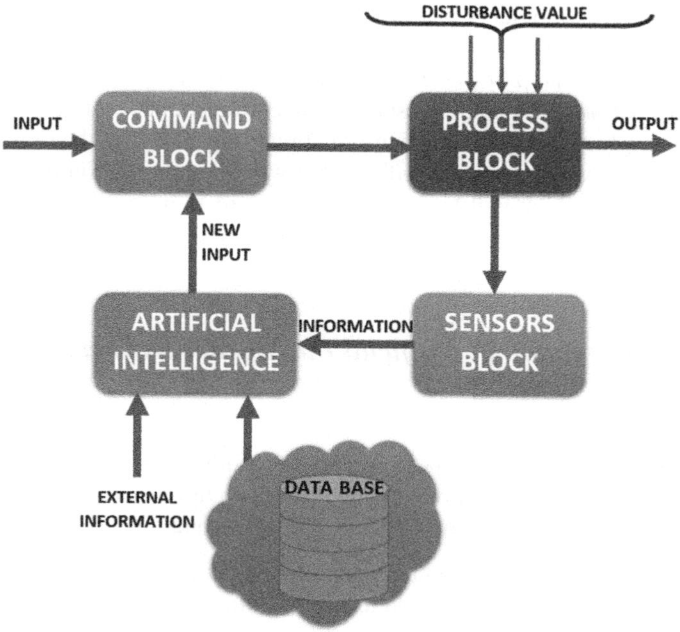


**Fig. 7** The block diagram of process using artificial intelligence

Adaptive management for manufacturing processes using the artificial intelligence systems, as structure and construction is based on parameters and values that define these processes, as they are shown in the Fig. 7, a block diagram whose functions are related to the collection, processing, storage, transmission, and comparison of information in the form of signals. The manufacturing processes are ensured by the control block through the parameters that define the input values, processes influenced by the disturbing values that define the phenomena that accompany these processes. Satisfying the optimal conditions imposed on a work process is determined by the parameters that characterize the output values, so that is required control and permanent supervision of its development, the function is provided by the sensor block that collects information about the process.

The calculation block in the structure of the artificial intelligence system is the one that processes the information received from the work process through the sizes of the parameters measured by the sensor block and certain information and signals received from the other production systems in the vicinity, as well as the information processed by artificial intelligence and compared to the database.

The new information, which have been achieved through calculations using artificial intelligence, is transmitted to the command block to direct the work process in the desired direction.

Thus, there are several parameters identified, depending on the purpose, because of the manufacturing process, which can be evaluated in size, such as the cutting forces, the process moments, the power consumed, the temperature, the process vibrations, the tool deformations etc. The existence and size of these parameters can be detected with the help of the sensors from the sensor block.

The information collected by the sensor block is transmitted to the artificial intelligence system which can also receive information from outside, in the form of correction information, which are analysed together with the information received by the sensor block, and which will be transmitted as correction signals to the command block to intervene on the parameters that characterize the process.

## 4.1 Implementation of Artificial Intelligence System on CNC Machine Tools

The increase of the degree of operability and flexibility when processing a part or the transition from one type of part to another, in conditions of increased efficiency, led to the development of a control system concept through which it can be ensured:

- Processing without direct human operator intervention;
- Manufacturing programs prepared using specialized software.;
- The control of the movement of the mobile subassemblies is performed by sensor systems;
- The movement is performed along several coordinate axes and with well-defined precision;
- The processing regime, the tools, change quickly.

  Increased flexibility of CNC machine tools is achieved by:
- Quick introduction of the dimensions for a new the part, respectively of the cutting regime;
- Control of all additional functions;
- Repeated use of NC programs that are optimized;
- Calling programmed correction values stored for different features: tools, devices etc.

The main disadvantage of CNC machine tools is related to the fact that in carrying out the process after an NC program, disturbing factors of any kind that occur in the process and influence working conditions cannot be considered, the program being a rigid control structure, which does not provide flexibility in the work process.

Starting from the generally valid tendency and specifying the work processes to ensure optimal development, in the case of machining processes on machine tools, the control systems are the ones that lead these processes, through them are transmitted to the machines all the options related to their development.

From the evolution of the control systems of the machine tools, it was noticed that they are characterized by the fact that:

- The work strategy is elaborated outside the process.

- The human operator is the one who is not directly involved in the management decision.
- The order is made by the direct and subjective action of the operator or indirectly through specific elements (stroke limiters, programming jacks, templates, or models).
- In all cases, the decision taken and launched can only be changed by a new intervention.

The size of the parameters that are introduced in these control systems, based on experience, or known empirical relationships, is not optimal, the processing being influenced by many factors that cannot be considered in the initial phase of making the NC program.

The main factors that introduce disruptions to the work process are:

- Diversity of cutting tools (materials, geometry, durability);
- Wide fields of variation of the processing addition and the nature of the semi-finished product;
- Large diversity of materials processing characteristics, diversity of static and dynamic characteristics of the machine tools, tools, and work devices.

All these aspects lead to the conclusion that an external optimization of the manufacturing processes on machines with numerical control is difficult to achieve, it introduces enough shortcomings.

The control systems with artificial intelligence represent progress compared to the control systems presented, by the fact that they can make decisions based on findings out what happens inside the process and ensure necessary corrections, without the need of human operator intervention during the process.

The control systems with artificial intelligence, through the concepts that underlie its realization, lead to the economic use of machine tools.

The concept underlying the artificial intelligence control systems of machine tools is based on the development of algorithms that derive from the conditions of optimization of the processing process imposed by technical and economic considerations. Depending on the purpose, several parameters that characterize the work process are constantly measured, the values of these measurements being the output values from the work process that is compared with reference values of the same nature, these are established based on mathematical models of optimizations, and from the result of the comparison derive values of the quantities that will be transmitted in the process as correction values, to adjust the technological parameters that would lead to the technical–economic effects initially imposed.

Two types of implementations of the artificial intelligence system are possible, depending on the output parameters that are to be obtained from the manufacturing process, namely:

a. Artificial intelligence systems designed to reduce processing costs (in-creasing productivity) This artificial intelligence system is based on regulating the cutting speeds, which ensure the regulation of some parameters of the processing which results in an economic effect, cost, productivity etc.;

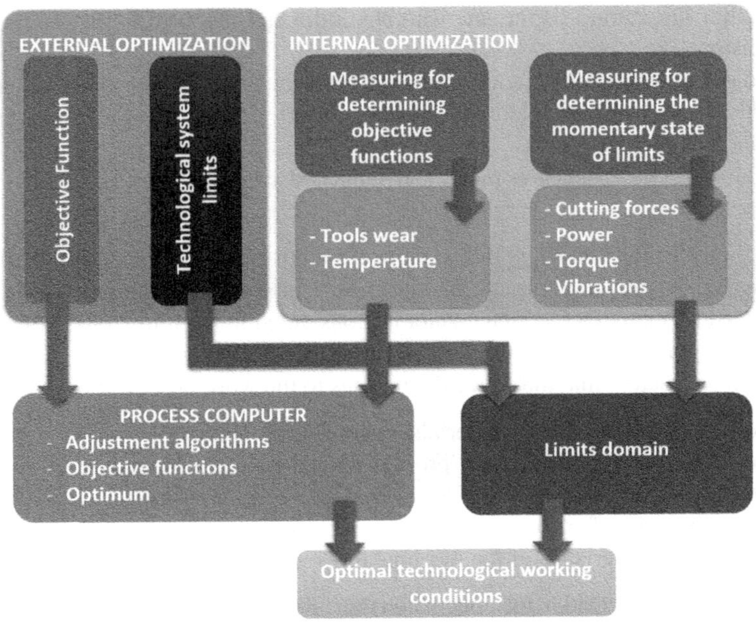


**Fig. 8** The values measured in the work process

b. Artificial intelligence systems to increase processing accuracy. This artificial intelligence systems are based on modifying the parameters of the cutting speed, so that a certain indicator of the manufacturing process (precision, productivity, cost price) is maintained in the field of optimum, considering the limit values. The optimization criterion is built on a mathematical model that includes in its structure the parameters of the cutting speed.

The principles scheme of the artificial intelligence system based on a block representation is presented in Fig. 8. The values measured in the work process and for the purpose pursued by their adjustments is presented in Fig. 8.

A principle functional diagram for a CNC machine tool in which an artificial intelligence system is implemented is shown in Fig. 9.

In such a system, the numerical control equipment is coupled to the artificial intelligence system by the numerical control itself. In this way, the usual programming of geometric, technological, and auxiliary functions is ensured, and through the artificial intelligence system with the help of sensors for measuring torque and for measuring deformations based on an initial strategy established the process can be driven and optimized.

By driving using the artificial intelligence system of the work process, the adopted strategy ensures the minimum cost of processing by measuring the torque and adjusting the feed speeds $V_{fn}$ (axial direction), $V_{fn}$ (radial direction), and the cutting speed $V_c$ (speed $S$), this being specific to roughing processes.

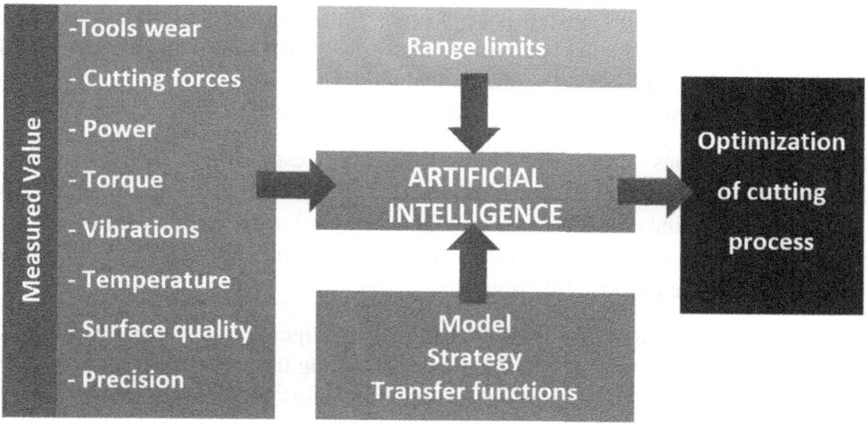


**Fig. 9** A principle functional diagram for a CNC machine tool in which an artificial intelligence system

In the case of finishing processes, the artificial intelligence system to increase processing accuracy comes into action as a priority for adjusting the processing regime ensuring the modification of feed rates and cutting speeds to ensure a required processing precision, by limiting the deformations of the main spindle.

In most cases, the artificial intelligence system to increase processing accuracy comes into action to regulate the processing mode, which measures a single output value in the working process and modifies one or two parameters of the cutting regime (most often the feed rate and cutting speed).

In the case of machining on a turning machining centers, the systems with artificial intelligence for adjusting the cutting regime of the rule systems measure: the main component $F_x$ of the cutting force, the torque at the main spindle $M_t$ or the motor power of the main drive and change the feed rate or speed of the workpiece, to ensure a maximum load of the machine. Artificial intelligence systems for changing the parameters of the cutting regime are based on measuring tool wear or temperature in the work area, changing the speed to obtain the minimum cost or optimal durability of the tool.

## 4.2 Mathematical Algorithm for Artificial Intelligence System

Mathematical modeling of a function with one or more variables to "build" an algorithm through which in a process of cutting is achieving the objective function (purpose), involves the development of mathematical ways to ensure the condition process optimization.

In general, variable parameters (one or more) that compose the function objective structure may have a domain of definition consisting of continuous and limited values, which obviously simplifies its mathematical model.

The function objective structure in their complexity can include:

1. Constant values - consecrated or experimentally established;
2. Stabilized values that are constant at a given time, such as the characteristics of the material, the characteristics of the cutting tool, the loading forces etc.;
3. Random values, to which certain values are assigned to achieve the optimum defined by the purpose function.

In the conception and design of an artificial intelligence system that would lead the work process on a machine tool, when establishing the objective (optimal) function, considered the algorithm based on which the work process is carried out, it is very important that in a first phase to defining the parameters and conditions with significant influences, in special the stabilized and random values.

It is very important to establish the field of definition of the objective function by specifying the technical limits of the variables and restrictions for which the process is possible and then establishing the decision parameters (independent variables) identified as having major influences in the process.

A cutting process led by an artificial intelligence system permanently involves adapting the process to an imposed condition, allowed as a reference quantity (parameter) with constant or variable value in time or space, according to a law that satisfies the objective function.

Driving a manufacturing process with an artificial intelligence system in which the volume of information can be very large (big data's), with a very short time to make a decision, is needed to involve different ways of processing from simple electronic devices to programmable controllers or PCs with high computing power, that can manage the information's and resolve the mathematical algorithm, but also to store all the information needed to be able to compare the information from next processes and have a base for comparing to make a correct decision.

Due to the permanent and rapid variation of the working conditions in the manufacturing processes, it is necessary that the information processing can be performed at high speeds. In this sense, we can exemplify the case of the variations of the input parameters and the characteristics of the "elastic" technological system, through which the production capacity or the processing precision can be limited.

It is known that the dimensions of the chip, in the case of machining on machine tools, are variable at the entry of the tool in the material, especially when machining a conical surface or when processing different surfaces (profiling), especially by varying the depth of cutting. The influence of the characteristics of the elastic technological system, related to the workpiece processing, the variation of the stresses on the workpiece, the variation of the elastic deformation is depending on the position of the tool on the coordinate axis, the tool holder construction etc.

It is known that the problems related to the characteristics of the elastic technological system are complex in the work process, the rigidity of the system is influenced by constructive factors, thermal phenomena, dynamic aspects, which are variable from one machine tool to another, even of the same type, but all with influences of variation of the thermal deformations of two specific subassemblies.

## 4.3 Artificial Intelligence Systems to Increase Processing Accuracy

As mentioned before, the artificial intelligence system to increase processing accuracy change the parameters of the processing model so that a certain output value that characterizes the process can be adjusted within the prescribed reference size limits that ensure an effect technically and economically favourable.

For such cases, NC programming of the machine tool ensures the cycle of dimensions with the initially programmed technological parameters (speeds, feeds, or the number of passes) and considered optimal for a certain requirement, accepted as an optimal condition (loading, accuracy etc.).

If the process is no longer carried out within the limits imposed by appropriate reference quantities (torque, power, drilling etc.) the artificial intelligence system inhibits the control circuits of technological parameters present by NC and switches the control system to the artificial intelligence system, thus ensuring the development of the process within the limits imposed by the reference input values.

In the case of numerical controlled machine tools, the moves on the coordinate axes are made by stepper motors, the artificial intelligence system monitors the working process by measuring the torque of the main spindle by torque sensors, as well as its speed. Thus, the artificial intelligence system modifies the initial values of the longitudinal advance and, if applicable, of the transverse one, established by NC programming, to maintain a maximum possible load corresponding to the power reference value. In establishing it, in addition to the nominal power of the main drive motor, the restrictions introduced by the tool and the part were considered.

## 4.4 Artificial Intelligence Systems Designed to Reduce Processing Costs (Increasing Productivity)

The artificial intelligence system designed to reduce processing costs (increasing productivity) ensure the management of cutting processes, so that by changing the parameters of the cutting regime to keep an indicator expressed by an objective function (cost price, productivity etc.) in an optimal field, considering the restrictions imposed.

In such cases, connecting the artificial intelligence system designed to reduce processing costs with the numerical controller of the machine tools can be achieved higher performance indicators, but with the observation, the artificial intelligence system designed to reduce processing costs are more complex and more difficult to achieve.

Thus, for such a case as an objective function was considered the F function, of the form:

$$F = \frac{V_m}{C_m + \frac{C_m t_s + C_s}{\mu_a}\mu} \quad (1)$$

For which the notations were introduced:

$C_m$ the cost of using the machine tool for one minute (operators remuneration + the cost of depreciating the machine tool);
$V_m$ the volume of material, removed in the unit of time;
$C_s$ Expenses related to the tool exploitation between two resharpening (change of cutting insert);
$t_s$ tool change—tool adjustment;
$\mu_a$, $\mu_v$ maximum permissible tool wear and speed of tool wear.

The cost of removing a volume of material $V_o$ is expressed by the relation.

$$C'_{m_o} V_o = C_m(T + t_s) + C_s \quad (2)$$

Were:

$$C'_{m_o} = \frac{1}{T}\int_0^T C_{m_0} dt \quad (3)$$

And:

$$V_0 = \int_0^T Btv_f df \quad (4)$$

where the above relation can be written:

$$\frac{1}{T}\int_0^T C_{m_0} dt = \frac{C_m + \frac{1}{T}(C_m t_S + C_s)}{\frac{1}{T}\int_0^T BtV_f da_p} \quad (5)$$

The notations that were used in (5) are:

$C_{m_o}$ the cost of one mm of material manufactured;
$C'_{m_o}$ the average value of $C_{m_o}$ during the durability T of the cutting tool;
$T$ durability of the cutting tool;
$B$ the width of the cutting tool;

$V_f$ cutting speed;
$a_p$ depth of cut.

Admitting that for the durability of the cutting tool the chisel T corresponds to the wear $\mu_a$ and $\mu_v$ its allowable wear can be expressed by:

$$\mu_a = \int_0^T \mu_v da_p \tag{6}$$

Which leads to:

$$T = \frac{\mu_a}{\mu_v} \tag{7}$$

The relationship considering the above becomes:

$$C_{m_o} = \frac{C_m + \frac{C_m t_s + C_s}{\mu_a}}{Ba_p V_f} \tag{8}$$

Noting that this is exactly the inverse of the F function.
It is chosen that the objective function be:

$$F = \frac{1}{C_{m_0}} \tag{9}$$

Considered as a function of capacity, which is expressed by:

$$F = \frac{V_m}{C_m + \frac{C_m t_s + C_s k}{\mu_a} \mu_v} \tag{10}$$

For which $0 \leq k \leq 1$ which reaches the extremes of the interval, for $k = 1$, when aiming to minimize the cost and $k = 0$ when aiming to maximize productivity or for an intermediate situation.

The volume $V_m$ removed can be expressed by the relation:

$$V_m = Ba_p V_f \tag{11}$$

Is noted that the optimization function $F$ can be easily calculated if the momentary values of the quantities $V_m$ ($a_p$, $V_f$) and $\mu_v$ are known. This way of expressing the objective function was adopted to facilitate the measurement of values mentioned earlier.

The momentary tool wear rate was established indirectly according to the relationship:

$$\mu_v = \frac{du}{da_p} \tag{12}$$

$$\mu_v = k_1 V_m + k_2 U_\theta + k_3 V_{M_t} \tag{13}$$

where:

$U_\theta$ represents the electrical voltage through which the temperature in the work area is expressed;

$V_{M_t}$ the speed of variation of the torque, known as:

$$M_t = \frac{dM_t}{da_p} \tag{14}$$

where: $k_1$, $k_2$, $k_3$ are wear constants.

From this brief foray into the mathematical model adopted for the proposed purpose function, it is noted that the system must measure the temperature of the cutting tool insert and the torque at the main spindle, as well as the vibrations of the main spindle. Under these conditions, the system control was integrated with the process computer to perform the following functions:

a. The calculation of the indicators $V_m$ and $\mu_v$ according to the relations presented previously.
b. Calculation of the momentary value of the capacity function $F$.
c. In accordance with a certain strategy adopted, depending on the momentary value of $F$, the parameters are adjusted: $n$—spindle speed and $V_f$—cutting speed.

Optimization is achieved by searching for $n$ and $F$ values by artificial intelligence system, considered search optimization, for which the optimization function has analytical extremes (external artificial intelligence systems) and is known in mathematics as the gradient method.

The block diagram of a numerical controlled system of a machine tools, with the integration of an artificial intelligence system for reduce processing costs, which aims as the main function (purpose) to optimize processing costs in direct relation to productivity, is presented in the Fig. 10.

An artificial intelligence system for numerical controlled machine tools is considered an analytical passive. This system modifies the work advance according to the direct measurement of the power in the main drive:

$$P_c = UI\cos\varphi\eta_{AP} = \frac{F_t D_s n_{AP}}{612010^3} \tag{15}$$

For which the tangential component of the cutting force is obtained indirectly.
The meaning of the notations is as follows:

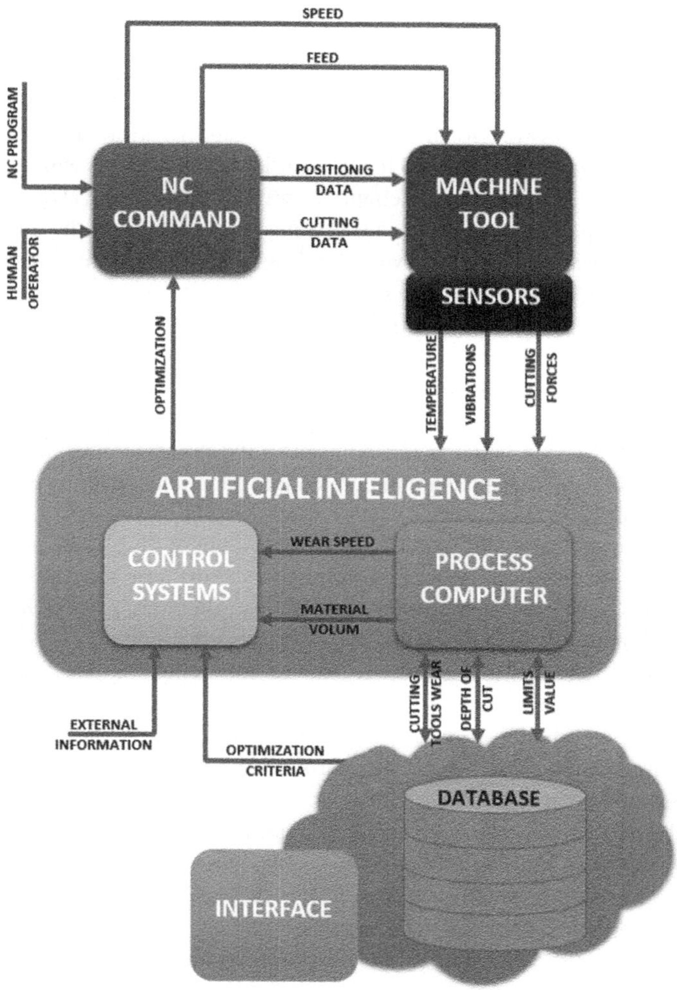


**Fig. 10** The block diagram of a numerical controlled system of a machine tool

| | |
|---|---|
| $U, I, \cos\varphi$ | are electrical values of voltages, current, and power factor; |
| $\eta_{\mathrm{AP}}$ | the yield of the main spindle; |
| $n_{\mathrm{AP}}$ | main spindle speed; |
| $D_s$ | diameter; |
| $F_t$ | the tangential component of the cutting force. |

## 5 Conclusions

In recent years, the industrial sector has been facing fierce competition in the market, imposed by the process of globalization. The Industry 4.0 system has grown with a high degree of expectation, due to the current IT infrastructure, which allows the industry to quick and efficiently implement the principles that stand at the base of this system. A challenge, in the Industry 4.0 systems, will be finding the human factor that has the talent and ability to develop and implement algorithms necessary for artificial intelligence, but also for the way of implementing human resource learning systems, taking advantage of the current infrastructure.

The evolution of industrial production to an industry 4.0 offers enormous opportunities for sustainable production, using the omnipresent infrastructure in the field of information and communication technology (IT).

Artificial intelligence systems can be used in the programming of numerically controlled machine tools, which are integrated into a manufacturing system, but also used as stand-alone.

Although the system is not so complicated, but a major disadvantage is the large volume of data that needs to be processed in a very short time for adjusting in real-time for each type of part. Determining the composition of the processing force that is established indirectly by measuring the power, introduces other errors, it is known that the variation of power is due to other than the component of tangential force and speed (efficiency for example).

In the situation when at a maximum load of the CNC machine tool it reaches the lower limit value of the feed rate, through the artificial intelligence system optimization will be achieved by dividing the cutting depth for several passes.

With all the obvious advantages that NC programming offers, they still have limits, dictated by the programming mode itself. It is known that this is an a priori stable external programming, which cannot consider the process in its intimacy, which limits the performance indices of machine tools.

The system has the disadvantage that the programming of the part is informative, and is made outside of the machine tools, in many situations the programmer is not on the same location whit the production layout, and he does not know the machine tool history. By this form of programming one of the first problem that appears is the incorrect choice of the cutting regimes, and this introduces errors.

The solution adopted and used on an increasing scale is to integrate the NC control system with data management systems (PLM). But by integrating them with artificial intelligence systems, the limits introduced by numerical programming are extended due to the advantages introduced by the artificial intelligence system, a system that introduces the possibility of internal modification of the NC program in real-time and considers the entire technological system of the CNC machine tool.

For such ways of driving the machining process on machine tools, the function of the NC control system has mainly the role of achieving the dimension cycle. Instead, the artificial intelligence system comes into action when disturbing factors appear in the process that tends to change the state of stability of the system. This action of taking over is being defined by the algorithms that underlie the unfolding of the process.

From simulations and experiments using machine tools with numerical control in which it was integrated, and artificial intelligence systems have proven effective in exploitation, they lead to increased productivity that can reach up to 50%. At the same time was observed that the cost of processing can be reduced by up to 33%.

## References


1. Acatech (2015) Umsetzungsstrategie Industrie 4.0 Ergebnisbericht der Plattform Industrie 4.0.
2. Barnett R (2011) Lifewide education: a new and transformative concept for higher education. In: Learning for a complex world: a life wide concept of learning, education, and personal development. s.l., s.n
3. G Chryssolouris 2004 A hybrid approach to the verification and analysis of assembly and maintenance process using virtual reality and digital Mannequin technologies Virtual reality and augmented reality applications in manufacturing Springer London 97 110
4. Eurostat (2019) Eurostat regional yearbook 2019. Luxembourg, Imprimerie Bietlot in Belgium
5. Kagermann H, Lukas W, Wahlster W (2011) Industrie 4.0—Mit dem Internet der Dinge auf dem Weg zur 4. industriellen Revolution. In: VDI Nachrichten. s.l., VDI
6. Kagermann H, Lukas W, Wahlster W (2015) Abschotten ist keine alternative. s.l., VDI Nachrichten
7. Ozkeser B (2018) Lean innovation approach in industry 5.0. s.l., s.n., pp 422–428
8. Perez-Lara M, et al (2018) Vertical and horizontal integration systems in industry 4.0. s.l., Springer Nature
9. Plattform Industry 4.0 (2015) Industrie 4.0 whitepaper FuE-Themen
10. Rada M (2017) Industry 5.0 definition. https://www.linkedin.com/pulse/indus-trial-upcycling-definition-michael-rada/. Accessed 21 Jan 2021
11. T Schuller R Desjardins 2007 Understanding the social outcomes of learning OECD Publication Paris
12. Skobelev P, Brovik SY (2017) On the way from industry 4.0 to industry 5.0: from digital manufacturing to digital society (6):307–311
13. Spath D, et al (2013) Produktionsarbeit der Zukunft – Industrie 4.0. s.l., Fraunhofer IAO, Fraunhofer Verlag
14. Stef D (2012) Dezvoltarea produsului in contextul fabricii digitale. Timisoara, Editura Politehnica Timisoara
15. Stock T, Seliger G (2016) Opportunities of sustainable manufacturing in industry 4.0. s.l., Procedia CIRP, pp 536–541
16. Swat M, Brünnet H, Bähre D (2014) Selecting manufacturing process chains in the early stage of the product engineering process with focus on energy consumption. In: Technology and manufacturing process selection: the product life cycle perspective
17. The Economist (2017) Lifelong learning—How to survive in the age of automation. https://www.econo-mist.com/special-report/2017/01/12/lifelong-learning-is-becoming-an-economic-imperative. Accessed 2 Feb 2021
18. Uhlmann E, et al (2014) Potential of reducing the energy consumption for machining TiAl6V4 by using innovative metal cutting processes. s.l., s.n
19. VDI/VDE-GMA (2015) Status report Referenz architekturmodell Industrie 4.0
20. World Economic-Forum (2018) *The* future of jobs report 2018, center of the new economy and society, Geneva: s.n

# Chapter 5
# Arguments for Emerging Technologies Applications to Improve Manufacturing Warehouse Ergonomics

**Anca Mocan, Alin Gaureanu, Gyula Szabó, and Beata Mrugalska**

**Abstract** Technology has disrupted each current industry, and supply chain is not going to be an exception. Businesses are already starting to establish interconnected global networks of Cyber-Physical Systems with the help of the Internet of Things and Cloud Computing. In this context, the chapter will debate aspects related to the new challenges of reducing ergonomics risks in manufacturing warehouse logistics by valorizing emerging technologies to create workplace wellbeing. After an extended literature review regarding the relevant ergonomics approaches in warehouse logistics, there will be presented some warehouse ergonomics solutions to be considered for the next generation of logistics system. The solutions described will refer to the monitoring and improvement of the ergonomic reality. Finally, conclusions and future trends will end the chapter.

**Keywords** Ergonomics · Industrial logistics · Innovation · Sustainability · Well-being · Emergent technologies · Innovation

A. Mocan (✉)
Amaris Consulting, Vernier, Switzerland

A. Gaureanu
Faculty of Management in Production and Transportation, Politehnica University of Timisoara, Timisoara, Romania
e-mail: alin.gaureanu@upt.ro

G. Szabó
Donát Bánki Faculty of Mechanical and Safety Engineering, Óbuda University, Budapest, Hungary
e-mail: szabo.gyula@uni-obuda.hu

B. Mrugalska
Faculty of Engineering Management, Poznan University of Technology, Poznan, Poland
e-mail: beata.mrugalska@put.poznan.pl

 
A. Draghici and L. Ivascu (eds.), *Sustainability and Innovation in Manufacturing Enterprises*, Advances in Sustainability Science and Technology,
https://doi.org/10.1007/978-981-16-7365-8_5

# 1 Introduction

The present chapter context is related to the industrial logistics system. Researchers and practitioners considered the Industry 4.0 Revolution is in the digital transformation phase which proves its maturity and announce the beginning of the Industry 5.0 Revolution. This will also impact the logistic systems and the supply chains. "Next generation supply chain optimization will pair both smart, interconnected digital environments with the human insight it takes to make the most out of them" [125, 135].

Industry 5.0 is perceived as a "next-generation auto manufacturing, armies of autonomous trucks or high-tech, robot-assisted surgical theatres" eliminating human operators from manufacturing plants and warehouses. Furthermore, it is promised (and already demonstrated in some industrial areas) that the applications and exploitation of new and emerging technologies will positively impact the comfort and well-being of our professional and personal lives, too. Also, there are expectations that digital transformation will be especially "powerful supporting logistics and supply chain management", referring to the whole processes and activities. "And while this next generation of supply chain solutions is still in the making, deploying Industry 5.0 technology to empower digital supply chain participants holds the promise of helping people across the supply chain fulfil their potential. This could mean [135]:

- Bringing more customization to a supply chain, improving not only customer satisfaction but also efficiency and margins.
- Reducing supply chain risks and waste based on more current information;
- Enabling supply chain and logistics functions to spend more time on strategic experimentation and less on fighting fires or matters of basic execution;
- Improving supply chain integration for more strategic partnerships;
- Getting more value from an organization's human capital, helping to retain and transfer knowledge around a particular supply chain's characteristic".

In this chapter, both theoretical and practical perspectives have been considered for increasing warehousing speed and accuracy to cope with ever-changing industry and customer requirements. During the literature review of different aspects considered relevant for the research, it has been discovered that ergonomic solutions for logistic operations were not considered to be the first area of interest in improving logistic systems, specifically the warehouses' workplaces and their general working conditions. Most researchers invested in how the Industry 4.0 paradigm would apply to processes automation and robotization and not how it could be applied to the human worker's activity optimization. Thus, the topic importance and relevance cannot be ignored in research anymore because the Industry 5.0 transition is already ongoing [125].

Some arguments of introducing the Industry 4.0 paradigm in the logistics activities are provided by facts that have been provided by important international organizations:

- A Price Waterhouse Cooper (PwC) survey [149] on Industry 4.0 implementation has shown that industrial leaders are digitizing essential functions of the enterprises, both "within their internal, vertical operations processes, as well as with their horizontal partners along the value chain. They are additionally enhancing their product portfolio with digital functionalities and introducing innovative data-based services". The study results showed an expectation of digitization expenses to increase from just 33% in 2017 to over 70% in 2020 [149].
- According to the World Economic Forum White Paper Digital Transformation of Industries in 2016, within SCM "key technologies will be included autonomous transport and drones, sensors for monitoring supply chains and 3D printing technology. Digitally enabled companies will incur procurement costs of 0.22% of net revenue, less than half of those of their peers (0.5%)". The value of digitization within the next decade will exceed €1333 billion [192].
- The supply chain will evolve and become smarter, more transparent, and more efficient because of this digital transformation and the subsequent use of intelligent cooperative systems. "There will be a particular focus in new models which will be more closely to individual customer needs, promoting a significantly increase of the decision-making quality and become more and more flexible and efficient in the near future" [13].
- Concerning Distribution Centers (DC), Industry 4.0 technologies can help enable automated systems to adapt to their environment to execute tasks more efficiently in collaboration with humans. Both from the technological and organizational point of view, equipments (such as low-cost sensors), computer vision, augmented reality (AR), wearables, Internet of Things (IoT), analytics and high-performance computing can be used to enhance existing automation and remove the existing historical situation of low automation in DCs without highly standardized products. The last years trend has shown that the ways in which DC facilities are being used is changing and there is a need for smarter, more adaptable systems services [36].
- Companies are aware of this change and are proceeding to adapt their investments accordingly. Based on a recent international study of DCs, "half of the surveyed IT and operations decision makers planned to move to a more modern, full-featured warehouse management system in 2015. By 2020 this number increases to 75%; 51% of those surveyed expected increased investment in real-time location systems that track inventory and assets throughout the warehouse as well as, equipping staff with technology (73%), bar code scanning (68%), tablets (66%) and IoT (62%)" which alone is predicted to connect 20 billion devices or "things" to the existing internet infrastructure by 2020 [197].

To reduce the productivity dips caused by poor ergonomic design of processes, that appears when ergonomics is seen as an afterthought rather than the first analysis that needs to be done, a solid ergonomic application framework is necessary to be developed. Besides the technological modifications that need to be ergonomically designed and integrated, Industry 4.0 also offers technological opportunities to support work-based learning, to reduce operators' physical loads and better monitor their health.

Although robotics and human–machine interaction are disciplines with an extended literature, but the actual complexity of professional human behavior opens new ways for research that allow creating new paradigms. The relative lack of understanding from the business world regarding emerging ergonomics developments in the research world also proves to be a fertile ground not only for improvements based on knowledge transfer, but for setting new and innovative developments.

Within the present chapter's research framework, the possibility of applying Industry 5.0 technology in the warehouse of a manufacturing facility will be analyzed. The existing market technology will be categorized, examined, and rated based on its capacity to improve the ergonomic efficiency within the warehouse, as part of the logistics system. Based on this rating and existing ergonomic framework, a model will be developed that can be used in a business environment, to track the current level of ergonomic applications, implementation practiced within the company and a development and to show the investment pathway needed to reach Industry 5.0 benchmarks.

The motivation of the research topic is determinate because ergonomics is becoming increasingly important in warehousing and logistics. Currently, due to European Union (EU) and national legislation, companies in Western Europe must reconsider their policies and focus more on individual rights, on implementing solutions of human-friendly logistics via excellent logistics ergonomics. The focus on reducing warehouse personnel turnover to reduce the unsafe environment that workers might experience. The companies with the most stable and productive warehouse working teams are the ones with the most developed programs for safety training, ergonomics, and housekeeping.

The increased interest in ergonomics, to reduce worker turnover, is coming at a time when companies have started to adopt Industry 5.0 techniques and technologies. "Scientific research will always be impeded if clear definitions are lacking and it has been shown so far that companies face difficulties when trying to develop ideas or act without understanding what to aim towards" [63, 141].

This proves that the current situation requires a structured analysis to define if ergonomic applications of Industry 5.0 technologies are suited for a specific company, turning the currently practiced empiric approaches into scientific ones. Companies need to understand what Industry 5.0 benefits they can reap and how to leverage those benefits into creating a safety, more issue-free work environment for their human operators. Due to the high interconnectivity and the digital transformation of Industry 5.0 technologies, companies also need to understand the legal limit of their application and how to approach their presentation to the operators that would be in direct contact with them.

Already some authors are introducing the Logistics 5.0 Revolution by considering or referring to the introduction of "smart era where robots, the cloud, and other technologies are obtaining, comparing, and analysing data to make processes more efficient, understand clients, and make better decisions" [167]. Top actors of the logistics practice recognized improvements of operational processes in terms of their speed, preciseness, and traceability. Thus, general perception is that Logistics 5.0 will be much more valuable for supporting Manufacturing 5.0.

"Whether companies implement Logistics 5.0 or not will greatly depend on how much they foster innovation. Usually, the companies that detect areas of opportunity and understand the benefits of their processes are the ones that implement tools like the Internet of Things (IoT) and robots. Because the cost of technology is much lower than before, small, and medium-sized companies can also make the best use of technologies like the cloud, which save them from having to invest in the construction of data centers and hardware" [167].

## 2 Relevant Ergonomics Approaches in Warehouse Logistics Research Areas (The Ergonomics Risks)

Ergonomics in some shape or form has existed ever since the beginning of time, as Australopithecus Prometheus started making pebble tools and scoops from antelope bones and selecting the best ones for use over and over, in a clear sign that the tools that made his tasks easier were preferable. From an industrial point of view, the solidifying of ergonomics as a science came along with the creation of scientific management, at the beginning of the nineteenth century.

Considering the historical perspective, the philosophical framework for the unique discipline of ergonomics (from the Greek "ergos" meaning work and "nomos" meaning natural law) has been first defined by W.B. Jastrzebowski in 1857. "Ergonomics was proposed as a scientific discipline with a very broad scope and a wide area of interests and applications, encompassing all aspects of human activity, including labour, entertainment, reasoning, and dedication" [82, 83].

In addition, the International Ergonomics Association supports the idea that "ergonomists contribute to the design and evaluation of tasks, jobs, products, environments, and systems to make them compatible with the needs, abilities, and limitations of people". General definition is graphically represented in Fig. 1. The subject of ergonomics is predominantly known in North America as "human factors", to emphasize the application of ergonomic methods to situations outside of work; the terms human factors and ergonomics are essentially synonymous (ISO 63785).

"Ergonomic problems at the workplace and bad work organization are part of the contributing risk factors to the abovementioned occupational safety and health problems. A number of situations within the workplace are conjectured to contribute to the increasing magnitude of musculoskeletal disorders (MSD) suffered by the workers, including postural stress from prolonged sitting, standing, or awkward position; stereotyped and repetitive tasks leading to chronic injury; peak overload injuries to the axial or peripheral skeleton; environmental factors; and psychosocial factors including psychological stresses, job dissatisfaction, and complex social issues, such as compensation laws and disability system" [136].

The MSD of the upper extremity have been discovered to be related to work for hundreds of years. Bernardini Ramazzini, considered the father of occupational medicine, was the first who developed a synthesis of working conditions and the

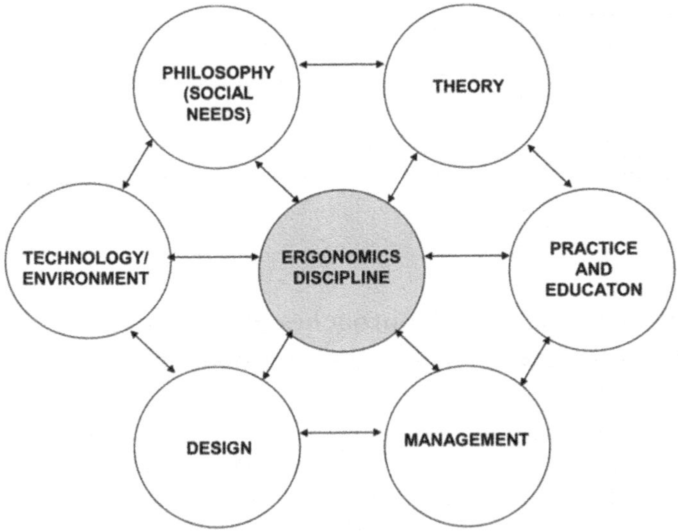


**Fig. 1** General dimensions of ergonomics discipline [83]

associated pathology, by considering the occupational health perspective. In his eighteenth century book "*De morbis artificum diatribe*", he said that these diseases "arise from three causes: first constant sitting, the perpetual motion of the hand in the same manner, and thirdly the attention and the application of the mind" [44].

During the age of industrialization, more attention has been given to the relation between work and health. In the 1830s, in the United Kingdom was recorded the first epidemic of work-related MSD for the case of the civil service processes; thus, the steel nib was introduced. It was suggested that a subsequent epidemic among telegraphists a few decades later, has led to the definition of the term "nervous breakdown" [105].

Later in "The Science of Labour and Its Organization" (1919), Józefa Joteyko introduced and discussed measurements of occupational fatigue and principles in the context of the scientific management wave [136]. Worker efficiency since then has been continuously improved upon, as ergonomics has gained more traction in the past few decades. However, the growth is not increasing as fast as would be expected, considering the new ground scientific ergonomics has recently covered [59].

In the mid-90s, a vision of the future of ergonomics was put forward by [161], when he argued that given the historical situation at the time, ergonomics was going to be looked at as a way to facilitate the improvement of the quality of life, as a partner in a multidisciplinary approach that would allow, entice, and enforce behavioral change that would benefit not only businesses, but the world at large. He mentioned that "ergonomics also, needed to accept the fact that few solutions are universal, as solutions need to acknowledge the morals and the ethics of place". With the rise in globalization and digitalization, new terms such as "situation awareness, mental workload and virtual reality" started to appear in practice and in the literature, along with complexity management and cognitive systems development. In his address in 1997, Helander considered "the 90s as the decade of cognitive and organizational ergonomics". The point was raised again in 2000, saying that "the need for a broader systems perspective in addressing ergonomics challenges has not yet been found" [127].

As can be seen in Fig. 2, there is a strong (0,83) correlation between the interest that is shown for the concept "Human Factors/Ergonomics" and the concept of "Industrial Revolution". It is also interesting to notice that even though there is a strong correlation between the two topics, there is also a downward trend throughout the years, showing that less interest has been given to the topics in the past decade and a half. The trend has however stabilized after the financial crash of 2008 and exists at the same level ever since. What is interesting to see however is the spread of this interest worldwide [125].

As shown in Fig. 3, industrialized countries, in their majority, are more interested in "Human Factors/Ergonomics", while mostly non-industrialized or lightly industrialized countries are interested in the new "Industrial Revolution". This spread is normal because ergonomics usually becomes enacted after the industrial revolution takes place. While the goal of ergonomics is to fix the issue before it ever occurs, it is not an easy task to achieve, leading it to mostly be a post-factum implementation in the industrial ecosystem [125].

Based on a relevant literature review, it has been shown that most decision support models that are designed by "considering the logic of order-picking as a business process concentrated only on the short-term economic impact" [57]. Thus, "decision-making processes are ignoring the influence of the order-picking processes on the human operator" [48].

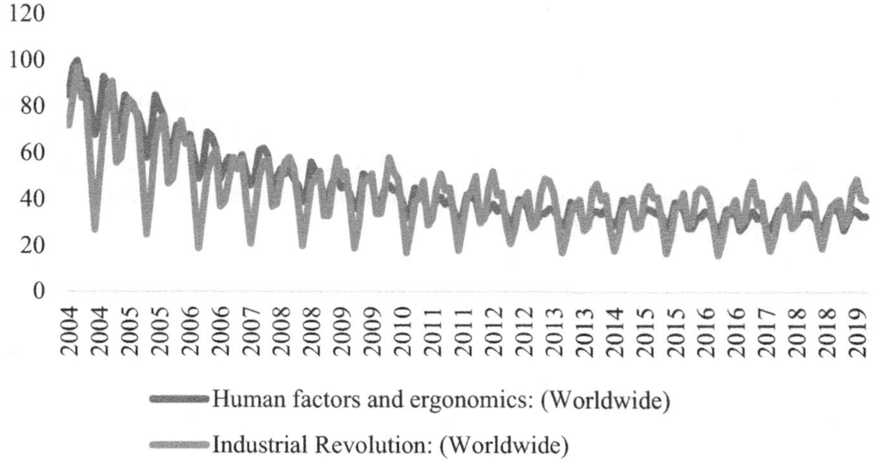


**Fig. 2** Correlation between "Ergonomics" and "Industrial revolution" (Google Trends Search, 27-05-2020)

**Fig. 3** Geographical areas of interest for ergonomics (Google Trends Search, 27-05-2020)

The question then stands, why have operational managers and engineers not kept up with the recent developments in research? The answers could lie in a lot of areas, such as ineffective terms of translating theory to practice, inadequate design of training programs, or even the communication style of these new discoveries not being suitable for corporate audiences [59]. Whatever the reason or combinations thereof, the result is a vast difference in ergonomics application maturity levels within the same industries or between similar sized firms of different industries.

In the following, we shall focus more on demonstrating the ergonomics implications in the case of the logistics systems started by analyzing the facts on the fatal and non-fatal accidents at work, Fig. 4 [45].

The context of "ergonomics in the warehouse logistics is often presented by its regulatory health and safety aspect (see the statistical data of Fig. 4), while not enough light has been shed on the competitive advantage opportunities that its applications can lead to" [124]. In most of the published studies related to this field, ergonomics principles are presented at the base level, oversimplifying the information and its implementation [93, 124, 170, 185].

The pressure felt by distribution center managers who say that safety is the number one driving concern in recent years, most likely causes this and that solutions are needed to increase it [171]. While also being forced to reduce costs and increase output speed, short-term vision leads them to apply superficial and non-integrated solutions to have a quick fix for the issue. This view of ergonomics and workers' safety is limiting both from an educational as well as a production perspective, as it keeps repeating the same basic information over and over, giving the reader the impression that as long as these sound bites are implemented in the overall production process nothing more needs to be done.

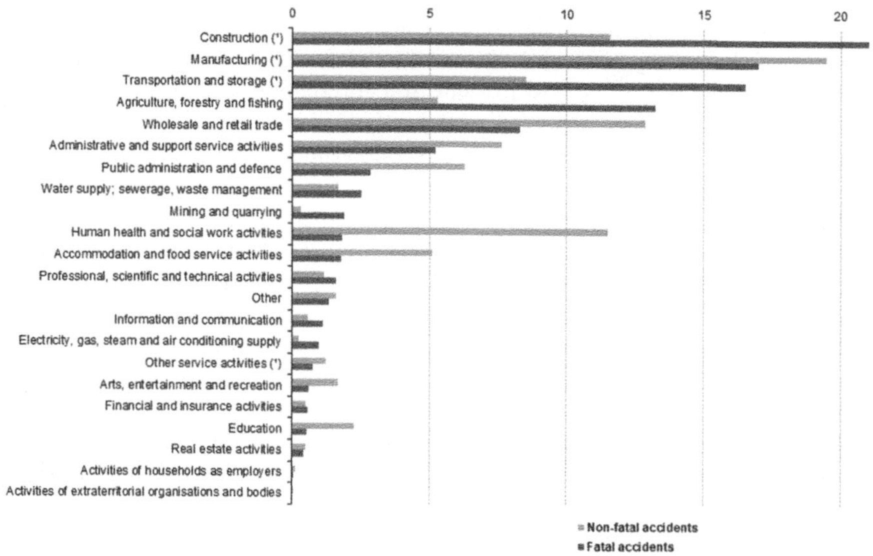


**Fig. 4** Percent of fatal and non-fatal accidents at work by NACE section, EU-28, 2020 (% of fatal and non-fatal accidents; Eurostat (hswn201) and (hswn202), 2020)

This is not the correct way of thinking, as ergonomics has a deeper well of solutions to offer to its users. Thankfully, the presentation of this topic has been changing in recent years along with the maturity of Industry 4.0 mind-set and the debate on the Industry 5.0 Revolution beginning. Here, ergonomic developments are seen less as being ergonomics for ergonomics' sake and more as belonging to the area of cost saving and operational streamlining facilitators. Within the logistics work areas, a large wave of technological advancements has been seen in warehousing, materials handling, and packaging. As mentioned previously, these are an integral part of the logistics process. Explaining it simply, inventory needs to be warehoused, material handling is necessary to move inventory between transportation vehicles and warehouses or within warehouses themselves, and packaging helps move individual products as efficiently as possible between two places. Firms can choose between operating their own warehousing facility and outsourcing the process to a third party. Regarding to this decision, activities that commonly happen in a warehouse are sorting, sequencing, order selection, transportation consolidation, and, in some cases, product modification and assembly [73]. Within these processes, handling or manipulation is an important activity that can be manual, fully automated, or somewhere in between.

EU-OSHA[1] and International Labor Organization (ILO) estimated that the rough annual cost related to work-related ill health and injury is costing the European Union 3.3% of its gross domestic product (GDP), or approximately €476 billion every

[1] Retrieved from: https://osha.europa.eu/en/about-eu-osha/press-room/eu-osha-presents-new-figures-costs-poor-workplace-safety-and-health-world (Access 17–04-2021).

year. Additional arguments on these aspects have been provided by [122]. "With an average warehouse size between 2,000 and 4,000 $m^2$, warehousing activities take up to between 2 and 5% of the cost of sales of a corporation meaning that improved efficiency and throughput time can lead to significant reduction of costs even in companies in which warehousing is not the core business" [71].

"The increase in productivity that ergonomic system implementation can lead to will also almost automatically lead to an improvement in throughput time which in turn leads to higher returns. This is achieved by reducing the number of errors and reducing the effort put into the picking and transportation. Reduced throughput also leads to less time in the warehouse, emptier warehouses as a result, therefore the realization that the need of space is not directly proportional to the number of items to be picked and shipped, but is a function of complexity, error and performance of the warehouse workers. It has been proven that investment in ergonomic improvements in the workplace can result on a return on investment ranging from 3:1 to 15:1" [61]. "Previous research has demonstrated that in the case of a workstation redesign in an assembly factory, settings made by ergonomists led to an increase of over 15% of the productivity and because of the higher quantity of the work output, the productivity per worker has been increased to €2000–2500" [62].

The proof therefore exists that cost reduction in ergonomics is possible, not just by reducing the cost of injury, but also by redesigning the way in which the work is done. This is being taken to the next level by the implementation of Industry 5.0 solutions within the supply chain.

## 2.1 Work-Related Physical Pain

"Within the management of the storage and movement of goods, a warehouse is a facility in the supply chain that is used to consolidate products, achieve economies of scale in manufacturing or in purchasing [14], or provide value added processes and shorten response time [55], being one of the main areas where logistics companies can gain competitive advantage by offering their clients tailored services" [122]. Stock management is at the core of a warehouse's activities. Within a warehouse, Manual Materials Handling (MMH) is defined as "any transporting or supporting of a load by one or more workers. It includes the following activities: lifting, holding, putting down, pushing, pulling, carrying, or moving of a load" (Council Directive 90/269/EEC).

Manual handling is developed in most of the working systems, but employees from the fields of construction, agriculture, and warehouse logistics are more exposed by frequent heavy loads processes. That implies that the worker is exposed to tasks that increase his ergonomic risk factors such as lifting, twisting, lateral bending, maintenance of static postures, heavy load carrying, or a combination of these. "Manual handling can result in fatigue, and lead to injuries of the back, neck, shoulders, arms, or other body parts. There are two types of injuries that can result from manual handling namely:

- *Acute* (e.g., cuts, bruises or fractures that are caused by sudden accidents);
- *Chronic* (e.g., the damage to the musculoskeletal system of the body due to gradual and cumulative wear caused by repetitive manual handling)".[2]

These chronic injuries are called *musculoskeletal disorders* (MSD, recognized by the EU-OHS). The most common MSD are caused by inappropriate material handling and often affect the neck, upper limbs, lower limbs, or the back. The following subchapters will delve deeper into the topics discussed relating to various work-related MSD (WRMSD) in warehouse logistics.

### 2.1.1 Back Pain and Work-Related Musculo-Skeletal Disorders

Research related to back pain in "WRMSD focuses mostly on low back pain. Low back pain is any back pain between the ribs and top of the leg. Work-related low back pain is any back pain originating in the context of work and is clinically considered to have been probably caused, at least in part, or exacerbated by the claimant's job. Usually, the origins of low back pain are grouped under four categories" [88]:

- Discogenic/neurological, meaning that one or more intervertebral discs are cause of the pain;
- Muscular/ligamentous, meaning that muscle sprains are the cause of the pain;
- Structural, meaning that bones and bone structure are the cause of the pain;
- Other disorders, meaning that anything from tumors to infections to genetic diseases are the cause of the pain.

"Within a warehouse environment, one of the most incorrectly done activities is the lifting and setting down of weights. Most people chose to pick up a weight by bending their back rather than squatting on their haunches" [122]. Furthermore, "when a load is held away from the body, the stress on the lower back increases substantially. The maximum weight one can lift safely is reduced dramatically the further away from the body that the load is handled" (Fig. 5).[3] "Long reaches increase the risk of a lower back injury. The altered mechanics of bending at the waist as opposed to the hips causes undue strain on the muscle ligaments and vertebrae that can lead to muscle pain, herniated disks, strains on the thoracolumbar fascia or latissimus dorsi muscles or sciatica" [122].

"Work-related low back pain and injuries are some of the most common musculoskeletal disorders that manual handling causes. About a quarter of European workers consider that work-related back pain injuries affect their health, and in areas such as construction, agriculture and transportation as many as 61% of people are suffering from some form of work-related musculo-skeletal disorders with the most

[2] As supported by EU-OHS publication available at: http://www.mtpinnacle.com/pdfs/E-fact_14_-_Hazards_and_risks_associated_with_manual_handling_in_the_workplace.pdf (Access 24–04-2021) and [42].

[3] Moving and Handling Techniques, Retrieved from: http://docplayer.net/16075167-Moving-and-handling-techniques.html (Access 20–04-2021).

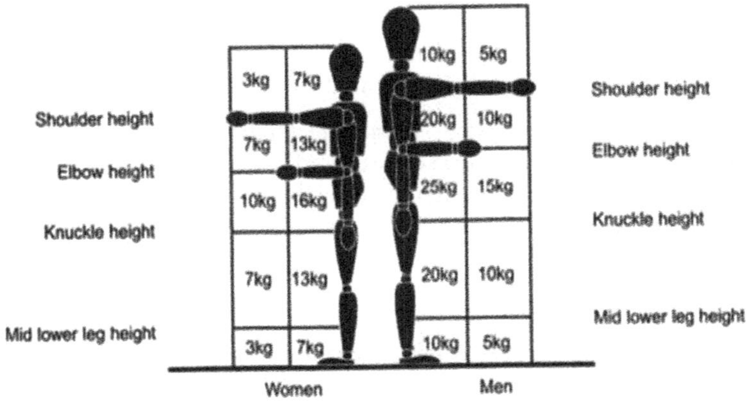


**Fig. 5** Numerical values for lifting loads. Retrieved from https://www.hsa.ie/eng/Publications_and_Forms/Publications/Retail/Management_Manual_Handling.pdf

reported health problem reported being backache at 43%" [122]. This was followed by "muscular pains in the neck or upper limbs (42%), headache and eyestrain, and overall fatigue (both 35%) and muscular pains in the hip or lower limbs (29%)".[4] According to the physical environment index, that "measures exposure to vibrations, tiring positions, lifting people, carrying heavy loads and repetitive movements, the most prevalent causes for MSD", the situation has been improving over the past 10 years in almost all countries except for France and the United Kingdom [45, 122, 125].

Manual Materials Handling (MMH) is a hazardous activity because it affects the lower back [94, 179]. Studies have shown that in the case of warehouse workers, they face significant risks for lower back pain, MMH being a dominant working style [37, 174]. "When the warehouse superstores subsectors are isolated from other subsectors, the proportion of injuries to the back ranges from 39 to 50% of all compensated injuries depending on the subsector, while the average proportion is 27.9% across all sectors in the province of Quebec" [174].

One of the most human-work intensive activity in warehousing is palletizing and depalletizing, which is defined as the transfer and stacking of material from or onto pallets. This activity's capability to cause MSD has caused researchers to analyze the ways in which spine loading can be reduced by means of package location within the pallet, pallet distance, operator position, pallet orientation, and palletizing condition.

Because human operators must often lift with the body in extended reach positions, studies have been conducted to analyze whether the location of the object on the pallet as well as the weight of the object may determine the risk of injury to the operator. One study found that "lifting a box from low and distant locations imposes nearly twice the risk probability as lifting the same weight box from a higher and closer

[4] Retrieved from: https://ajuntament.barcelona.cat/tempsicures/sites/default/files/6th_european_working_conditions_survey.pdf.

location, revealing that 97% of the lifts from the lower layer of the pallet resulted in spine compression values above 3400 N and would be expected to increase the risk of an occupationally related low back disorder". Increase in the weight of the box also increased spinal loading, but this impacted spinal loading greater when the position from which one lifts was taken into consideration as well. "Distant locations on a pallet required the worker to reach farther and increase the mechanical moment arm of the worker-environment system". Increasing the mechanical moment arm was equally significant between long vertical and long horizontal reach. The findings had "several implications for the design of the distribution workplace, such as how one would benefit from raising the pallets off the floor or adding handles to the boxes with the purpose of raising the height of the lift" [122]. The introduction of height increasing solutions was corroborated by other research, which proved that self-levelling carousels and adjustable carts were effective in reducing the spine loads when compared to the traditional pallet-cart condition and reduced loads, too [150].

The pallet distance from the operator is directly proportional to the torso kinematics (velocities and accelerations) and the risk of low back disorder. This kinematic also increases when packages are picked up at the lower end of the pallet [78]. "Lifting from the floor was reported to produce up to twice the amount of spinal loading as lifting from a higher up location, such as elbow height. It has been shown that lifting from below the knee or from the floor magnified the negative effects of MMH on workers with low-back pain symptoms, older workers, workers with knee osteoarthritis, and workers with a high body mass index" [137].

The study of [49] analyzed the design of operator workstations and their effect on reducing the worker injury risk. Its results clearly demonstrated "that positioning the pallet at the end of a conveyor belt, results in a significant reduction in loading on the lumbar spine compared to positioning the pallet on the side of the conveyor, most likely due to the ability of using the momentum of the bag as it comes off of the conveyor as opposed to having to forcefully redirect the bag from its course along the conveyor when the pallet is located on the side" [49]. Additionally, the study showed that controlled lower-level loading has a higher impact on the pressure exerted on the spine. The way to reduce this loading was either by raising the pallet level, or by loading the pallet with the help of uncontrolled drop stacking. It was posited that lumbar compression for the drop technique was approximately 600–800 N lower than those for controlled placement at the lower levels [49].

### 2.1.2 Upper Limb Disorders Generated by the Manual Materials Handling

"Holding and carrying involve static endurance, which can be determined by the length of time a limb can maintain a certain position. The amount of muscular strength is the maximum amount of force that a muscle can exert under maximum contraction. It has been shown that isometric (or static) activities cause greater levels of exhaustion than isotonic ones, meaning that holding a weight for a period is more straining that moving it" [153]. "Holding a weight also changes the point of gravity, thus putting a strain on posture muscles such as the trapezius and the erector spinae muscles to maintain a proper upright position" [122].

Work-related Upper Limb Disorders (WRULD) are related to the injuries of different body segment as the neck and the upper limb segment (shoulders, arms, forearms, wrists, and hand) because of physical work exposure that involves application of force, either to move objects or to keep them steady. Some of WRULDs have clearly defined signs and symptoms (such as tendonitis, carpal tunnel syndrome, or osteoarthritis), while others do not have a clearly defined symptom list, presenting as pain, discomfort, or numbness [183]. The most common WRULDs are (Petreanu and Seracin, 2018)[5]:

- *"Neck*: Tension Neck Syndrome, Cervical Spine Syndrome;
- *Shoulder*: Shoulder Tendonitis, Shoulder Bursitis, Thoracic Outlet Syndrome;
- *Elbow*: Epicondylitis, Olecranon Bursitis, Radial Tunnel Syndrome, Cubital Tunnel Syndrome;
- *Wrist/Hand*: De Quervain Disease, Synovial Cyst, Trigger Finger, Carpal Tunnel Syndrome, Guyon's Canal Syndrome, Hand-Arm Vibration Syndrome, Hypothenar Hammer Syndrome".

According to "results from the sixth European Working Conditions Survey 2015, 44.4% of workers cited problems with muscular pains in the shoulders, neck, and/or upper limbs, thus making WRULDs the most common form of occupational disease in Europe". The "carpal tunnel syndrome studies have found prevalence rates of 7–14,5%". According to Eurostat data, "upper limb disorders are more self-reported than lower limb disorders" and these seem to affect women more than men. While men report problems more often, "women reported upper limb disorders as the most serious work-related health problem" [45].[6] The main WRULD factors are [4]:

- "Force application resulting in heavy mechanical loads on the neck, shoulders, and upper limbs;
- Working in awkward positions-muscles must contract, and greater mechanical loads are placed on the body;
- Repetitive movements, especially if they involve the same joints and muscle groups, and if there is an interaction between forceful activities and repetitive movements;
- Prolonged work without the opportunity to rest and recover from the load;
- Local compression of tools and surfaces;
- Hand/arm vibration, causing numbness, tingling or loss of sensation, and requiring greater force when gripping".

[5] Presented in the OSH WIKI. Retrieved from: https://oshwiki.eu/wiki/Risk_factors_for_musculoskeletal_disorders_development:_hand-arm_tasks,_repetitive_work (Access 18–04-2021).

[6] Eurostat (2018). Statistics Explained, Accidents at work statistics. Retrieved from: https://enuveprod-universitatpolit.netdna-ssl.com/php_prevencionintegral/sites/default/files/noticia/45658/field_adjuntos/11539.pdf (Access 24–04-2021).

These work-related musculo-skeletal disorders (WRULD) factors are exacerbated by work environment such as poor workspace layout, or poor lighting practices; individual factors such as the physical capability of the worker, or level of experience; and organizational and psychosocial factors such as lack of control over the performed tasks, or time pressure. Within the European context, there are a series of main directives related at preventing WRULDs, namely, 89/391/EEC,[7] "which covers the measures to encourage improvements in the safety and health of workers, 90/270/EEC,[8] which, covers the minimum health and safety requirements for work with display screen equipment and 90/269/EEC,[9] which covers the identification and prevention of manual handling risks".

Within a warehouse environment, the activity that causes the highest incidence of WRULD is order picking, due to the "manual carrying and lifting of bulky and/or heavy items. Order picking is the process of retrieving items from their storage locations in a warehouse to fulfil customers' orders", it is the most labor-intensive (manual and manual-mechanical process), and thus high time-consuming in manufacturing warehousing, accounting for more than 50% of warehouse operating costs [48, 57, 178].

The way in which WRULDs are assessed is by using the Rapid Upper Limb Assessment (RULA) methodology. This is an assessment that is used to analyze the ergonomic risk that workers face in relation to upper extremity musculoskeletal disorder [116]. "A single page worksheet is used to evaluate required body posture, force, and repetition. Based on the evaluations of the movements, scores are entered for each body region in section A for the arm and wrist, and section B for the neck and trunk. After the data for each region is collected and scored then the tables on the form are then used to compile the risk factor variables, generating a score that represents the level of MSD risk", which ranges from negligible to very high [116].

The findings of a WRMSD study presented by [15] "showed that the heart rate was significantly changed from rest level during performing both lifting and pulling products tasks. On the other hand, the pulling task showed a greater total wrist and arm score than the lifting task, that may be due to the range of heavy loads in the pulling task. In the pulling task, the wrist pain score (4,78) was the highest among other body regions. Following the wrist pain score was the prevalence of lower arm disorders (4,16). According to the total score of the RULA method, the lifting product task was a greater ergonomic hazard than the pulling products task in term of MSDs as well as being more physically demanding, since the score of the lifting task (5,76) was significantly greater than the score of the pulling task (4,88)" [15]. This type of pain can lead to severe productivity loss, as one study found that 56% of analyzed

---

[7] Retrieved from: https://eur-lex.europa.eu/legal-content/EN/ALL/?uri=celex%3A31989L0391 (Access 24–04-2021).

[8] Retrieved from: https://eur-lex.europa.eu/legal-content/EN/TXT/?uri=CELEX%3A31990L0270 (Access 24–04-2021).

[9] Retrieved from: https://eur-lex.europa.eu/legal-content/EN/TXT/?uri=celex%3A31990L0269 (Access 24–04-2021).

workers reported a productivity loss associated with pain intensity, pain interference with work, and fear avoidance [111, 125].

### 2.1.3 Lower Limb Disorders During the Manual Materials Handling

The lower extremity skeletal system consists of the pelvic girdle segments, thigh, lower leg, foot and the joints, cartilages tendons and muscles that connect the regions with each other. The largest articulations in the area are the hip, knee, and ankle, other articulations including that between the tibia and fibula, the tarsals, the tarso-metatarsal regions, the metatarsals, the metatarso-phalangeal segments, and the phalanges [169]. The most common risk factors associated with the Work-Related Lower Limb Disorders (WRLLD) are the repetitive kneeling or squatting, fixed postures, or frequent jumping from a height.

The biggest issue related to the study of lower limb disorders is the underestimation of the problem. At the European level, "a data trends analysis on muscular pains of the lower limbs is not possible because the indicators were introduced only in the 2000 and there is no universally recognized list of occupational musculoskeletal diseases. Some national data can provide more extensive information on lower limb disorders, distinguishing by body part affected" [43].

Local statistics show that there are significant gender differences between the type of WRLLD women and men must work through. "Women are significantly exposed to prolonged standing and walking, reporting more problems in hips, legs, and feet, while men are more affected by knee problems. Prolonged standing and walking are notable risk factor in the "traditional" sectors such as agriculture and construction, but also greatly affects workers in service professions, above all in hospitality and retail, a fact that is, as mentioned before, barely reflected in monitoring and recognition of lower-limb disorders"[10] [43].

Even in smaller scale studies, "the course of lower limb pain is studied less frequently than back and upper limb pain. In a study population of 139 patients with hip complaints, 24% reported recovery after three months, increasing to 37% after 12 months. A study of 251 patients done in relation to knee pain resulted in 25% of respondents reporting recovery after three months, increasing to 44% after 12 months" [41].

The same lack of interest issue appears when analyzing the risk assessment models and tools that can be used for lower extremity regions. There is no WRLLD-specific assessments and the tools that can be used are intended to be used for whole body use. Within the logistics systems, the lack of information is even more jarring. From statistics data available, it could be known that back and lower limb disorders can occur in the case of truck drivers and warehouse employees, among others such as parcel handlers, or operators of cranes and other large vehicles [146]. But overall, there seems to be very little warehouse specific literature in relation to WRLLD.

---

[10] Retrieved from: https://osha.europa.eu/en/publications/osh-figures-work-related-musculoskeletal-disorders-eu-facts-and-figures.

The few papers found are not warehouse industry-specific, related to issues such as activities involving pulling and pushing. "The amount of force that can be exerted by your limbs depends on body posture and the direction of force. For example, when standing, one can exert more force when pulling backwards than when pushing forwards. Pushing is preferable to pulling for several reasons such as the awkward positioning of the arm stretched behind the body during pulling while facing in the direction of the walk that places the shoulder joint in a posture that can increase pain and possible injuries. Similarly, pulling while walking backwards can lead to accidents very easily as there is no line of view to see the travel path" [29, 122]. "Further, research demonstrates that people can usually exert higher push forces than pull forces. In some situations, pulling may be the only viable means of movement, but such situations should be avoided wherever possible, and minimized when pulling is necessary. It has been shown [67] that pushing and pulling lead to an increase in shoulder aches on a dose response relation and that sometimes lower back issues can also occur" [122].

### 2.1.4 Prevention of Work-Related MSD

As already presented in the previous sub-chapters, the health of a workplace environment is directly affected by [162], as shown in Fig. 6:

- Things that employees bring with them to the workplace: personal experiences, health practices, genetic makeup, and attitudes;
- What the workplace does to and for employees once they are there: the steps that an organization takes in creating its work culture and climate (the physical and psychosocial organization of work).

To remove or diminish the effects of WRMSD at work, a series of preventive strategies can be applied. Prevention is characterized by considering three different levels:

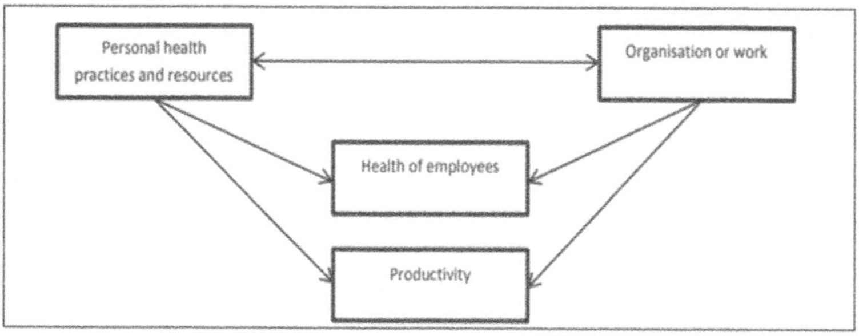


**Fig. 6** Interrelation of health and productivity

- *Primary prevention*, in which the efforts are focused on discouraging people from using unhealthy behavior, thus applied prior to the onset of diseases. Within an organization, it includes risk assessment of processes and ergonomic trainings;
- *Secondary prevention*, in which the efforts are focused in stopping the development of a disease and treating it before irreversible pathological changes take place. In a work context, it involves the identification and health monitoring of workers at risks;
- *Tertiary prevention*, in which the efforts are focused on applying all available measures to reduce impairment or disability and to promote adjustments to irremediable conditions. Within a work environment, this implies return-to-work actions.

While companies are working mostly in applying primary and secondary prevention within their local policies, the international administrative organizations develop legislative contexts for the application of tertiary prevention. The EU-OSH Agency (2007b) even recommends employers to act "to protect workers from the risks of manual handling" by managing risk assessment activities as well as setting up prevention campaigns. Their recommended prevention measures included are (EU-OSH 2007b):

- "Designing and organizing tasks to avoid manual handling completely, or at least restrict it";
- "Using automation and lifting equipment";
- "Organizing manual handling tasks in a safe way, with loads split into smaller ones, and proper rest periods provided";
- "Providing information and training to workers on tasks, and the use of equipment and correct handling techniques".[11]

The current methods of treating WRULD currently imply active symptom surveillance, which can lead to early assessment and treatment. To assess the workplace risk, education and involvement of workers and line managers is needed to create a collaborative and non-adversarial approach. External help should also be used to ergonomically assess the jobs being done and re-engineer the working stations to avoid unsafe repetition, force, and prolonged abnormal postures [60]. Treatment and recovery for chronic work illness often prove to be unsatisfactory with the result in that case being permanent disability and loss of employment. According to the present literature study on effectiveness of work-related intervention and workers rehabilitation, we conclude that it is scarce.

The impact of training advice and working techniques for lifting equipment in a way that prevents back pain is not yet clear. Currently, due to the prevalence of back pain and its resulting economic consequences, "employers must ensure that workers receive proper training and information on how to handle loads correctly" [110], with the help of specific techniques that have been presented by government agencies to reduce the load on the back. "Minimum health and safety requirements

[11] As mention in the EU-OSHA document: http://www.mtpinnacle.com/pdfs/E-fact_14_-_Hazards_and_risks_associated_with_manual_handling_in_the_workplace.pdf (Access 04–05-2021).

for the manual handling of loads where there is a risk particularly of back injury to workers" (Council Directive 90/269/EEC[12] of 29 May 1990; [110]).

The study of [110], found that training did not lead to a positive outcome concerning diminishing the risk of back pain. With a sample size of just over 2500 people, the resulting confidence intervals show that the possibility of the review being too small to detect differences in incidence cannot be excluded. However, almost all studies analyzed showed almost non-significant differences between training and not training. It could be that the reason "for the lack of an effect was because the intervention was not appropriate", but the studies were classified based on the learners' participation and studies that involved more intense training methods did not show outcomes that are more positive [110].

The review did not show significant capability of prevention of back pain related to workers being trained and advised by health professional on the correct lifting and handling procedures [110]. Given this situation, it could be posited that perhaps current training techniques are not the way in which to reduce the incidence of back pain. Either new training techniques must be developed, or tools must be used to help practitioners reduce the number of tasks that could cause back pain injuries in the first place.

It has been shown that direct observation methods (e. g., Ovako Working posture Assessment System, OWAS) are not replicable in dynamic work conditions and are subject to intra- and inter-observer variability [15]. Colombini [30] edited a consensus document in association with the International Ergonomics Association technical group for work-related musculoskeletal disorders, which included checklists and models that describe and evaluate each of the principal risk factors. In this paper, he defined the generally accepted requirements such as tasks, cycles, and technical actions. Even though the authors provided an exposure assessment method that takes awkward positions and repeatability frequency, the authors also acknowledged the difficulties involved in validating methods where many interactions occur. The need for standardization of assessment methods and methodologies is still present in today's environment for complex ergonomic activities.

As with most chronic diseases, MSDs have both occupational as well as non-occupational risk factors. As the body does not stop being in use once office hours are over, physical stresses to musculoskeletal tissues continue. In addition to occupational demands, other daily life aspects (e.g., physical activity, sports, housework) could generate physical stresses to the musculoskeletal system [148].

Within a warehouse environment, one of the most important factors that influences the pervasiveness of risk factors is the relation between stock volume and available storage space. "An imbalance between the amount of stock and the available storage space has three general types of consequences: the increase of risk factors related to the development of musculoskeletal disorders caused by increased MMH operations, an increased risk of accidents, such as falls or loss of balance and impacts on

[12] https://eur-lex.europa.eu/legal-content/EN/TXT/?uri=celex%3A31990L0269 (Access 04–05-2021).

productivity and quality of service due to issues such as stock loss or time wasted that ultimately lead to customer dissatisfaction" [37].

## 2.2 Work-Related Stress

According to Council Directive 89/391/EEC,[13] work-related stress is considered part of the legal domain of occupational safety and health within the European Union. According to the EU-OSHA, "work-related stress could be defined as a pattern of emotional, cognitive, behavioral, and physiological reactions to adverse and harmful aspects of work content, work organization and the working environment. It is a state characterized by high levels of agitation and distress and often feelings of not coping". Work-related stress is experienced "when the demands of the work environment exceed the workers" ability to cope with (or control) them.

According to the EU Labor Force Survey, between the years of 1999 and 2007, approximately "28% of respondents, roughly corresponding to approximately 55.6 million European workers, have reported that their mental well-being had been impacted by exposure to psychosocial risks", with the most common selected risk factor being too much work and too little time to do it in. In addition, the EU-OSHA have underlined that "the cost to Europe of work-related depression was approximately €617 billion annually, made up of costs to employers resulting from absenteeism and presentism (€272 billion), loss of productivity (€242 billion), health care costs of €63 billion and social welfare costs in the form of disability benefit payments (€39 billion)". This is a topic that managers are aware of, with 79% of them reporting high concerned level of workplace stress. Also, managers considered the main reasons for work-related stress to be "time pressure, difficult customers and poor communication between management and employees" (EU-OSHA 2014, completed by the OSH Wiki article: "Psychosocial risks and workers health"[14]).

A 2015 study from Harvard and Stanford University business schools found that job-related stress problems have a monetary impact on the US economy of approximately €170 billion a year in health care costs and contributed to almost 120.000 deaths a year. By looking at more than 200 other studies, the researchers have found that job insecurity increased the odds of reporting poor health by 50% and long work hours increased mortality by almost 20%.

While they have been shown to be ripe in environments that pervade the physically unsafe work environment [144], warehouse jobs are not exempt from work-related mental health issues either. In some trucking companies, 200–300% employee turnover is the norm [85], while retention rates for the warehousing industry range from 80 to 25%. The cost of replacing an employee in this sector is estimated to be of thousands of dollars [56, 131]. While not all departures are directly related to mental

[13] Retrieved from: https://eur-lex.europa.eu/legal-content/EN/ALL/?uri=celex%3A31989L0391.

[14] Retrieved from: https://oshwiki.eu/wiki/Psychosocial_risks_and_workers_health (Access on 23-04-2021).

health, it is widely acknowledged that big warehousing companies are not highly receptive to employee mental distress. Research related to this topic shows both workers and team leaders presenting with different stress symptoms such as work overinvestment, passive behavior, and a reduction in self-esteem [175]. Besides the pure link with mental stress, the primary or basic needs of safety, resting, and food motivates employees [177], reminding that work-related stress can be the cause of workers' health problems.

Amazon workers, for example, are facing increased risk of "mental and physical illness", due to their exhausting (11 miles of walking per shift) and demanding (one order picked every 33 s) jobs [77]. Other companies such as Blue Apron had warehouses ripe with violence.

Estimates of employee turnover within the warehousing industry range from 25 to 75% per year, with costs of replacement ranging in the thousands of euros [56, 113]. In the current economic environment where warehouse workers are being pushed to fulfil delivery quotas daily, a lot of unethical and unsustainable work practices in warehousing environments have come out into the public view with the help of investigative journalism reports.

The following paragraphs will dive deeper into the ways in which warehouse workers are impacted by these stress factors and the issues that the pervasiveness of the risks is causing.

1. *Job content*

Job content depends on job characteristics that can lead to poor mental health for workers, such as the limited variety, fragmented and/or meaningless work, short work cycles, and low use of employees' skills. As research has shown, "job security turned out to be one of the most important factors for recruiting and retaining warehouse employees, whereas monetary incentives have little or nothing to do with warehouse employee turnover". The warehouse size was "directly related to the employee turnover due to the lack of personal attention paid to individual employees. This may have had the adverse impact on their retention" [117].

Warehouse workers are asked to do repetitive and monotonous jobs by the very design of their work [114]. The purpose of a warehouse is to stock items. These items are delivered to the warehouse via another type of transportation. The job of a warehouse worker involves removing the items from the transportation means they arrived at and storing them at a predefined location. After a specific amount of storage time (dependent on the type of material being stored and the type of warehouse being used), the items must be shipped to another location. Depending on the type of warehouse, orders are either large quantities of a limited number of products, such as in the automotive business [14], or small-batch wide-range orders for the warehouses delivering directly to consumers. To achieve this, the items must be picked up from their previous designated storage location, bundled up, packaged, and shipped out. Between these steps in the process, various number of required scans must be done to alert the warehouse IT system of the location of the goods at any given moment. These scans can happen automatically on location or must be performed manually by operators. These repetitive motions not only cause physical pain and injury—women in the United States, for example, are absent from work for due to repetitive motion injuries twice as often as men—but also lead to psychosomatic and musculoskeletal complaints and absenteeism. Underuse of skills is also an issue to be mentioned, as a study using data from employees of Oregon manufacturing and warehouse firms showed. Overeducated workers are less satisfied with their job and more likely to resign than workers of an adequate education are [1, 11, 64].

Monotonous job content is also what can lead to mental retirement. "Employees who are mentally retired are disconnected from their work and from the organization. Compared to others, they invest less in their work, their employability and development, and they have gradually lost their connections with their job, their colleagues, and the organization" [69]. Given the aging population that has led to the implementation of a higher retirement age throughout many countries, it is important to find ways in which to keep workers engaged with the work that they are doing for as long as possible.

2. *Work overload and workplace*

Workload can be defined as the amount of work that needs to be completed by an employee at a given time. The over or under loading of an employee is a psychosocial risk factor, together with time pressure (or the speed in which the tasks assigned must be completed). "Mental workload can be described as the relationship between the cognitive resources that are necessary to fulfil a specific task and the operator's cognitive resources that are available"[15] [189].

Mental and physical load can be expressed with the same components, namely, stress, represented by "task demands and strain, represented by the impact on the human" [194]. This similarity is used even in the international standard on mental workload (ISO 10075[16]). Demands "can have multiple facets, such as time pressure or task complexity and they can be overcome with different support", such as team members or resources available to the worker [194].

Workload problems receive a lot of attention, as research has proven that both overload and underload can be problematic [2, 20, 46, 107]. Another stress factor described by studies has been the difference between quantitative and qualitative workload [22, 108]. Quantitative workload refers to the amount of work that must be done while qualitative workload refers to the difficulty of that work. These two factors are independent of one another.

One of the reasons necessary for studying mental workload is to establish the relationship this has with operator performance (and subsequent errors coming from suboptimal workload conditions). Suboptimal workload in this case "can mean either overload or underload [23]. Overload occurs, when the operator is faced with more stimuli than they can handle" [194], and this is common occurrence for warehouse workers to have 12 h shifts. During the end of the year holiday season, warehouse managers often apply mandatory 12 h shifts or mandatory overtime to be able to

[15] Retrieved from: https://pure.rug.nl/ws/files/60962213/Proceedings_of_the_Human_Factors_and.pdf (Access on 20-03-2021).

[16] As mentioned at: https://www.iso.org/standard/66900.html (Access on 20-03-2021).

cope with the influx of orders that must be processed. This causes work overload, as most of the time the same number of people are expected to handle a higher number of orders. In some cases, like Walmart, store shelves are not being replenished at the needed rate because of chronic understaffing utilized as a cost saving mechanism. As Sanford C. Bernstein analyst explained in an interview, there are only so many ways to cut costs before the decisions become detrimental to the work process (confirmed by the studies of [39, 51, 181]).

The high picking targets that must be achieved by workers cause work-related pain and strain, such as carpal tunnel or Work-Related Lower Limb Disorders (WRLLD). In retail warehouses, such as those of Amazon or Walmart, workers are expected to pick upwards of 1000 packages per day [114]. In some locations, workers are not even paid per hour, but per number of packages moved, meaning that to receive a proper paycheck at the end of the day the pressure to load/unload trucks coming into the warehouse is high [75]. This pressure is increased even more by the fact that in companies' search for more efficiency they end up being "*monitored and supervised by robots*". The system goes so far as to track "time off task". If workers break from scanning packages for too long, the system automatically generates warnings which can eventually lead to the employee being fired [100, 125].

3. *Control*

Job control implies the degree of involvement of an employee in the decision-making process, relative to her/his job role. Within warehouses, mainly in the retail areas, the control of the pickers is slowly being relinquished to automated processed. Some studies underlined that job demands are significantly directly related to emotional exhaustion and depersonalization and in addition to this, as self-efficacy beliefs concerning coping with aggressive behavior diminished so did personal accomplishment feelings [24, 35].

The job strain model developed by Robert [81] has been used countless times ever since its inception. According to his model, the combination of high psychological demands and low control over work can turn into both a risk factor for cardiovascular diseases [17], as well as for the elevation of the arterial blood pressure [9, 96].

This means that in an industry where job control is constantly diminished by the implementation of automated alternatives, workers will find it increasingly harder to get involved in their work or exhibit any decision-making initiative. This increases the level of stress that the workers are under and creates a vicious cycle regarding worker turnover and employee retention. Without increase of job control, warehouse workers will not have any internal motivation for the job that they are doing and will rely exclusively on external, mostly financial compensation for their loyalty, which, in an industry that is focused on cost cutting through any means necessary, will not be a long-term solution.

4. *Work schedule*

Work schedule can be defined as the amount of time a person has to work during a day and the placement of these hours during the day.

Within the warehousing industry, work schedules imply long shifts and few breaks with badge in and badge out times. The metrics being used to calculate the work schedule and attendance caused some workers a "*state of constant anxiety that we could be fired at any moment for not meeting metrics*" [51].

Long weekly work hours and overtime have been associated with shorter sleep duration or sleep disturbances in several studies [155]. Shift work (any work schedule not scheduled between 7 AM and 6 PM) in and of itself is detrimental to healthy sleep patterns [38].

Within Europe, a national study done in the Netherlands shows that in the transport and communication sector 21% of work time was covered via shifts, with, 37% was evening or night shift work and 41% was weekend work. According to the fifth European Working Conditions Survey, a large part of the workers in the research sample recognized that they work outside the normal work schedule of 8 h/day during daytime, during weekdays. The survey shows that "around 17% of employees do shift work, while 21% work on call and on call work is carried out mainly in the transport sector (30%), followed by construction (27%) and public administration and defense (24%)" [45].[17]

5. *Work environment and equipment*

Problems related to these two areas could occur because of inadequate equipment availability, suitability, or maintenance and poor conditions such as lack of space, poor lighting, and excessive noise. These improper conditions can lead to both physical as well as mental health impairments [186]. In the last years, because of high degrees of temperature generated by the climate change, companies decide to support their employees by providing water and flexible work schedule some time. For example, Amazon has assured paramedics in ambulances outside their warehouses, ready to act in the case of dehydrated or heat stress affected employees.

Some former employees of the retail giant have come forward to say that the isolation and constant surveillance that they had experienced in the Amazon warehouses led to a toxic work environment that they claim was worse than being homeless [54]. Besides the equipment that should be used to make someone's job more efficient, there is equipment that is used to track and monitor a worker's actions. In a piece of investigative reporting, Amazon pickers described how the company equips them with handheld scanners that is used both to scan the items they retrieve as well as monitors time between scans. "Pickers must hit a certain number of scans per hour and if they start missing their targets, a manager will show up to see what they are doing" [51]. The same report mentioned how the working conditions caused one

[17] Retrieved from: https://www.eurofound.europa.eu/ (Access on 22–04-2021).

worker to have an "asthma attack during his night shift"[18] which led to his hospitalization. The most shocking of all claims are that the toilets are few and far between, causing workers to have to pee in bottles. "*As for bathrooms being a short distance from anyone in the fulfilment centre, that's just not true. There aren't enough of them, and they are always a good distance from you*" [114].

6. *The employees' role in organization*

The effect of role clarity on its performance consequences can be explained by two cognitive theories, namely:

- Expectancy theory of motivation (the likelihood that focusing their effort on a specific service dimension will increase the level of performance on that dimension) and
- Attribution theory (the belief that employees do not only want to maximize their rewards but also attain cognitive knowledge about the process of the environment).

Role clarity therefore aids in the creation of the desired work focus and offers the clarity necessary to master one's job. Job clarity therefore leads to job satisfaction, organizational commitment, and increased job performance [129]. The lack of role clarity or ambiguity has been shown to lead to a three times higher absence risk [182]. In addition to clarity, aspects such as role overload or responsibility can negatively impact individual health.

Ambiguity role happens when an employee has insufficient information about their work role. As defined by [187], it happens when, "the individual just doesn't know how he or she fits into the organization and is unsure of any rewards no matter how well he or she may perform". Role ambiguity is manifested at moments of change in the organization, such as restructuring or streamlining of human resource [74]. This can lead to confusion about objectives, expectations, and generally about the responsibilities of the job. Early studies have shown that [80] workers who dealt with role ambiguity were at a higher risk of experiencing work-related tension, lower levels of self-confidence and lower levels of job satisfaction.

In the case of warehouse workers, results reveal similar psychosocial stressors. While culture can provide a small differentiation between the ways in which role stress is perceived [66], it was clear that role ambiguity compromised mental health by increasing sleep disturbances, job dissatisfaction and desire to leave the company. On the other side role clarity increased job satisfaction [8, 180].

7. *The organizational culture and functions*

The organizational culture is related to the management styles and leadership that can cause negative consequences for employees [103]. Studies in the literature have recognized that employees are socializing in the organizational environment

---

[18] Retrieved from: https://www.businessinsider.in/peeing-in-trash-cans-constant-surveillance-and-asthma-attacks-on-the-job-amazon-workers-tell-us-their-warehouse-horror-stories/articleshow/64015295.cms.

context, using a set of three forces: (1) the hierarchical communication channels (with their managers or leaders), (2) in inter-group and inter-departmental communication channels (with other employee), and (3) adapting themselves to the organizational specifics (e. g., policies, structures, procedures, methods and tools implemented for collaboration, co-operation and communication).

Person-organization fit refers to the match between the needs and preferences of the employee with the ones of the company they work in; it can be split into two categories:

- Needs-supplies which considers person-organization fit to occur only when the organization, manage to satisfy individual's needs, desires, or preferences, (focus on the individual occupational needs);
- Demands-abilities "looks at person-organization fit from the perspective of the individual's ability required to meet organizational demands, therefore the organization's happiness with the hire" [8].

It has been shown that "individuals make job choices based on perceived organizational fit with the company's personality, as they want to feel that they are becoming part of a group" [8]. While this is true for office workers, "warehouse employees may be more interested in immediate rewards", such as salary and benefits, rather than considering how they aligned to the organizational culture [8]. The lack of capability could lead to the increase level of occupational stress. As managers are responsible for passing on the culture and function of a company to their workers their behavior has been shown having a major impact on employees' emotional well-being [33, 95].

Furthermore, organizational culture and functions are sources of job satisfaction. If a warehouse employee is satisfied with their job, they will put more effort into improving themselves and the work environment around them [104, 143], whereas if they are dissatisfied, they intent to behave in an unproductive manner (having tardiness, workday skipping, complete quitting) [119, 120]. Sometime, within some retail warehouses, workers feel as they are terrorized by their team leaders [114].

Research shows "that employees who have realistic expectations about their company are more likely to be satisfied" [8], therefore the best way to maximize benefits for companies is to increase the hiring and retaining of "employees with expectations that can be met by the company". As this is a hard thing to assess before the hiring [157], managers should want to invest in improving the human resource practices of the company "to ensure that the best-matched employees are identified" [8].

Besides the pre-hiring screening, which should focus on person-organization fit warehouse management should also analyze the "candidates' compatibility with the leadership style of the supervisory staff and the cultural characteristics of the company" [8]. To do these, human resources personnel should spend time to better understand the supervisors' leadership style and the work environment culture. While this kind of activity leads to a desired high fit, it is true that the current shortage of warehouse workers has left many positions unfilled and thus put more pressure on managers to hire workers who do not possess optimal levels of person-organization fit. In this case, the solution relates to creating a person-organization fit post hiring, by implementing long-term solutions such as mentorship or "big brother" socialization programs to guide new hires about existing company norms on the one hand and to listen to their perceptions of incompatibility or dissatisfaction on the other. The purpose is "to encourage active solicitation of employee ideas and opinions, increasing the sense of employee empowerment as well as creating a source of information for managers to improve working conditions in a timely manner" [8].

8. *Interpersonal relationships*

They are related to social or physical isolation, lack of support from peers and superiors, interpersonal conflict, or sexual harassment. Bad or negative interpersonal relationships could generate unsatisfaction at work, anxiety, and stress; positive interpersonal relationships and a good climate of work relation in the organization could positively affect work productivity and employees' morale. "Social context such as colleagues and the atmosphere at work can be one big engagement factor. Colleagues give the employee good resource of knowledge and emotional support. This can decrease work-related stress, which is one big factor in employee engagement" [10].

Within the fast-paced environment of retail warehousing, a lot of stress is caused by the lack of inter-personal relationship that are allowed, for the sake of efficiency [114].

Violence in the workplace is another factor that causes psychological damage [98]. Across the EU "3 million workers had reported being subjected to sexual harassment, 6 million to physical violence, and 12 million to intimidation and psychological violence," leading to the publication of the "guidance on the prevention of violence at work." Women have been shown to be at a higher risk than men of being sexually harassed or assaulted at work. With some reports indicating that that 90% of women face sexual harassment in the workplace and, over 11,000 people filling complaints about sexual harassment with the [40], thus it is clear that the problem is pervasive. Sexual harassment has severe consequences for both victims and employers, creating problems of absenteeism, low productivity, and increased turnover [40, 193]. With already existing pay gaps between men and women in the warehousing industry it is sometimes hard for women who rely on their salary in these sectors to come forward about their harassment for fear of losing their jobs. This creates a toxic environment for many women's mental health and forces them to prioritize their job to their wellbeing [176].

9. *Career development*

Career development risks are related to career stagnation, poor pay, job insecurity, and uncertainty.

The lack of career development capabilities could turn into a source of stress determined by its relationship with the competency's development. Job insecurity and status incongruity (over or under promotion, career stagnation or career ceiling) have been shown to be both sources of stress as well as having a direct impact on physical health [109]. These stressors primarily apply to older employees, who have reached their career ceiling and place a high value on stability and do not want to experience the erosion of status related to retirement. Job insecurity and redundancy related stress is even increased when the companies in which people work expect commitment [147]. Poor pay, or poor payment schedule (payment/piece instead of payment/hour) is a stress source that causes negative effects in the rate of working [84]. Due to the routine of the job and getting used to the demands that it has on workers, "more experienced warehouse workers are less inclined to give up on their current jobs than less experienced warehouse workers" [117]. Status incongruity linked either to promotional lag or social class of the worker while growing up has been shown to cause psychiatric illness, mental health issues and put a person at a higher risk of coronary heart disease [5, 163].

10. *Homework interface*

Home-work interface is related to the demand conflicts between work like and home life, such as the stress and conflicts relating to dual careers or low support at home. While it does not only relate to domestic life or family life, but most research has also focused on the relationship between work and spouses or leisure time use [31, 50]. Another strong focus was put on women workers, especially when the family unit has young children and a way to resolve work life-family life conflicts must enhance. While most studies presented the combination of "thrusting male-caring female" dynamic. In more recent years more emphasis has been put on the dual career couple. The combination lends itself to the dissolution of traditional role expectations as both partners can express negative feelings as threat and/or anxiety. The "wasted leisure time syndrome" [50] describes the frustration people with full time jobs feel when coming back home from work and not finding the time to do more than non-involving light leisure activities. Lundahl [106] observed that workers with very high level of fatigue in their jobs are demonstrating less involvement in their social lives. Many studies have shown that this lack of capability to engage in more demanding leisure activities caused a negative psychological effect on the average worker [34, 50]. The general realization was that "the unsatisfactory mental health of working people consists in no small measure of their dwarfed desires and deadened initiative, reduction of their goals and restriction of their efforts to a point where life is relatively empty and only half meaningful".

"Unpredictable scheduling practices lead to a worker's inability to plan any other aspects of life—which ranges from everything from childcare to attending school or taking a second job—which end up leading to overall negative effects on family life and their children's outcomes". "While employers' use of unpredictable schedules or just-in-time schedules is seen to boost profits by cutting labor costs, taking a more holistic view reveals these practices can cause harmful ripple effects for firms, families, and the economy". Local optimization, as is practiced in this case, does overall more harm than good. Irregular "schedules transfer the risk of doing business to workers and can harm a company's productivity as well" [21].

Already from research from previous decades it was shown that while "almost 73% of rotating shift workers were satisfied with their work-life balance the least satisfied were those with split or irregular shifts (about 65% were satisfied), on call or casual (62%), or with other shifts (63%)" [54, 190]. In other words, the people with the least say in what their schedules were going to be. "For families with children where both spouses worked full time shift work could exacerbate the balance needed to raise children in a good environment". The survey "shows that about 75% of full-time day workers whose spouse also worked full time were satisfied with their work life balance. When their spouse worked part time or was not in the labor force, about 77% were satisfied. While the proportion of full-time workers unhappy with their work-life balance varied, the main reasons for dissatisfaction were similar and were related to similar work-life balance issues. The main reasons for full time worker dissatisfaction were not enough time for family and too much time spent on the job". Not enough time for other activities was also a main trigger point. "Shift workers were slightly more likely than their day worker counterparts to worry about not spending enough time with family or friends (56 vs. 51%). Those working irregular schedules seemed the most affected by role overload and would sacrifice sleeping time to try to maintain enough family time" [190]. "They would also feel constantly stressed trying to accomplish more than they could handle" [190]. In other cases, workers do not have enough breaks during the day to allow them to talk to their family. This affects single mothers most of all.

In the past decades, the standard industry practice to increase workplace productivity has been the lean production standard work. On the one hand work has become faster and more efficient, but on the other hand more stressful for the workers. It has been shown in the past that issues such as job tension and fatigue were significantly greater in lean production environments than in traditional companies [101, 102]. Standardized work also increased the level of stress that workers felt [140]. While the mental toll was mostly left invisible, the incident rate has clearly increased [47, 118] while the reporting of illness has decreased a result of managerial pressures to work under pain [18, 139].

While the problems of these cases cannot be unidirectional linked to lean and might be a compound issue of managerial decisions that have as a result cost cutting, they ultimately lead to understaffing and overtime work. It is clear from the studies presented thus far that on the one had warehouse specific injuries are wide in scope and high in quantity, and on the other that the management of most warehouses did not understand the impact of illness related costs on their bottom lines [125].

Within the emergence of the Industry 4.0 industrialization model and the transfer to Industry 5.0 Revolution, there is a clear opportunity to use emerging technologies not only in monitoring the wellbeing of our workers, but also in supporting them do their jobs in a more ergonomic fashion. The following subchapter will present the ways in which the Industry 4.0 emerging solutions can be integrated into daily worker life, the theoretical and practical challenges and opportunities these implementations raise and the ways in which management can assess which tools would best suit their working environment [125].

# 3 Ergonomics Solutions in the Industry 5.0 Context

By 2025, 10% of people are expected to be wearing clothes connected to the internet and the first implantable mobile phone is expected to be sold [70]. As the Industry 4.0 paradigm shifts forward to Industry 5.0, it will bring along with it a supply chain "disruption with long-term gains in efficiency and productivity". This will cause "transportation and communication costs to drop, logistics and global supply chains to become more effective, and will ultimately diminish the cost of trade which will open new markets and drive economic growth" [159]. At the same time, there is a chance that the increase in automation will cause unrest in the labor markets. In the march towards lower costs, it is unclear if the automation will substitute labor across all the economy and create an even larger inequality gap or if automation will increase the possibility for a removing low pay low reward jobs and change the employment landscape for everyone [159]. With the help of the Internet of (every)Things, Cyber-Physical Systems communicate and co-operate with each other and with humans in real-time, with the goal of personalizing mass production. As manufacturers focus on the impact these new technologies will have on the production it is also important to assess the areas of impact that ergonomics will have from these new technologies. Industry 5.0 will offer ergonomists both the possibility to reduce workload strain, both physically and mentally as well as the possibility to better monitor workers' biological reactions to assigned tasks. This would allow them to act preventively and step in before it is too late.

The following paragraphs will discuss the two branches and the technologies already being used within them, while focusing particularly on Wearable Computers and Smart textiles. This decision is related to the fact that these technologies have the human at their core, while virtual reality/augment reality (VR/AR) and 3D printing, even though supporting the work that a warehouse worker would do daily and exist independently of the person using them. They are modern solutions for training workers regarding the products that they are handling.

## *3.1 Monitor Ergonomic Reality*

a. **Wearable computers**

"With the decreasing size and increasing power of most of the components of a modern computer, it was only a matter of time until computers that could be worn on the body began appearing" [16]. The social implications of a given technology are not always clear during the technology's introduction and it takes even more time for scientists to start analyzing the effects said technology has on the world around [145]. While still in its infancy, wearable computing is graining steady ground both in the private as well as corporate life. Although its advances are still slow, wearable computing has the potential to have one of the biggest efficiencies improving impacts to date in society. Research is currently focused on using wearable computers to get this kind of technology in places where it previously was not available before, due to space or activity restrictions, such as for medical monitoring, or mechanical inspection [121, 138, 173].

A wearable computer implies a computer that can be worn or transported and on top of this, a wearable computer is a computer that is controlled by the user and has operational and interactional consistency. They should be designed in such a way that the user sees them as part of themselves. As such, for differentiating wearable technology from those being or considering portable, three criteria can be proposed" [90–92]:

- "The device is attached to the body and does not require muscular effort to remain in contact with the body (i.e., you do not have to hold it);
- The device remains attached to the body during the body's orientation or activity (i.e., you do not have to take it off to perform a task specific action)"[19];
- "The device does not have to be detached to be interacted with (i.e., the first two criteria are not violated when the device is in use)".[20]

Given the above-mentioned criteria, one can separate between three different types of wearable computers that can assist with warehouse operations, as following:

- Head-mounted computers (eyeglass-like);
- Arm-mounted computers (watch-like);
- Hand-molded computers (glove-like).

While the usage of head mounted and arm molded computers are mostly used to improve the ergonomic reality by removing the need for old versions of handheld tools, the arm mounted computers can be most easily used for worker monitoring. Within a logistical environment the number of steps being taken, the paths being followed, and the amount of effort needed to move packages are pieces of information that can give insight into the efficiency of the tasks laid out to the workers and their reaction to it.

It has been shown that wearable devices and smartwatches were accurate for tracking step counts (with smartphones being more accurate than wearable devices) [27]. With the addition of GPS tracking capabilities this opens the possibility to assess worker spaghetti charts in real time, measuring the difference between a computationally devised optimal walking path and a the one most preferred by the worker. This can also support in facility reorganization, by providing the data regarding most frequent path combination [121].

While research has shown that the difference between medical grade devices and those available in retail is still to be properly bridged, the capabilities and accuracy of these devices are increasing from one release to another [151]. For example, some [58] have mentioned that the improved quality of the sensors in

---

[19] As mention also in: https://cmst.be/publi/doctv.pdf.

[20] Supported by the research retrieved from: http://oa.upm.es/43010/1/JORGE_CANCELA_GONZALEZ.pdf.

newer releases is capable of better tracking movement. The combination of sensors, such as the primary 3-axis accelerometers with secondary sensors such as magnetometers, and gyroscopes can compensate for the lack of accuracy obtained in data for motion tracking. While the tracking motion is reliable with the help of wearable computers the researchers suggested that smart textiles would be preferable for heartrate monitoring as the measurements would be more precise.

Research into the capabilities of wearable hand mounted computers to convey more information regarding worker mental workload is already underway, with mixed results related to the current technology [12, 158]. While the capability to assess long term changes in biological output based on mental workload modification might be possible, the research does not show confidence in their capabilities to measure it short term.

b. **Smart textiles**

The "field of ergonomics can be split into two major areas: the conduction of basic ergonomic research that improves the body of ergonomic knowledge, and the assistance in product development and design. To conduct ergonomic research, a lot of emphasis is put on finding models of interaction between humans and their respectively analyzed work environments. These models involve an ergonomics expert that analyzes the current work environment and acts as a change agent, by modelling the perceived actions of the subject and analyzing their impact. The problem with this approach is that the perceived actions and the actions done are never 1:1. Monitoring sensors are often uncomfortable and movement restraining, causing the analyzed person to change their movements to be able to accommodate the new machinery. This leads to wrong assumptions and measurements and reduces the added benefit that an ergonomics assessment can bring. Participatory ergonomics seeks to reduce this misinterpretation. It is a branch of ergonomics that emphasizes employees' self-potential for conducting ergonomic improvements at work, specifying that end-users should be actively involved in planning and implementing ergonomics solutions [128]. The challenge in this case is to create an environment that is auspicious both to scientific enquiry as well as worker input. One of the ways in which both can be accommodated is with the help of wearable technologies" [121].

"Wearable technology is a term that refers to clothing or accessories that are created or enhanced using embedded electronics [86], while others posit that it can be used to aid their users by monitoring information about the user themselves or the surroundings they interact with on a regular basis. While currently the general market for wearable technologies is small, due to the high cost of manufacturing on the one hand and the sense of intrusion of privacy on the other, innovations in the mobile and electronic healthcare area are already providing doctors and patients with expanded capabilities of physiological monitoring. Smart sensors are being used for perioperative monitoring and rehabilitation medicine allowing physicians to monitor patients in home and in community settings, which lead to a better understanding of the impact clinical interventions have on the level of mobility and the quality of life of the patient [3]. The creation of effective and unobtrusive wearable devices is one of the basic applications of pervasive computing, and that this creation can be used to improve the quality of ergonomic research by providing both the means of seamless user analysis as well as the solution to specific ergonomic issues that arise in warehouse logistics" [121, 154].

"It is important to first make the distinction within wearable technologies between wearable computers and smart textiles. Wearable computers imply electronics that are housed within a fashion accessory, and which allow the consumer to carry out their tasks without being obstructed. Smart textiles are products were using either the physical properties of the material, or electronics woven into the fabric can measure and/or react to stimuli from the user or environment [65]. They have a smaller range than wearable computers but allow the comfortable wearing of sensors for longer periods of time, making long term monitoring studies easier to do. This paper will further present the benefits of smart textiles in ergonomics research and work design applications" [121].

"Smart textiles are defined as textile products such as fibers and filaments, yarns together with woven, knitted, or non-woven structures, which can interact with the environment/user. Smart textiles are divided into three subgroups" [121, 172]:

- "Passive smart textiles: only able to sense the environment/user, based on sensors";
- "Active smart textiles: reactive sensing to stimuli from the environment, integrating an actuator function and a sensing device";
- "Very smart textiles: able to sense, react and adapt their behavior to the given circumstances".

"Passive smart textiles can help researchers via fabric sensors which can offer access to information such as body temperature [165], heart rate [32], movement and muscle tension, amongst others. At the same time carbon electrodes integrated into fabrics allow for the reading of environmental features such as moisture, salinity, and contaminants" [121, 195].

"Ergonomics analysis often relies on models of human movement. The human modelling tool used when showing and visually evaluating results makes a difference, in that there is a bias that leads to a more thorough analysis of *human looking* models and their postures than that of manikins or enhanced stick figures [97]. The issue with this is the fact that the more humanoid a model looks like the more time and effort must be spent to add the extra layers of information and design. It's also safe to say that regardless of the amount of time spent improving the model, by its very definition a model is a representation of a human and not an actual human, thus there is a compromise being made between the number of characteristics a human model retains and which information is eliminated in the process of digitalization and abstraction of the real-life information" [121].

"Due to the cost efficiency of smart textiles in contrast with the combination of motion capture suits and cameras, analysis on actual humans during the work that they do in their actual working environments becomes a possibility, especially given that traditional warehouse logistics work involves a wide range of movements and positions that would be difficult to catch on a static camera" [121]. Furthermore, due to the influence of the Hawthorne effect, the knowingly observed subject behaves differently under scrutiny [112]. "As represented in Fig. 7. textiles can gather

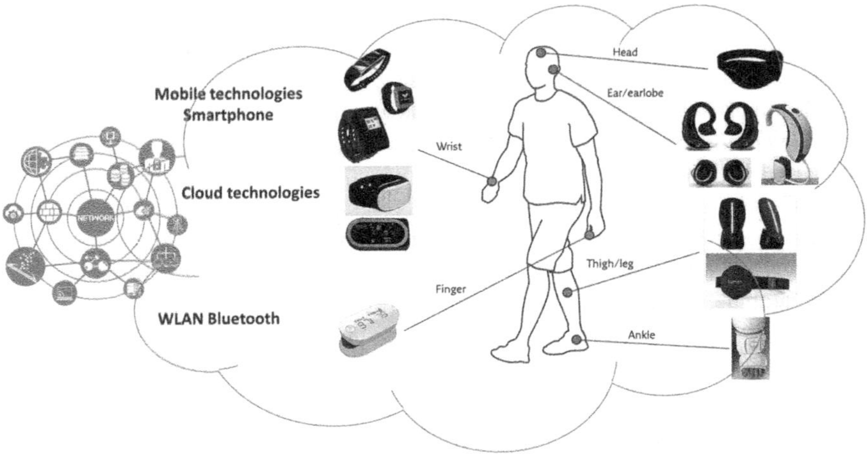


**Fig. 7** Remotely collecting user data

information remotely and repeatedly over the course of a study, therefore creating an environment where the information can be extracted easily while limiting any possible observational biases would decrease the quality of the raw data thus leading to a better understanding of the underlining issues" [121].

"As a second step, after a baseline has been established, the possibility of using active smart textiles comes into play. These react to stimuli from the environment, by integrating both actuators and sensors. These textile products can react automatically to the stimuli they receive, by becoming, among other things thermo-regulated or water resistant. At this point the same type of analyzes can be done as for the baseline, but with modified parameters allowing to see how the change in environment impacts the workers and what kind of changes can improve or worsen the working situation" [121].

"The third and final step would be to implement a study with the worker wearing very smart textiles, where the textiles can react in a personalized way with the wearer, adapting themselves based on previous experience and learning to react better to the wearer's movements so that they lessen ergonomic strain whenever possible. Depending on the type of material used and the strength of the fabric, the possibility of a movement training harness arises, which would teach the wearer how to do correctly do their daily movements and offer support when the movement done is straining. By changing the fabric of the textile to a material that offers more support, one can effectively create an exoskeleton that could take on part of the physical strain, removing it from the wearer. An experimental analysis should therefore have a minimum of four settings: normal textiles; baseline/passive smart textiles; activated smart textiles; activated very smart textiles" [121].

"These settings would allow the researcher to gain a deeper knowledge into the means through which ergonomic improvements can be brought to the analyzed workspace" [121].

"New fibers and textile materials are being discovered and improved upon every day, making intelligent clothing not only a possible future for ergonomics research, but a plausible one, where intelligent clothing can be worn like ordinary clothes and ensure that the wearer is protected from strain and discomfort when working. Wearable technologies could make ergonomics analysis more reliable and easier to undertake" [121]. This can only be done accurately by customizing the smart clothing to the applicant's body shape and data that needs to be gathered [28]. The information being gathered is only as good as the information that can be transmitted to a processing center, meaning that interconnectivity of smart clothing is paramount to its successful use.

"Whether it is the use of smart textiles to perform a long-term analysis of the day-to-day work of a warehouse employee or to actually offer them a tool through which their job becomes easier, there is clear merit in the analysis of the applicability of emerging technologies in aiding participatory ergonomics. More analysis is needed into the applicability of smart textiles in a warehouse environment particularly the type and number of sensors necessary to collect relevant and qualitative data while not constraining the wearer by creating an environment where their moves would not be natural. Similar attention should be paid to the interaction between the electro-magnetic fields created by wearing electrically charged clothing and the human body. Previous studies have shown that there can be serious health consequences because of electromagnetic interference within the human body, so the development of smart textiles should not be prioritized at the expense of human health" [121].

## *3.2 Improve Ergonomic Reality*

### a. **A debate on using wearable computers**

Due to the ubiquitous use of barcoding for product identification, the current process of warehouse item scanning involves some type of handheld barcode scanner, which, as can be seen in Fig. 8 is composed of a scanner window, a trigger switch, and a cable interface port. This can be held in a gun-like manner, i.e., the scanner is held with the dominant hand, gripped between the index finger and thumb, with the index finger on the trigger switch and the middle, ring, and pinkie finger around the base of the grip, between the trigger switch and the cable interface port.

Given the fact that the human arm has a tremor even in a static position that influences accuracy in a negative, depending on the width and breadth of the scanner window the time it takes to find the accurate position for proper scanning is not optimal. The rapid and repetitive pressing of the trigger switch can also lead to loss of pressure sensitivity in the index finger. With the introduction of tablet-like scanners or tablet integrated scanners, the holding position has changed towards a cell phone-like manner, with the palm turned upwards, the wrist flexed laterally to allow the scanner window proper access. Depending on the weight of the scanner and

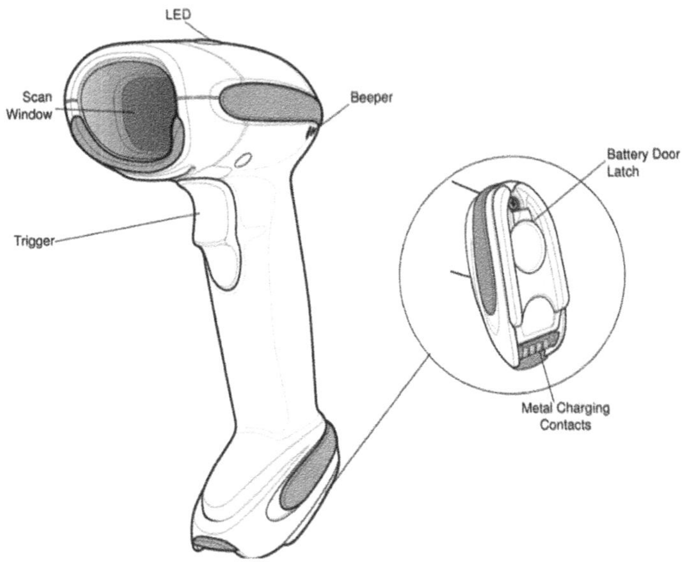


**Fig. 8** Standard components of a handheld scanner (As mentioned in the User Manual: https://fccid.io/UZ7DS6878/User-Manual/User-Manual-I-1246139 (Access on 24–10-2018)

the positioning of the keyboard/buttons on its surface, prolonged use of this position can induce muscle fatigue [164] or even enlarges the median nerve, causing pain in the thumb and decreasing hand functions [72, 142].

Ever since the 1960s, barcodes have been used to scan products. They are standardized, readable from almost any angle and at a wide range of distances, cheap to produce, and easy to print [52, 160]. In warehouse management, barcodes have been used for everything from goods storage to re-parcellation to movements inside as well as outside the storage facility.

While easiness of reading and processing barcodes has been an issue that has been solved in the retail business, the standard countertop laser scanner being equipped with mirrors that can obtain signals from multiple angles, it is a fact that handheld laser scanners are less powerful and less accurate than the countertop ones. They are not capable of gathering multiple signals and the necessity of operating smaller optical devices leads to worse readings or longer reading times [191]. Due to this, there is a clear need to analyze different scanning solutions that might limit the amount of time spent per scanning either via integrating the scanning within the already necessary manipulation movements, improving the scanning technology, or both. Wearing technology instead of holding has the advantage of freeing up hands for this function [19].

Head-mounted scanners provide the operator freedom to move their arms unimpeded to perform the task the way they usually do, thus causing the slightest amount of inconvenience. When designing a head-mounted scanning system, one needs to take into consideration the location of the scanner and how this will be activated during regular operations. Because of this point of attention, scanners attached to eyeglasses can directly take advantage of the line of sight and of the natural tendency to look at what we are manipulating. Therefore, this limits the need to create a special "to scan" movement as the scanning takes place as soon as your line of sight aims towards the package if continuous scanning is applied. When designing such scanners as presented in Fig. 9, the mass, the bulk, the heat dissipation, and optical performance [168] of the equipment must be taken into consideration. Loading of the ear and nose should be tolerable for several hours. Scanning should be easy to stop when encountering another person, in order not to cause visual inconveniences [123].

Arm-mounted scanners take advantage of the hand placement during manipulation movements. It eliminates the need for a grip of the scanner, as it is attached to the wrist, and in some cases can extend to the finger, like in the case of Fig. 10. For visual scanning, the device places the scanner on the front facing side of the intermediary phalange of the index finger and the trigger button on the side of the index, for easy push access from the thumb. This offers more movement control, as the scanning happens both at a closer distance and that the user's fingertip [123].

The biggest downside to the optical scanning is the need to press a trigger button and the instability of the finger scanner on the finger during strenuous manipulation. This can be fixed by the usage of radio-frequency identification (RFID) technologies [133]. These were first used in the 1940s as a method to identify allied airplanes,

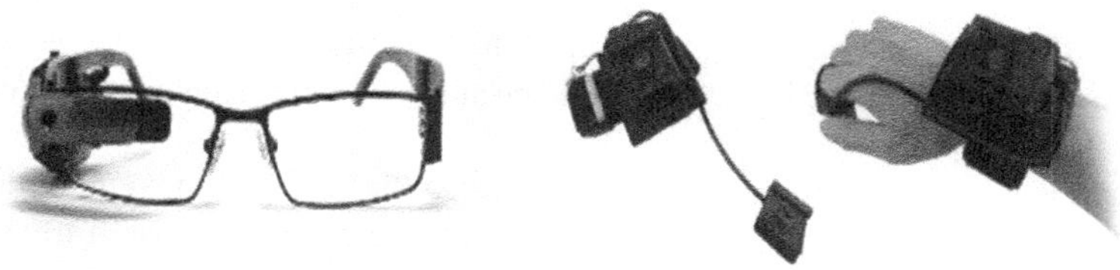

**Fig. 9** Eyeglass mounted scanner (used for practical tests as designed by: https://www.google.com/glass/start/) and arm mounted scanners (the one use for the practical tests has been provided by: As seen on: http://www.tuvie.com/wt4000-wearable-computer-for-workers/)

**Fig. 10** Arm-molded scanner (as used in the practical test in according to the descriptions at: https://m.iotone.com/hardware/proglove-mark/h5903)

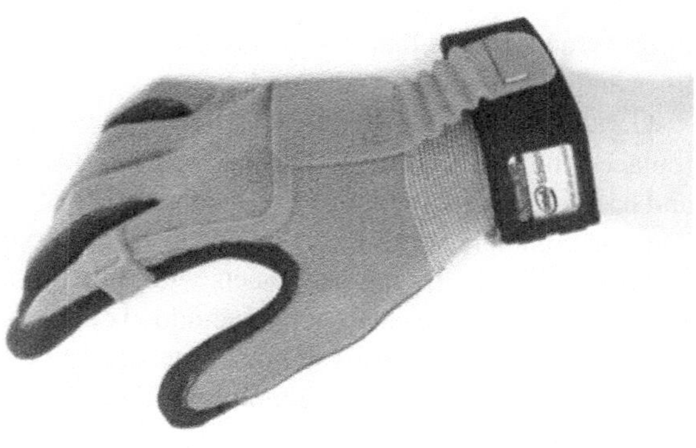

but have current frequent applications in transportation, services, supply chain, and manufacturing. During recent years, an interest in RFID technology has arisen with the goal of increasing readers' mobility as something transparent for users, and their combined integration into wireless communications systems [130]. Passive RFID tags are small electronic components with an integrated circuit and a small antenna usually sealed in one small package. The tags do not need a battery, the reader via electromagnetic induction energizes them during access. In addition, "each tag contains a numeric code that uniquely identifies the object and can be queried by a wireless reader" [184]. While the price of this technology is slightly higher than barcodes, the ability to scan an item without having as specific target the tag itself, can lead to easier storage solutions and less manipulation of a package to be processed.

If the RFID reader can be placed within the wristband while also being screened from accidentally scanning nearby items, it would lead to reducing the amount of movement necessary and thus creating an uninhibited work experience for the operator.

The arm-molded scanner takes the wristband scanner concept and integrates the industry mandated hand protection that workers always must wear during their operation. As can be seen from Fig. 10, it leads to less slippage of the finger-mounted grip as this feature is fully incorporated into the glove itself, minimizing unwanted movements. The implementation of a glove-mounted scanner is in RFID glove has been extensively researched [99, 130]. While the proprietary version is still currently used exclusively for optical barcode and QR scanning (based on Quick Response Code), prototypes are already being developed to include shock resistant and water-resistant RFID capabilities. This would lead the operator to not have to rely on any extra movement to scan the item, as it would be automatically scanned upon touching it.

Wearable computers are an easy way to reduce unnecessary movement that can cause strain or injury. While there may currently be a barrier related to the cost of equipping workers with such tools, during the next few years, as technology advances, the costs will decrease, as it has already been shown for other technologies [26]. These mean that the incentive to use this equipment must be clearly stated and disseminated within the business decision-making community to ensure the quickest possible adoption of the ideas within the non-academic environment. Currently, the cost of applying smart technologies to existing jobs is most related to cost-efficient, improvement means that does not require the restructuring that automated scanning lines and robotization would imply.

Certain warehouses are already starting to use Google Glasses' functionalities to replace handheld scanners as they have been proven to pick-time reducing out an item and pack it for shipping by 25%. The possibility of eliminating the scanning work by equipping cameras with RFID means to record images related to the identities and locations of all RFID-tagged objects within that image. This can lead to workplace reorganization where customer value-added operations are being applied in the place of essential, but repetitive and non-value adding tasks such as scanning [188].

Besides the reduction of ergonomic strain alongside reduction of throughput times, the application of wearable computers has the advantage of creating an innovation mind-set within the organization where the objective of doing more efficient work in a healthier way can be easily understood by the workers as they are directly involved in the usage of the new technology. As of now, the topic offers fertile ground for more investigation.

b. **Virtual reality, augment reality, and 3D printing**

With the creation of virtual reality (VR) in the second half of the twentieth century and its increased availability in consumer markets at the beginning of the twenty-first, the manufacturing industry's interest has been raised regarding virtual training for manual assembly tasks. Virtual environments have the capability to deliver cost-efficient, safe, and potentially effective training. Their unique selling point is that they can allow full operator training prior to them starting at their assigned workstation in an environment that would almost fully resemble their actual work one. This leads to the possibility for an accelerated end-to-end manufacturing process and a directly proportional efficiency increase. Studies have shown the benefits of VR applications and serious games in training and education both towards the public [53, 115] as well as specifically in an industry and warehouse environment [132, 196]. In the case of a forklift training application, the researchers have found that the virtual course was more time efficient than the traditional one and, enthusiastically perceived by users from the target group. In the case of bimanual assembly tasks, the virtual training was proven to be promising and validated the effectiveness of virtual training in these conditions.

Besides the training capability of the VR/AR technologies, 3D printing brings about a new change in the way in which supply chains operate [6]. From the first prohibitively expensive 3-D printing device, created in 1984 by Charles W. Hull, the variety of additive manufacturing printers on the market has led to a drop-in hardware cost [7]. Thanks to this, the way in which current supply chains operate is on the brink of being disrupted. Firstly, the technology has the potential to support reshoring and local sourcing, as well as lead to a decentralizing of the means of production [126].

The increase in low volume high variability demands from customers can also be satisfied with the help of the Rapid Prototyping capabilities that Additive Manufacturing (AM) is creating. It is recognized that Rapid Prototyping provides manufacturing cost and time reductions, by allowing a more in-depth product analysis from the early innovation stages, by by-passing the costly one-off traditional manufacturing. Additionally, AM is also faster compared to traditional manufacturing, as shown in the analysis done by Siemens, and briefly presented in Fig. 11.

Besides the possibility for prototype manufacturing, additive manufacturing (AM) can be used in component manufacturing for pieces that require low quantities or parts with low tolerance limits. Considering that "20% of the 3D printing market is made up of component part production for the aerospace and automotive industries" [6],[21] it is clear to see that the technology is not a long shot, but something that is already being applied in this day. Even if the pieces manufactured with the help of

[21] Retrieved from: https://file.scirp.org/Html/1-9201960_75953.htm (Access on 23–04-2019).

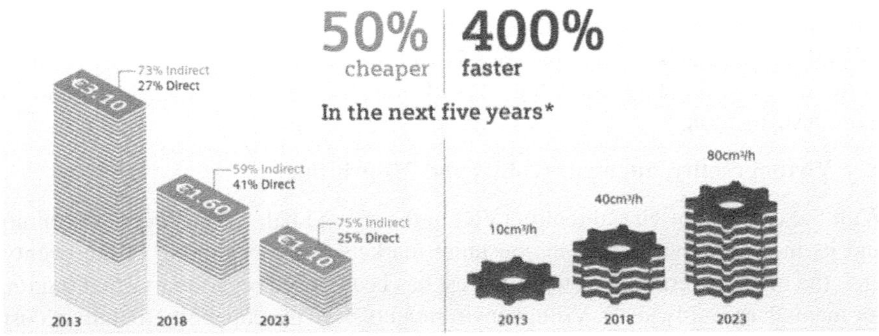


**Fig. 11** Cost and time of additive manufacturing [166]

"AM won't replace existing conventional subtractive production methods", they are expected to be applied successfully in a series of niche areas [6, 87, 152, 156].

Given the information presented thus far a gap in the research has been found regarding the usage of wearable computers in factory warehouse logistics. While the technology is being used for retail warehouses, where the products picked are varied and in low volume, industry warehouses, where variability is lower and volume higher, do not seem to be given the same level of attention. There is also no literature regarding the theoretical model to be used in applying Industry 5.0 solutions. Currently, managers are inundated with new technology without having a proper guideline as to how all of this should be best applied. Given the high cost of the technology a trial-and-error way of working is not sustainable and even counterproductive, as reluctance to invest in machines that do not prove useless might decrease over time.

## 4 Conclusions and Future Trends

The chapter's content described the application of ergonomics approach and principles within the field of warehouse logistics. The split was made between physical and mental workloads, and it concluded that the current level of ergonomic knowledge and application in warehouses is minimal, that worker health and safety are not seen as anything more than legal compliance topics. Furthermore, by providing ergonomic solutions that could be adopted by companies in actual context of the transition from Industry 4.0 to Industry 5.0, there were presented and categorized solutions that monitor the worker and solutions that improve the worker's job by making it more ergonomic.

Despite the difficulties of the pandemic period 2019–2020 (2.2 million individuals have officially died from COVID-19 and more than 100 million people were infected[22]), there have been registered a high rate of digital transformation in many industrial areas, as the logistics is, too. In this case, professional life has been fast pushed forward to the Industry 5.0 context (e.g., processes related to how the distribution, acquisitions and delivery, purchase etc.), while other economic areas stop existing (e.g., the hospitality industry) or some others took a step back. Recently, the following five trends "Global Supply Chain and Logistics"[23] has been announced:

- *Green logistics*—refers to the emergent of the sustainable development values, principles, and practices in the supply chain management. "With the increased awareness that transportation is a large contributor to greenhouse gas emissions, green logistics has gained significant traction over the years. Designing value chain towards advanced energy management systems and electric and solar-powered vehicles to lower their overall carbon footprint will emerge in a rapid way" [76]. In addition to these, circular logistics concept (opposite to the linear logistics) is more often used in the practice implying costs and waste reductions in the long term, and reduced impact on the environment. This modern and actual practice includes reverse logistics and it depends on the local regulations, requirements on recycling, and waste disposal [76, 79],
- *Supply Chain Resilience* has arisen during the pandemic of COVID-19 and important changes are related to the sourcing processes from- and offshoring to low labor cost markets lost their motivation in the conditions of green supply chain; labor costs decrease in the condition of high automation process mainly in the warehouses and there are rather the energy costs, taxes, ease of doing business, together with fiscal and political stability factors that are considered critical factors or criteria in the decision-making process [76]. Related to the use of emergent technologies, there must be mentioned the use of 3D printing that will boost the resilience capabilities. In addition, local products and local produce products and services will be preferred and will impact the shortness of the supply chain that will be more adaptive and personalized, customized to their clients. In the nearest future, Blockchain and Cloud technologies, Internet of (every)Thing, Big data, predictive analytics, and Artificial Intelligence solutions will be used to optimize the logistics systems [76, 79, 134],
- *Shipment of Refrigerated Goods*—It is related to the increase of the population and the increased need for shipping refrigerated goods, including foods, beverages, and pharmaceuticals [76]. A first challenge confronted the drug industry for the global distribution of COVID-19 vaccine. Thus, it is estimated that "shipping companies will need to expand their fleets of trucks and shipping containers as well as climate-controlled warehouses" [76],
- *Omnichannel Services*—These refer to the achieving of high degree of flexibility on the supply chain and maximum customer satisfaction by using omnichannel approach. The focus will be to offer personalized experiences for customers

---

[22] According to the analysis available at: https://blog.greencarrier.com/5-global-supply-chain-and-logistics-trends-2021/ (Access on 08–05-2021).

[23] Retrieved from: https://blog.greencarrier.com/5-global-supply-chain-and-logistics-trends-2021/ (Access on 08–05-2021).

through a synergy created in the supply chain (integrating the efforts of marketing, sales, and logistics) [25, 68];

- *Technology and 5G*—referring to the full potential use of the IoT to track inventory in the supply chain and thus, creating a trusted record of an item's journey. "One of the most disruptive uses for 5G networks is driverless transportation. 5G is likely to make this possibility a reality, which will have a huge impact on the logistics industry" [76, 89].

Finally, researchers and practitioners have recognized the high impact of the digital transformation that will metamorphosize the future work process in warehouse logistics. In addition, it is expected from the "supply chain leaders to foster an environment that attract individuals with an infinite mindset, where learn, re-learn and learn again is a natural" [76]. Occupational health and safety will be a priority and thus, ergonomics, sustainable development management, and human resources management will need to be integrated to assure occupational wellbeing.

## References

1. Acosta-Ballesteros J, del Pilar Osorno-del Rosal M, Rodríguez-Rodríguez OM (2018) Overeducation of Young workers in Spain: how much does the first job matter? Soc Indic Res 138(1):109–139
2. Aldrich TB, Szabo SM, Bierbaum CR (1989) The development and application of models to predict operator workload during system design. In: Applications of human performance models to system design. Springer, Boston, MA, pp 65–80
3. Appelboom G, Camacho E, Abraham ME, Bruce SS, Dumont EL, Zacharia BE, et al (2014) Smart wearable body sensors for patient self-assessment and monitoring. Arch Publ Health 72(1):28
4. Arezes P, Carvalho P (eds) (2014) Advances in safety management and human factors. In: Proceedings of the AHFE 2017 international conference on safety management and human factors, Los Angeles, California, USA. Accessed 17–21 July 2017; Advances in intelligent systems and computing book series, AISC, vol 604, Spinger Cham
5. Arthur RJ, Gunderson EK (1965) Promotion and mental illness in the Navy (No. 65–8). Navy medical neuropsychiatric research unit San Diego CA
6. Attaran M (2017) Additive manufacturing: the most promising technology to alter the supply chain and logistics. J Serv Sci Manag 10(3):189–205
7. Attaran M (2017) The rise of 3-D printing: the advantages of additive manufacturing over traditional manufacturing. Bus Horiz 60(5):677–688
8. Autry CW, Daugherty PJ (2003) Warehouse operations employees: linking person-organization fit, job satisfaction, and coping responses. J Bus Logist 24(1):171–197
9. Babu GR, Jotheeswaran AT, Mahapatra T, Mahapatra S, Kumar A, Detels R, Pearce N (2014) Republished: is hypertension associated with job strain? A meta-analysis of observational studies. Postgrad Med J 90(1065):402–409
10. Bakker AB, Leiter MP (2010) Work engagement: a handbook of essential theory and research. Psychology press
11. Baran J (2018) A side effect of a university boom: rising incidence of overeducation among tertiary educated workers in Poland. Econ Bus Rev 4(2):41–63
12. Barber D, Carter A, Harris J, Reinerman-Jones L (2017) Feasibility of wearable fitness trackers for adapting multimodal communication. In: International conference on human interface and the management of information. Springer, Cham, pp 504–516

13. Barreto L, Amaral A, Pereira T (2017) Industry 4.0 implications in logistics: an overview. Procedia Manuf 13:1245–1252
14. Bartholdi JJ III, Hackman ST (2008) Allocating space in a forward pick area of a distribution center for small parts. IIE Trans 40(11):1046–1053
15. Basahel AM (2015) Investigation of work-related Musculoskeletal disorders (MSDs) in warehouse workers in Saudi Arabia. Procedia Manuf 3:4643–4649
16. Bass L (1995) Is there a wearable computer in your future? In: IFIP International conference on engineering for human-computer interaction. Springer, Boston, MA, pp 3–16. https://dl.gi.de/bitstream/handle/20.500.12116/6499/Bass_1995.pdf?isAllowed=n&sequence=2. Accessed 06 Feb 2021
17. Belkic KL, Landsbergis PA, Schnall PL, Baker D (2004) Is job strain a major source of cardiovascular disease risk? Scand J Work Environ Health 30(2):85–128
18. Berggren C (1993) Lean production—The end of history? Work Employ Soc 7(2):163–188
19. Bhutani A, Bhardwaj P (2018) Barcode printers market share—Industry size forecast report 2024. https://www.gminsights.com/industry-analysis/barcode-printers-market-size
20. Bommer SC, Fendley M (2018) A theoretical framework for evaluating mental workload resources in human systems design for manufacturing operations. Int J Ind Ergon 63:7–17
21. Boushey H, Ansel B (2016) Working by the hour: the economic consequences of unpredictable scheduling practices. Washington Center for Equitable Growth, Washington, DC
22. Bowling NA, Alarcon GM, Bragg CB, Hartman MJ (2015) A meta-analytic examination of the potential correlates and consequences of workload. Work Stress 29(2):95–113
23. Brookhuis KA, De Waard D (2010) Monitoring drivers' mental workload in driving simulators using physiological measures. Accid Anal Prev 42(3):898–903
24. Brouwers A, Tomic W (2016) Job-demands, job control, social support, self-efficacy, and burnout of staff of residential children's homes. Educ Pract Theory 38(1):89–107
25. Buldeo Rai H, Verlinde S, Macharis C, Schoutteet P, Vanhaverbeke L (2019) Logistics outsourcing in omnichannel retail: state of practice and service recommendations. Int J Phys Distrib Logist Manag 49(3):267–286
26. Bureau of Labor Statistics (2015) U.S. Department of Labor, the economics daily, Long-term price trends for computers, TVs, and related items. https://www.bls.gov/opub/ted/2015/long-term-price-trends-for-computers-tvs-and-related-items.htm. Accessed 16 Feb 2021
27. Case MA, Burwick HA, Volpp KG, Patel MS (2015) Accuracy of smartphone applications and wearable devices for tracking physical activity data. JAMA 313(6):625–626
28. Chen M, Ma Y, Song J, Lai CF, Hu B (2016) Smart clothing: connecting human with clouds and big data for sustainable health monitoring. Mob Netw Appl 21(5):825–845
29. Cheung Z, Feletto M, Galante J, Waters T, Centers for Disease Control and Prevention (2007) Ergonomic guidelines for manual material handling. In NIOSH Publication (No. 2007–131). NIOSH
30. Colombini D (1998) An observational method for classifying exposure to repetitive movements of the upper limbs. Ergonomics 41(9):1261–1289
31. Cooper CL (1981) The stress check: coping with the stresses of life and work. Prentice Hall
32. Coosemans J, Hermans B, Puers R (2006) Integrating wireless ECG monitoring in textiles. Sens Actuat A 130:48–53
33. Corey DM, Wolf GD (1992) An integrated approach to reducing stress injuries
34. Cox T, Howarth I (1990) Organizational health, culture and helping. Work Stress Int J Work Health Organiz 4:107–110
35. De Jonge J, Dollard MF, Dormann C, Le Blanc PM, Houtman IL (2000) The demand-control model: specific demands, specific control, and well-defined groups. Int J Stress Manag 7(4):269–287
36. Deloitte, Taliaferro A, Guenette CA, Agarwal A, Pochon M (2016) Industry 4.0 and distribution centers Transforming distribution operations through innovation, https://www2.deloitte.com/content/dam/insights/us/articles/3294_industry-4-0-distribution-centers/DUP_Industry-4-0-distribution-centers.pdf. Accessed 20 Mar 2021

37. Denis D, St-Vincent M, Imbeau D, Trudeau R (2006) Stock management influence on manual materials handling in two warehouse superstores. Int J Ind Ergon 36(3):191–201
38. Drake CL, Roehrs T, Richardson G, Walsh JK, Roth T (2004) Shift work sleep disorder: prevalence and consequences beyond that of symptomatic day workers. Sleep 27(8):1453–1462
39. Dudley R (2013) Customers flee wal-mart empty shelves for target, Costco. Bloomberg
40. Equal Employment Opportunity Commission (2011) Sexual harassment charges EEOC and FEPAs combined: FY 1997-FY 2011. https://www.eeoc.gov/enforcement/sexual-harassment-chargeseeoc-fepas-combined-fy-1997-fy-2011. EU. Accessed 23 Mar 2021
41. EU-OSHA (2007a) Work-related musculoskeletal disorders: back to work report. https://ketlib.lib.unipi.gr/xmlui/bitstream/handle/ket/970/TE7807300ENC-Work-related-musculoskeletal-disorders-Back-to-work.pdf?isAllowed=y&sequence=2. Accessed 20 Mar 2021
42. EU-OSHA (2007b) E-fact 14—Hazards and risks associated with manual handling in the workplace. http://www.osha.mddsz.gov.si/resources/files/pdf/E-fact_14_-_Hazards_and_risks_associated_with_manual_handling_in_the_workplace.pdf. Accessed 19 Mar 2021
43. EU-OSHA (2010) OSH in figures: work-related musculoskeletal disorders in the EU—Facts and figures. https://osha.europa.eu/en/publications/osh-figures-work-related-musculoskeletal-disorders-eu-facts-and-figures. Accessed 19 Feb 2021
44. European Foundation for the Improvement of Living and Working Conditions (2006) Fourth European working conditions survey. European Foundation for the Improvement of Living and Working Conditions. https://www.eurofound.europa.eu/. Accessed 20 Mar 2021
45. European Foundation for the Improvement of Living and Working Conditions (2014) Fourth European working conditions survey. European Foundation for the Improvement of Living and Working Conditions. https://www.eurofound.europa.eu/Eurostat. (2018) Statistics Explained, Accidents at work statistics. https://enuveprod-universitatpolit.netdna-ssl.com/php_prevencionintegral/sites/default/files/noticia/45658/field_adjuntos/11539.pdf. Eurostat. Accessed 08 Feb 2021
46. Frankenhaeuser M, Johansson G (1976) Task demand as reflected in catecholamine excretion and heart rate. J Hum Stress 2(1):15–23
47. Fucini JJ, Fucini S (2008) Working for the Japanese—Insides the Mazda's American auto plant. The Free Press, New York
48. Gajšek B, Đukić G, Opetuk T, Cajner H (2017) Human in manual order picking systems. In: Management of technology—Step to sustainable production. https://bib.irb.hr/datoteka/877213.Human_in_Manual_Order_Picking_Systems.pdf. Accessed 20 Mar 2021
49. Gallagher S, Heberger JR (2015) The effects of operator position, pallet orientation, and palletizing condition on low back loads in manual bag palletizing operations. Int J Ind Ergon 47:84–92
50. Gardell B (1973) Quality of work and non-work activities and rewards in affluent societies
51. Ghosh S (2018) Peeing in trash cans, constant surveillance, and asthma attacks on the job: Amazon workers tell us their warehouse horror stories. https://www.businessinsider.com/amazon-warehouse-workers-share-their-horror-stories-2018-4?r=nordic. Accessed 22 Feb 2021
52. Global Market Insights Inc. (2016) Barcode printers market size by technology (Laser, Ink-Jet, Impact, Direct Thermal, Thermal Transfer), by product (Desktop, Mobile, Tabletop/Industrial), by application (Retail, Manufacturing, Shipping, Government, Healthcare). In: Industry analysis report, regional outlook, application potential, competitive market share & forecast (2015–2022). https://www.gminsights.com/industry-analysis/barcode-printers-market-size. Accessed 18 Mar 2021
53. Girard C, Ecalle J, Magnan A (2013) Serious games as new educational tools: how effective are they? A meta-analysis of recent studies. J Comput Assist Learn 29(3):207–219
54. Gracely N (2014) Being homeless is better than working for Amazon. The guardian. http://www.theguardian.com/money/2014/nov/28/being-homeless-isbetter-than-working-for-amazonGolden,2015. Accessed 20 Mar 2021
55. Gong Y, De Koster RB (2011) A review on stochastic models and analysis of warehouse operations. Logist Res 3(4):191–205

56. Gooley TB (2001) How to keep good people. Logist Manag Distrib Report Radnor 40:55–60
57. Grosse EH, Glock CH, Jaber MY, Neumann WP (2015) Incorporating human factors in order picking planning models: framework and research opportunities. Int J Prod Res 53(3):695–717
58. Haghi M, Thurow K, Stoll R (2017) Wearable devices in medical internet of things: scientific research and commercially available devices. Healthc Inform Res 23(1):4–15
59. Harrison PD, Martins MR, Tsai LW (2006) An application of the PM book maturity model. In: Proceedings of the XXVI Brazialian congress on production engineering-fortaleza, Brazil
60. Helliwell PS (1999) The elbow, forearm, wrist and hand. Best Pract Res Clin Rheumatol 13(2):311–328
61. Heller-Ono A (2014) A prospective study of a macroergonomics process over five years demonstrates significant prevention of workers' compensation claims resulting in projected savings. Evaluation 30:90
62. Hendrick HW (2003) Determining the cost–benefits of ergonomics projects and factors that lead to their success. Appl Ergon 34(5):419–427
63. Hermann M, Pentek T, Otto B (2015) Design principles for Industrie 4.0 scenarios: a literature review. Technische Universität Dortmund, Dortmund
64. Hersch J (1991) Education match and job match. Rev Econ Stat 140–144
65. Hertleer C, Schwarz A, Van Langenhove L (2012) Smart textiles: an overview. In Intelligent textiles and clothing for ballistic and NBC protection. Springer, Dordrecht, pp 119–136
66. Hoppe A, Heaney CA, Fujishiro K (2010) Stressors, resources, and well-being among Latino and white warehouse workers in the United States. Am J Ind Med 53(3):252–263
67. Hoozemans MJM, Van der Beek AJ, Frings-Dresen MHW, Van der Woude LHV, Van Dijk FJH (2002) Pushing and pulling in association with low back and shoulder complaints. Occup Environ Med 59(10):696–702
68. Hossain TMT, Akter S, Kattiyapornpong U, Dwivedi Y (2020) Reconceptualizing integration quality dynamics for omnichannel marketing. Ind Mark Manage 87:225–241
69. Huijs JJ, Houtman IL, Taris TW, Blonk RW (2019) Effect of a participative action intervention program on reducing mental retirement. BMC Public Health 19(1):194
70. Hutt R (2016) What are the 10 biggest global challenges? In: World economic forum, vol 21
71. Hwang HS, Cho GS (2006) A performance evaluation model for order picking warehouse design. Comput Ind Eng 51(2):335–342
72. İnal EE, Demirci K, Çetİntürk A, Akgönül M, Savaş S (2015) Effects of smartphone overuse on hand function, pinch strength, and the median nerve. Muscle Nerve 52(2):183–188
73. Ismail R (2008) Logistics management. Excel Books India
74. Ivancevich JM, Matteson MT (1980) Stress and work: a managerial perspective. Scott Foresman
75. Jamieson D (2011) The new blue collar: temporary work, lasting poverty and the American warehouse. Huffington Post
76. Jemdahl J (2021) 5 global supply chain and logistics trends 2021. https://blog.greencarrier.com/5-global-supply-chain-and-logistics-trends-2021/. Accessed 26 Apr 2021
77. Jivanda T (2013) Amazon workers have 'increased risk of mental and physical illness'. https://www.independent.co.uk/news/uk/home-news/amazon-workers-have-increased-risk-of-mental-and-physical-illness-expert-claims-8962304.html. Accessed 22 Feb 2021
78. Jorgensen MJ, Handa A, Veluswamy P, Bhatt M (2005) The effect of pallet distance on torso kinematics and low back disorder risk. Ergonomics 48(8):949–963
79. Julianelli V, Caiado RGG, Scavarda LF, Cruz SPDMF (2020) Interplay between reverse logistics and circular economy: critical success factors-based taxonomy and framework. Resour Conser Recycl 158:104784
80. Kahn RL (1973) Conflict, ambiguity, and overload: three elements in job stress. Occup Mental Health
81. Karasek Jr, RA (1979) Job demands, job decision latitude, and mental strain: implications for job redesign. Admin Sci Qu 285–308

82. Karwowski W (1991) Complexity, fuzziness, and ergonomic incompatibility issues in the control of dynamic work environments. Ergonomics 34(6):671–686
83. Karwowski W (2005) Ergonomics and human factors: the paradigms for science, engineering, design, technology and management of human-compatible systems. Ergonomics 48(5):436–463
84. Kasl SV (1992) Surveillance of psychological disorders in the workplace
85. Keller SB, Ozment J (1999) Managing driver retention: effects of the dispatcher. J Bus Logist 20(2):97
86. King MF (2011) Fashion, the body and technology: tracing early 20th century techno-utopian ideas, aesthetics and impulses in 21st century wearable technology (Doctoral dissertation, Queensland University of Technology)
87. King R (2012) 3D printing coming to the manufacturing space and outer space. Bloomberg LP, Bloomberg
88. Khalil TM, Rosomoff RS, Abdel-Moty EM (1993) Ergonomics in back pain: a guide to prevention and rehabilitation. Van Nostrand Reinhold, New York, pp 174–185
89. Khatib EJ, Barco R (2021) Optimization of 5G networks for smart logistics. Energies 14(6):1758
90. Knight JF (2002) The ergonomics of wearable computers: implications for musculoskeletal loading (Doctoral dissertation, University of Birmingham)
91. Knight JF, Baber C, Schwirtz A, Bristow HW (2002) The comfort assessment of wearable computers. In: iswc, vol 2, pp 65–74
92. Knight JF, Deen-Williams D, Arvanitis TN, Baber C, Sotiriou S, Anastopoulou S, Gargalakos M (2006) Assessing the wearability of wearable computers. In: 2006 10th IEEE international symposium on wearable computers. IEEE, pp 75–82
93. Knill B (2002) Practical ergonomics for plant people. https://www.mhlnews.com/labor-management/article/22032920/practical-ergonomics-for-plant-people
94. Kumar S (1994) The epidemiology and functional evaluation of low-back pain: a literature review. Eur J Phys Med Rehabil 4(1):15–27
95. Landy FJ (1992) Work design and stress panel. In: Washington DC (ed) Work and well being. American Psychological Association, pp 115–158
96. Landsbergis PA, Dobson M, Koutsouras G, Schnall P (2013) Job strain and ambulatory blood pressure: a meta-analysis and systematic review. Am J Public Health 103(3):e61–e71
97. Lämkull D, Hanson L, Örtengren R (2007) The influence of virtual human model appearance on visual ergonomics posture evaluation. Appl Ergon 38(6):713–722
98. Leather P, Brady C, Lawrence C, Cox T (eds) (1999) Work-related violence: assessment and intervention. Psychology Press
99. Lee C, Kim M, Park J, Oh J, Eom K (2010) Design and implementation of the wireless RFID Glove for life applications. Int J Grid Distrib Comput 3(3):41–52
100. Lecher C (2019) How Amazon automatically tracks and fires warehouse workers for 'productivity'. https://www.theverge.com/2019/4/25/18516004/amazon-warehouse-fulfillment-centers-productivity-firing-terminations. Accessed 12 Feb 2021
101. Lewchuk W, Robertson D (1996) Working conditions under lean production: a worker-based benchmarking study. Asia Pac Bus Rev 2(4):60–81
102. Lewchuk W, Robertson D (1997) Production without empowerment: work reorganization from the perspective of motor vehicle workers. Cap Class 21(3):37–64
103. Ližbetinová L, Lorincová S, Caha Z (2016) The application of the organizational culture assessment instrument (OCAI) to logistics enterprises. NAŠE MORE: znanstveno-stručni časopis za more i pomorstvo, 63(3 Special Issue):170–176
104. Locke EA (1969) What is job satisfaction? Organ Behav Hum Perform 4(4):309–336
105. Lucire Y (1986) Neurosis in the workplace. Med J Aust 145(7):323–327
106. Lundahl A (1971) Fritid och rekreation: Utkast till kapitel 11 i betänkande om svenska folkets levnadsförhållanden. Allmänna förlaget
107. Lundberg U, Forsman L (1979) Adrenal-medullary and adrenal-cortical responses to understimulation and overstimulation: comparison between type A and type B persons. Biol Psychol 9(2):79–89

108. Malhotra S, Chadha O (2012) Stress in the context of job satisfaction: an empirical study of BPO sector. Int J Res IT Manag 2(1):24–38; Mann (1998)
109. Margolis BL, Kroes WH, Quinn RP (1974) Job stress: an unlisted occupational hazard. J Occup Environ Med 16(10):659–661
110. Martimo KP, Verbeek J, Karppinen J, Furlan AD, Takala EP, Kuijer PPF, et al (2008) Effect of training and lifting equipment for preventing back pain in lifting and handling: systematic review. BMJ 336(7641):429–431
111. Martimo KP, Shiri R, Miranda H, Ketola R, Varonen H, Viikari-Juntura E (2009) Self-reported productivity loss among workers with upper extremity disorders. Scand J Work Environ Health 301–308
112. McCambridge J, Witton J, Elbourne DR (2014) Systematic review of the Hawthorne effect: new concepts are needed to study research participation effects. J Clin Epidemiol 67(3):267–277
113. McElroy JC, Rodriguez JM, Griffin GC, Morrow PC, Wilson MG (1993) Career stage, time spent on the road, and truckload driver attitudes. Transp J 5–14
114. McClelland M (2012) I was a warehouse wage slave. Mother Jones. https://www.motherjones.com/politics/2012/02/mac-mcclelland-free-online-shipping-warehouses-labor/4/Messing
115. Merchant Z, Goetz ET, Cifuentes L, Keeney-Kennicutt W, Davis TJ (2014) Effectiveness of virtual reality-based instruction on students' learning outcomes in K-12 and higher education: a meta-analysis. Comput Educ 70:29–40
116. Middlesworth M (2007) A step-by-step guide to the RULA assessment tool. https://ergo-plus.com/rula-assessment-tool-guide/Milczarek
117. Min H (2007) Examining sources of warehouse employee turnover. Int J Phys Distrib Logist Manag 37(5):375–388
118. Mirer FE (1989) Worker participation in health and safety: lessons from joint programs in the American automobile industry. Am Ind Hyg Assoc J 50(8):A598-603
119. Mobley WH (1977) Intermediate linkages in the relationship between job satisfaction and employee turnover. J Appl Psychol 62(2):237
120. Mobley WH, Horner SO, Hollingsworth AT (1978) An evaluation of precursors of hospital employee turnover. J Appl psychol 63(4):408
121. Mocan A, Daniel C, Draghici A (2017a) Analyzing warehouse ergonomics using smart textiles. In: Hajek P, Vit O, Basova P, Krijt M, Paszekova H, Souckova O, Mudrik R (eds) Proceedings of the international conference of Central-Bohemia-University, CBUIC, vol 5. Prague, Czech Republic, pp 1181–1184
122. Mocan A, Draghici A, Mocan M (2017b) A way of gaining competitive advantage through ergonomics improvements in warehouse logistics. Res Sci Today 13:7–15. https://www.rstjournal.com/wp-content/uploads/2017/10/RST-Supplement-2-2017.pdf. Accessed 22 Feb 2021
123. Mocan A, Draghici A (2018)
124. Mocan A, Draghici A (2019) Maximizing data gathering opportunities by using fitness trackers during ergonomic assessments a case study. In: Thriving on future education, industry, business and society; Proceedings of the make learn and TIIM international conference 2019. ToKnowPress, pp 95–102. http://www.toknowpress.net/ISBN/978-961-6914-25-3/papers/ML19-012.pdf. Accessed 25 Feb 2021
125. Mocan A (2020) Industry 4.0 in warehouse ergonomics: possible applications of emerging technology. Ph.D. thesis, Politehnic University of Timisoara, Romania
126. Mohr S, Khan O, Kersten W, Blecker T, Ringle CM (2015) 3D printing and supply chains of the future. In: Innovations and strategies for logistics and supply chains, epubli GmbH, Berlin, pp 147–174
127. Moray N (2000) Culture, politics and ergonomics. Ergonomics 43(7):858–868
128. Motamedzade M, Shahnavaz H, Kazemnejad A, Azar A, Karimi H (2003) The impact of participatory ergonomics on working conditions, quality, and productivity. Int J Occup Saf Ergon 9(2):135–147
129. Mukherjee A, Malhotra N (2005) Antecedents and consequences of role clarity in explaining employee-perceived service quality in call centers. Mark Theory Appl 7:15

130. Muguira L, Vazquez JI, Arruti A, Ruiz-de-Garibay J, Renteira S, Mendia I (2009) RFIDGlove: a wearable RFID reader. In: IEEE international conference on eBusiness engineering, 2009. ICEBE'09. IEEE, pp 475–480
131. Mullins LJ (2007) Management and organizational behaviour. Pearson education
132. Murcia-Lopez M, Steed A (2018) A comparison of virtual and physical training transfer of bimanual assembly tasks. IEEE Trans Visual Comput Graphics 24(4):1574–1583
133. Nambiar AN (2009) RFID technology: a review of its applications. In: Proceedings of the world congress on engineering and computer science, vol 2, pp 20–22
134. Nandi S, Sarkis J, Hervani AA, Helms MM (2021) Redesigning supply chains using blockchain-enabled circular economy and COVID-19 experiences. Sustain Prod Consum 27:10–22
135. Neights G (2021) Industry 5.0 and the supply chain. https://talkinglogistics.com/2020/08/11/industry-5-0-supply-chain/. Accessed 12 Mar 2021
136. Niu S (2010) Ergonomics and occupational safety and health: an ILO perspective. Appl Ergon 41(6):744–753
137. Ngo BP, Yazdani A, Carlan N, Wells R (2017) Lifting height as the dominant risk factor for low-back pain and loading during manual materials handling: a scoping review. IISE Trans Occup Ergon Hum Fact 5(3–4):158–171
138. Ockerman JJ, Pritchett AR (1998) Preliminary investigation of wearable computers for task guidance in aircraft inspection. In: Digest of papers. second international symposium on wearable computers (Cat. No. 98EX215). IEEE, pp 33–40
139. Parker M, Slaughter J (1994) Beware! TQM is coming to your campus. Thought Action 10(1):5–30
140. Parker SK, Jackson PR, Sprigg CA, Whybrow AC (1998) Organizational interventions to reduce the impact of poor work design. HSE Books
141. Paschek D, Mocan A, Draghici A (2019) Industry 5.0—The expected impact of next industrial revolution. In: Thriving on future education, industry, business and society; Proceedings of the make learn and TIIM international conference 2019, pp 125–132. https://ideas.repec.org/h/tkp/mklp19/125-132.html. Accessed 21 Feb 2021
142. Patel S, Park H, Bonato P, Chan L, Rodgers M (2012) A review of wearable sensors and systems with application in rehabilitation. J Neuroeng Rehabil 9(1):21
143. Petty MM, McGee GW, Cavender JW (1984) A meta-analysis of the relationships between individual job satisfaction and individual performance. Acad Manag Rev 9(4):712–721
144. Phipps E (2012) Temporary labor: issues of health and safety experienced by warehouse workers in the Inland empire
145. Pool I (1977) The social impact of the telephone. J Soc Hist 12(3):480–482
146. Pope MH, Andersson GB, Frymoyer JW, Chaffin DB (1991) Occupational low back pain: assessment, treatment, and prevention
147. Porter LW, Steers RM, Mowday RT, Boulian PV (1974) Organizational commitment, job satisfaction, and turnover among psychiatric technicians. J Appl Psychol 59(5):603
148. Punnett L, Wegman DH (2004) Work-related musculoskeletal disorders: the epidemiologic evidence and the debate. J Electromyogr Kinesiol 14(1):13–23
149. PwC (2016) 2016 global industry 4.0 survey industry 4.0: building the digital enterprise. https://www.pwc.com/gx/en/industries/industries-4.0/landing-page/industry-4.0-building-your-digital-enterprise-april-2016.pdf. Accessed 15 Feb 2021
150. Ramsey T, Davis KG, Kotowski SE, Anderson VP, Waters T (2014) Reduction of spinal loads through adjustable interventions at the origin and destination of palletizing tasks. Hum Factors 56(7):1222–1234
151. Rosenberger ME, Buman MP, Haskell WL, McConnell MV, Carstensen LL (2016) 24 hours of sleep, sedentary behavior, and physical activity with nine wearable devices. Med Sci Sports Exerc 48(3):457
152. Royte E (2013) What lies ahead for 3-D printing. Smithsonian Institute, Washington DC
153. Salter N (1955) The effect on muscle strength of maximum isometric and isotonic contractions at different repetition rates. J Physiol 130(1):109

154. Sanchez BB, de Rivera DS, Sánchez-Picot A (2016) Building unobtrusive wearable devices: an ergonomic cybernetic glove. J Internet Serv Inf Secur 6(2):37–52
155. Sasaki T, Iwasaki K, Oka T, Hisanaga N, Ueda T, Takada Y, Fujiki Y (1999) Effect of working hours on cardiovascular-autonomic nervous functions in engineers in an electronics manufacturing company. Ind Health 37(1):55–61
156. Sasson A, Johnson JC (2016) The 3D printing order: variability, supercenters and supply chain reconfigurations. Int J Phys Distrib Logist Manag 46(1):82–94
157. Schein EH (1991) What is culture. Sage Publicatios, Newbury Park, CA, pp 243–253
158. Schmalfuß F, Mach S, Klüber K, Habelt B, Beggiato M, Körner A, Krems JF (2018) Potential of wearable devices for mental workload detection in different physiological activity conditions. In: Proceedings of the human factors and ergonomics society Europe, pp 179–191
159. Schwab K (2016) The fourth industrial revolution: what it means, how to respond (2016). World Econ Forum. https://www.weforum.org/agenda/2016/01/the-fourth-industrial-revolution-what-it-means-and-how-to-respond/. Accessed 15 Feb 2021
160. Seideman T (1993) Barcodes sweep the world. Am Herit Invent Technol 8(4):56–63
161. Senders JW, Moray NP (1995) Human error: cause, prediction, and reduction
162. Shain M, Kramer DM (2004) Health promotion in the workplace: framing the concept; reviewing the evidence. Occup Environ Med 61(7):643–648
163. Shekelle RB, Ostfeld AM, Paul O (1969) Social status and incidence of coronary heart disease. J Chronic Dis 22(6–7):381–394
164. Shim JM (2012) The effect of carpal tunnel changes on smartphone users. J Phys Ther Sci 24(12):1251–1253
165. Sibinski M, Jakubowska M, Sloma M (2010) Flexible temperature sensors on fibers. Sensors, 10(9):7934–7946
166. Siemens (2017) Additive manufacturing: facts and forecasts. https://www.siemens.com/innovation/en/home/pictures-of-the-future/industry-and-automation/Additive-manufacturing-facts-and-forecasts.html
167. Solistica (2021) Emerging technologies transforming the logistics industry. https://blog.solistica.com/en/emerging-technologies-transforming-the-logistics-industry. Accessed 25 Apr 2021
168. Spitzer MB, Rensing NM, McClelland R, Aquilino P (1997) Eyeglass-based systems for wearable computing. In: Digest of papers. First international symposium on wearable computers. IEEE, pp 48–51
169. Standring S (ed) (2015) Gray's anatomy e-book: the anatomical basis of clinical practice. Elsevier Health Sciences
170. Stone S (2015) Ergonomic safety tips for the warehouse. https://www.cisco-eagle.com/blog/2015/04/02/ergonomic-safety-tips-for-the-warehouse/. Accessed 15 Mar 2021
171. Stone S (2017) Oversimplification: four approaches to guarantee limited safety improvement. https://www.ehstoday.com/safety-leadership/oversimplification-four-approaches-guarantee-limited-safety-improvement. Accessed 12 Mar 2021
172. Stoppa M, Chiolerio A (2014) Wearable electronics and smart textiles: a critical review. Sensors 14(7):11957–11992
173. Sunkpho J, Garrett JH, Smailagic A, Siewiorek DP (1998) MIA: a wearable computer for bridge inspectors. In: Digest of papers. Second international symposium on wearable computers (Cat. No. 98EX215). IEEE, pp 160–161
174. St-Vincent M, Denis D, Imbeau D, Laberge M (2005) Work factors affecting manual materials handling in a warehouse superstore. Int J Ind Ergon 35(1):33–46
175. St-Vincent M, Denis D, Imbeau D, Trudeau R (2006) Symptoms of stress related to the characteristics of customer service in warehouse superstores. Int J Ind Ergon 36(4):313–321
176. Telegraph Report (2018) Tesco 'facing record £4bn equal pay claim' after 'hiding in plain sight for years'. https://www.telegraph.co.uk/business/2018/02/07/tesco-facing-record-4bn-equal-pay-claim-hiding-plain-sight-years/
177. Thomas AB (2005) Controversies in management: issues, debates, answers. Routledge
178. Tompkins JA, White JA, Bozer YA, Tanchoco JMA (2010) Facilities planning. Wiley

179. Troup JDG, Davies JC, Manning DP (1988) A model for the investigation of back injuries and manual handling problems at work. J Occup Accid 10(2):107–119
180. Tucker MK, Jimmieson NL, Jamieson JE (2018) Role stressors in australian transport and logistics workers: psychosocial implications. Saf Sci 109:12–19
181. Tuttle B (2016) Hey Walmart, it's hard to make sales when store shelves are empty. https://business.time.com/2013/03/27/hey-walmart-its-hard-to-make-sales-when-store-shelves-are-empty/
182. Väänänen A, Kalimo R, Toppinen-Tanner S, Mutanen P, Peiró JM, Kivimäki M, Vahtera J (2004) Role clarity, fairness, and organizational climate as predictors of sickness absence: a prospective study in the private sector. Scand J Publ Health 32(6):426–434
183. Van Tulder M, Malmivaara A, Koes B (2007) Repetitive strain injury. The Lancet 369(9575):1815–1822
184. Want R (2006) An introduction to RFID technology. IEEE Pervasive Comput 1:25–33
185. Ware BF, Fernandez JE (2014) Warehouse ergonomics/tips and techniques to decrease injury risk. https://www.ehstoday.com/industrial-hygiene/warehouse-ergonomicstips-and-techniques-decrease-injury-risk
186. Warr PB (1992) Job features and excessive stress. Prevention of mental Ill health at work. HMSO, London
187. Warshaw LJ (1979) Managing stress, vol 8299. Addison-Wesley, Reading, MA
188. Whelan R (2015) DHL unit plans google glass experiment in U.S. Warehouses, https://www.wsj.com/articles/dhl-unit-plans-google-glass-experiment-in-us-warehouses-1439568950
189. Wickens CD (2008) Multiple resources and mental workload. Hum Factors 50(3):449–455
190. Williams C (2008) Work-life balance of shift workers. Ottawa, Ontario, Canada, Statistics Canada, pp 75–001
191. Wittman T (2004) Lost in the supermarket: decoding blurry barcodes. SIAM News 37(7):16
192. World Economic Forum (2016) Digital transformation of industries: digital enterprise. http://reports.weforum.org/digital-transformation/wp-content/blogs.dir/94/mp/files/pages/files/digital-enterprise-narrative-final-january-2016.pdf
193. WWJ report (2012) Women at work. Women at risk. http://www.ww4j.org/uploads/7/0/0/6/7006481 3/atworkatrisk.pdf
194. Young MS, Brookhuis KA, Wickens CD, Hancock PA (2015) State of science: mental workload in ergonomics. Ergonomics 58(1):1–17
195. Zadeh EG, Rajagopalan S, Sawan M (2006) Flexible biochemical sensor array for laboratory-on-chip applications. In: 2006 International workshop on computer architecture for machine perception and sensing. IEEE, pp 65–66
196. Zawadzki P, Buń P, Górski F (2018) Virtual reality training of practical skills in industry on example of forklift operation. In: International conference on innovation, engineering and entrepreneurship. Springer, Cham, pp 46–52
197. Zebra (2016) Building the smarter warehouse: warehousing 2020 redefining supply chain automation in the age of digital technology: North America report, https://www.prnewswire.com/news-releases/zebra-technologies-study-finds-seven-in-ten-decision-makers-to-accelerate-warehouse-technology-investment-by-2020-300257135.html

# Chapter 6
# New Approaches to Product Development in the Current Industrial and Economic Context

**Adrian Ciprian Firu, Anamaria Feier, Felicia Banciu, Alin Ion Tapirdea, and George Draghici**

**Abstract** An important aspect that defines the product design and development process relates to the identification and collection of market needs or the definition by the customer of a set of specifications transformed into design and functional parameters. In most cases, new product development in the industrial environment involves creating an updated version of an existing product. The current economic context is playing a major role in the trend towards individualized mass production. There is a shift from a linear economy to a circular economy, with a focus on optimizing how resources are used to reduce emissions over the lifetime of products or services. The ability to adapt even the entire design of a product to meet customer needs is essential in the manufacture of individualized products and in the product development process; databases prove to be useful translating a part of the routine decision-making process to a computer program. This study presents a model of product development process including a technical configurator with adaptable interface and an e-service web application, which involve customers directly in the design of product development.

**Keywords** Product design · Industry 4.0 · Product configurator · Mass production · Individualization · Customization · Production · Competitiveness

---

A. C. Firu · A. Feier · F. Banciu · A. I. Tapirdea · G. Draghici (✉)
Politehnica University of Timisoara, Faculty of Mechanical Engineering, Timisoara, Romania
e-mail: george.draghici@upt.ro

A. C. Firu
e-mail: ciprian.firu@student.upt.ro

A. Feier
e-mail: ana-maria.feier@upt.ro

F. Banciu
e-mail: felicia.banciu@upt.ro

A. I. Tapirdea
e-mail: alin.tapirdea@student.upt.ro

 
A. Draghici and L. Ivascu (eds.), *Sustainability and Innovation in Manufacturing Enterprises*, Advances in Sustainability Science and Technology,
https://doi.org/10.1007/978-981-16-7365-8_6

# 1 Introduction

The chapter aims to present a conceptual framework to understand how the features of Industry 4.0 and the generated technologies can supports the transition to Industry 5.0, the role that co-creation of/and added value has supporting open and sustainable innovation in this transition by presenting a product configurator for the automotive industry. Thus, a presentation of the limitations that currently exist in product development, a brief review of the concepts of mass customization product and mass individualization product is made to highlight the need and increase flexibility and innovation in new product development. Mass individualization is based on an open architecture of new products and emphasizes open innovation and customer involvement in the design process, open architecture concept and customer's role in design of individualized products, the role of databases in the product development process.

A different "society should arise from Industry 4.0 to a proposed Industry 5.0 (proposed in Japan) which strategic concept is to guide technological innovation processes and put technological innovation in the spotlight because it can be used as a tool for social innovation and not just a factor leading to changes in companies and business processes. In this way, it had been clearly stated that technological innovation plays an essential role in all the transition to Industry 5.0" [3].

"The Japanese Government had not adopted the term Industry 4.0 due to consideration that these new technologies are too simplistic to impact the pillars of Japanese society but has chosen the term "super-smart society" and stated what that Industry 5.0 will be achieved through his concept" [3, 21].

The concept of Industry 5.0 was first time used in the Fifth Science and Technology Basic Plan [8], "which is a program whose aim was to connect physical space (the real world) and cyberspace, taking advantage of all the potential of new technologies in order to achieve an ideal future society" [3].

The concept of co-creation is an Open Innovation processes and is the pillar that supports the "transition from Industry 4.0 to Industry 5.0, or rather the essential 'enabling mechanisms.' Open Innovation processes have already evolved, beginning with the most studied and used inbound ones [7, 23]. Open Innovation differs greatly from the closed innovation model, which was the only one known about 15 years ago" [3, 6, 7].

# 2 Limitations of Product Development

With the help of advanced technologies associated with Industry 4.0, integrated into the developing process of intelligent products, with communication functions, information gathering during operation, it is possible to redefine the current way of designing a new product.

In the view of the expert consultants at Berger Strategy Consultants [5], four pillars underpin the digitization of industry. They are the use of digital data through the ability to collect, manage and analyse it, automation to achieve cognitive products that interact with the environment and can self-organize, and connectivity to synchronize supply chains with the aim of shortening innovation cycles and the appropriation of new business models that ensure the sustainability of the innovation process.

By adopting and integrating the Industry 4.0 concept, companies can overcome the limitations of the current business model, starting with the development of new products in a much shorter time, which implies cost reduction. By developing smart products, capable of providing information throughout the entire life cycle, on one hand, quality issues in operation are solved, and on the other hand, innovation opportunities for the next generations of products are identified [17].

An important aspect that defines the product design and development process relates to the identification and collection of market needs or the definition by the customer of a set of specifications transformed into design and functional parameters so that the finished product satisfies as much as possible of the initial set of specifications.

In most cases, new product development in the industrial environment involves creating an updated version of an existing product. A new product version often involves elements of novelty in terms of functionality or design, but at the same time, a large part of the specifications or technology is taken from the previous product version, a process known as reconfiguration [1]. In this case, companies usually are in competition to develop new products, based on their previous models, but with new or innovative elements, which can assure their portfolio adaptation on the existing competitive markets.

The process of designing and developing new products entails high costs and quality risks, particularly through the need to invest in new equipment and production lines. These high costs are sustained only through high production and sales volumes, making it extremely difficult to sustain new product development for small, customized or individualized production runs.

In the current context, the new product development process must respond to the needs of an increasingly competitive market, which requires a wide variety of products to be developed in the shortest possible time at the lowest possible cost. At the same time, a standardized, regularly reviewed and verified process is the guarantee of product quality, safety and security.

The product development process is a chain of steps needed to take a product from idea or concept to market. Given the increasing complexity of products, in most cases, we are talking about multidisciplinary teams involved in the development of a product. Therefore, achieving a product that meets the required quality standards and getting it to market as quickly as possible depends on effective collaboration between those involved in the development process.

Broadly speaking, the product development process follows the same steps related to clarify the tasks, conceptual design, embodiment design, detail design and then manufacturing [18].

The result obtained from the execution of a process step is analysed and if it is not satisfactory, that step must be repeated. The iteration principle is a basic principle of the traditional new product development process. Iterations involve repeating steps until the desired result is achieved. The iteration method must be applied constantly and in small "loops" to be effective. Otherwise, when a problem occurs, the project team must start the development process from the beginning.

According to Ref. [4] "the challenge is to analyse in real time the impact of a changed requirement and to adapt the product and supply chain with the same speed of reaction. The advantages of increased speed of reaction are reduced product development time to market launch and reduced total cost of intellectual ownership". These advantages can be exploited if methods are found to eliminate downtime both within the company between departments and between phases in the product life cycle, and along the supply chain between customer and supplier.

In actual context of the transition from Industry 4.0 to Industry 5.0, the new product's development process is of particular importance in the success of an organization, in increasing the competitiveness of that organization and in ensuring product quality. Therefore, in recent years, the focus has been on improving and streamlining processes by eliminating waste, unnecessary steps that do not add value and optimizing the resources used.

## 3 A Debate on Mass Production, Mass Customization and Mass Individualization

In the last years, there has been a shift from **mass and customized** production to **individualized production** [13]. Complex products with multiple and individualized functions require a considerable degree of innovation in the design phase and a significant degree of flexibility in production.

"In mass production, the product architecture is unified. In mass customization, the product architecture is modular. In mass individualization, the sequence of the three operations is more complex" [13] as described in Fig. 1.

In addition, Table 1 described the concepts paradigms with brief summarization of their characteristics "corresponding to product architecture, the type of product built and the role of the customer" [13].

The **mass customization** paradigm was introduced in the 1980s and has brought with it important changes in the range of options, the customer can choose from Ref. [12], which identifies the factors that made the concept of mass customization possible as: Architecture composed of product families, Reconfigurable production systems, delaying the point at which different features emerge. While custom mass production has clear advantages in increasing customer engagement, the customer has a choice of pre-configured modules, and the result may not fully meet their wishes.

Figure 2 shows a shift from mass production characterized by large volumes of a single product pushed to market, to mass customization characterized by module-based products that can be configured differently to achieve multiple variants, with the trend moving towards mass individualization achieved through open-architecture

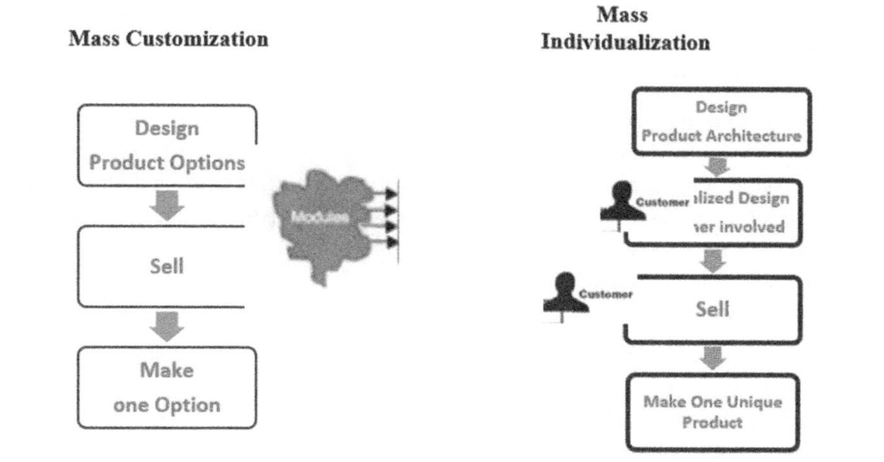


**Fig. 1** The mass customization and individualization organizational chart. Adapted from Ref. [13]

**Table 1** Classes of product architectures [13]

| Paradigm | Product architecture | Product built | Customer's role |
|---|---|---|---|
| Mass production | Unified | Identical product | Choose a product |
| Mass customization | Modular | Product built with one option | Choose among offered options |
| Mass individualization | Open-platform | Customer designed | Involved in final product design |

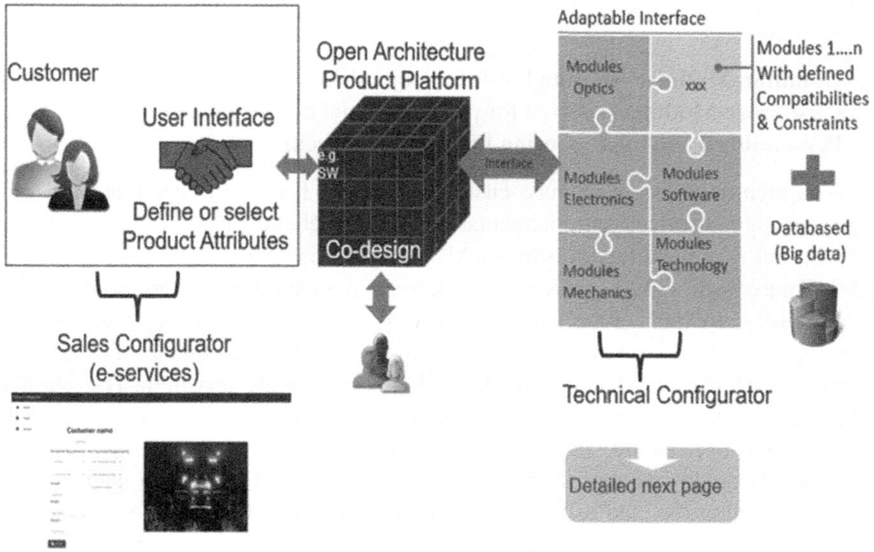


**Fig. 2** Transitions between the three major manufacturing paradigms

products that are developed at the request of and with the involvement of an individual customer using Smart Product-Service Systems.

**A Smart Product-Service System (PSS) is defined as**:

- Integrated **smart products and e-services** into single solutions. Smart products make use of information and communication technology (ICT) to collect, process and produce information, while **e-services** are web portals, apps and means alike, which facilitate the communication between service providers and consumers.
- **Smart PSSs** are relatively novel in the market, and there is limited available information that can help designers be prepared for this new task of **integrating products and services**.

Web-based and e-Commerce systems have become more common and have proved to be very effective in capturing the pulse of the market.

These web-based toolkits aim to provide a set of user-friendly design tools that allow trial-and-error experimentation processes and deliver immediate simulated feedback on the outcome of design ideas.

Once a satisfactory design is found, the product specifications can be transferred into development process.

Figure 2 presents a model of a new concept of product configurator (configure-to-order and engineering-to-order) and an e-service web application, which involve customers directly in the design of product development. The customer can define product attributes in the configurator, and with this tool, the configurator is very friendly to use.

The product configurator has an adaptable interface using different types of modules, which are specific for each case of product development, and which depend on the technical requirements of the customer. The various types of modules that can be used are:

- **Common Modules**—design for different product families;
- **Customize Modules**—design for product or function specific;
- **Personalized Modules**—design for customer specific.

Although the concept of mass customization has been introduced as early as a few years ago, its effective implementation is a great challenge in terms of flexibility and the ability of manufacturers to quickly change variants of manufactured products [29]. Figure 3 shows how mass customization is realized in the smart factory.

In mass individualization case, the customer is directly involved in the product development process, so the chances required for achieving the final product characteristics and functionalities meeting the customer's needs are very high. The relationship with the customer is also strengthened, as there is direct communication with the design team from the early stages of the product design process. Thus, customer understands exactly what product features is paying for and in this way his/her attention is no longer directed exclusively to the product price but to the benefits of his choices (product functionalities, characteristics, impact on the environment etc.). In addition to these benefits, this product design approach reduces/eliminates the working task repetition and increases product innovation.

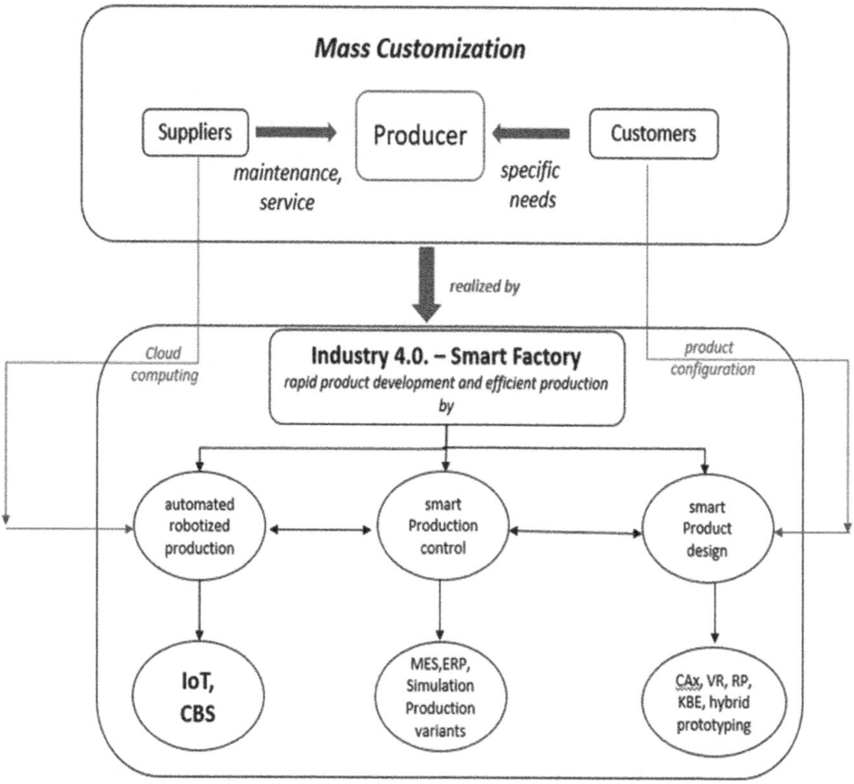


**Fig. 3** Mass customization in the smart factory. Adapted from Ref. [29]

The perspective presented by [14] states that there are three possible product individualization strategies from which a company can choose:

1. The manufacturer tailors the product for a specific customer;
2. Customer tailored or self-individualized products;
3. Individualization strategy without tailoring individualized products or services.

## 4 Open architecture concept and customer's role in the design of individualized products

The customer's role in the design of individualized products is increasing exponentially as they have active participation in their design. It is again [12] who reveals the elements that make co-design between customer and manufacturer possible:

- Open architecture products are based on an open platform that allows the integration of standard, custom and individual modules;

- Custom design that allows customer involvement at different stages of the development process. In this approach, the flexibility of the process must allow collaboration between actors with different degrees of expertise and experience;
- On-demand manufacturing systems that respond quickly to changes brought about by customer wishes (e.g., additive manufacturing);
- Cyber-physical systems through user-friendly interfaces support participatory design.

The role of open architecture products is to create a bridge between the developers and the users of the product. The three elements that make up a modular structure are architecture, modules and interfaces [19].

Open architecture products and adaptable module-based design support and lead innovation in integration technologies. In [19] there are presented different solutions that could be adopted in the case of an open architecture product development approach such as:

- Improved functionality by incorporating new advanced technologies into product design;
- High-quality standards because validated modules are used, as well as the knowledge and processes associated with them;
- Reduced costs because it relies on the reuse of common modules and components;
- Fast delivery times due to reuse of design elements and manufacturing processes;
- Individualization using customized modules to meet changing requirements;
- Environmentally friendly products as their lifetime are increased, reducing waste.

As can be observed, there is a need to introduce into the product development process, a methodology dedicated to planning the architecture of the future product. Such a methodology oriented towards product individualization is proposed by [11] and is presented as a five-step process:

1. Analysis of the initial situation;
2. Analysis of customer wishes;
3. Analysis of potential conflicts;
4. Modification of the product architecture;
5. Validation of the architecture change.

Regarding customers' involvement in the product design process, in the research described by [24] it is defined as a co-creation "process that allows individual customers to influence the design of future services or products, developing experiential marketing". The concept of co-creation of value can reflect the fact that "value is not created exclusively by companies but from their interaction with different customers" [10]. The concept of co-creation is sustaining the Open Innovation approach, and in most cases, the customers' input ideas could be found original, new and/or feasible and can be considered valuable for companies' innovation processes [9].

End users are involved in the development of individualized products from the early stages of the creative process. To make this possible, companies provide customers with co-design equipment such as product configurators [15].

In the current economic climate, it has been demonstrated time and again that a company's performance is directly proportional to the satisfaction of its customers. As a result, there is a migration towards individualized products. The individualization strategy chosen by each company aims to strike a balance between increasing the variety of its products and the costs generated by the higher degree of complexity. According to [26], product configurators have the following fundamental functions: presentation of the product offer to customers, real-time information on price, delivery time and product specifications, verification of the new product variant in terms of its integrity and validity and the compilation of the bill of materials.

In general, the product configurators used so far are not customer-oriented and do not help to identify requirements, because they do not consider the customer's knowledge about a particular product, which leads to frustration when the customer has to use such a tool. Although essential features of a product configurator and methods to improve such a platform have been presented in the literature, no references to improve the product configuration process were found.

IT and new manufacturing techniques are tools through which a new customer-oriented product development process can be realized. These include product configurators with which the customer can choose from a list of existing features resulting in a recommended product, which meets the customer's desires [30].

In the view of [2], the future of manufacturing lies in modular and efficient manufacturing systems, which allow the production of individualized products while maintaining the economics of mass production. One of the technologies that lend itself to a high degree of product customization is the 'Digital Twin', which can be a digital replica of a product, features, functionality, processes or systems. It is also based on other Industry 4.0 specific technologies such as the Internet of Things (IoT), Artificial Intelligence (AI), autonomous robots, Augmented Reality (AR) and Additive Manufacturing (AM).

Disadvantages of individualized products include high initial investment costs, long design and development time, low volumes and low productivity compared with mass and large series production. At the same time, the market is shrinking because customized products meet the specific needs of a particular customer.

As highlighted in this chapter, the trend is towards migration to mass individualization. In contrast to customized products, individualized products aim to satisfy a single customer. It is, therefore, a customer-centric approach. It is the customer who has the power to radically change the product design or even create new products within budget and lead time. Mass customization is based on modules that can be configured to achieve the final product. This mechanism allows the reuse of basic design parameters within a product family [27]. On the other hand, the ability to adapt even the entire design of a product to meet customer needs is essential in the manufacture of individualized products.

The current economic context is playing a major role in the trend towards mass individualization. There is a shift from a linear economy to a circular economy, with

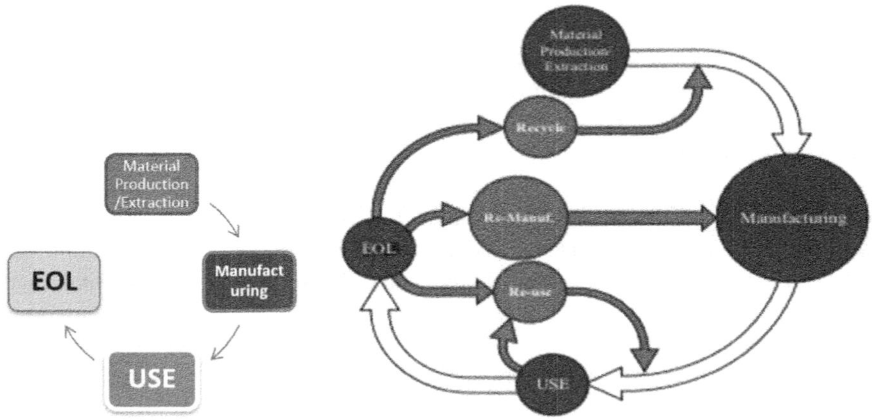


**a**. Linear Economy **b.** Circular Economy

**Fig. 4** Economics models (rebuilt after [16])

a focus on optimizing how resources are used to reduce emissions over the lifetime of products or services. Figure 4a, b shows a comparison between the two economic models, linear economy and circular economy (rebuilt after [16]).

Product modularity has clear advantages for the circular economy model. Among the most important are the possibilities to improve products by upgrading functionality with a module, thus avoiding the need to replace an entire product; modularity allows the replacement of components and facilitates the maintenance and repair process.

Product reconfiguration is the modification of structural and functional components to adapt products to new operational requirements. Reconfiguration can be considered as a way of creating new variants of a product. According to [16], the development of families composed of different variants of the same product is associated with a more efficient use of resources, and thus occupies an important place in the current circular economy model.

"New technologies are clearly essential within this evolution and will continue to play an increasingly important role. However, the true innovation of co-creation, as well as Industry 4.0 and Industry 5.0, is the different general approaches to individuals and their true value, both for companies and society. Value co-creation, on the one hand, has made the role of individuals very clear. It is now evident that individuals can not only contribute to innovation in companies but also to the society in increasingly innovative ways. The value co-creation, contribute to the transition from Industry 4.0 to Industry 5.0 in an almost indirect way through innovations realized" by companies considering these processes (as debated by [3]).

## 5 The Role of Databases in the Product Development Process

Developing smart products in the context of Industry 4.0 means adapting the way products are designed. Intelligent design is characterized by speed, efficiency and elimination of errors.

The design process using databases meets these criteria [29]. Databases store information about process steps, their timeline and execution methods. The main reason why databases should be integrated into the product development and design process is to reduce the resources involved. Up to 80% of the time spent designing a product is consumed with the execution of routine tasks [20]. If repetitive tasks are executed faster, with the help of Knowledge-Based Engineering (KBE) systems, it means that the time saved can be redirected to creative activities, thus increasing the number of variants brought to market in a shorter time and facilitating the **customized products' production** (Fig. 5).

Database systems have the role or simulate the knowledge of an expert in a particular field (expertise, high qualify competences in a specific field, activity, or process). In the research described in [22] is define the ontological approach using a database

**Fig. 5** Design time structure according to the method used (rebuilt after [20])

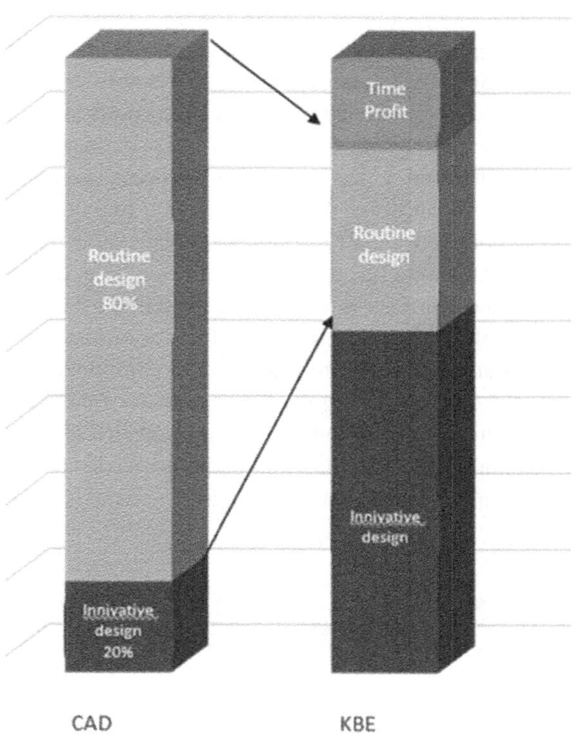

system (similar to a complex knowledge repository) as "a way to collect knowledge, find and record solutions or reuse knowledge later".

Through a critical analysis of several works dealing with the topic of using systems designed for the purpose of developing products in the context of the fourth industrial revolution, Ref. [28] identified problems in the following areas:

- Knowledge management—database systems are not yet reliable enough to become a widely used tool in the industrial domain.
- Cost estimation—computer-aided design tools currently in use cannot provide sufficient cost forecasting information in the early stages of the development process;
- Product development—intelligent systems are used for the efficient processing of large amounts of data and information. Therefore, in the future, the knowledge of experts and the experience of development teams must be redirected to creative tasks, with repetitive tasks being taken over by various systems integrated into the development process.

## 6 Product Configurators

Routine and repetitive tasks in the traditional design of new product development processes are resources and time consuming. The results obtained in design step execution should be analysed and if it is not satisfactory, that step must be repeated, the iterations are done until the desired results are achieved.

In the product development process, databases prove to be useful translating a part of the routine decision-making process to a computer program. In this way, the engineers are relieved of routine decision-making work, which is time consuming, by being able to use their resources to create value through innovation and non-routine tasks that cannot be automated. An example in this sense, are product configurators where the customer is directly involved in the development of the product, involving the development of the same product for different customers (Fig. 6).

A Lamp open architecture product configurator from the automotive industry is presented in the following paragraphs. As functions to be offered, the Lamp open architecture product configurator presents to customers the products offer, information about price, specifications and time needed to delivery, bill of materials, product variant all according to input requirements. Also, a lamp product is a complex one

Fig. 6 Customer input—product configurator

and integrates interdisciplinary teams and knowledge. New or existing products with individualized functions require a high degree of innovation in design and to be able to manufacture the products with a high degree of flexibility in production.

This configurator is based on a model presented in Ref. [30], which delineates two stages of open architecture product configurator development:

- Modular design stage with macro-level impact in the performance of an entire product family;
- The scalable design stage, which focuses at the micro-level on optimising design specifications.

To make the transition from requirements to physical products, the Lamp open architecture product configurator has as basis the axiomatic design model of Nam Pyo Suh [25]. According to the model (Fig. 7), the first stage will collect the requirements and needs of the industrial environment, from the perspective of Industry 4.0. The requirements and needs (CAs) of the industrial environment will be translated into functional requirements (FRs) and constraints (Cs). Defining functional requirements (FRs) will determine the level of digitization and the degree of automation required throughout the product lifecycle.

For each functional requirement (FRs), the required design parameters (PRs) will be nominated, which will contribute to the fulfilment of the functional requirements. The design parameters will define the new stages of design and development of new products, the integration of the design process with manufacturing processes

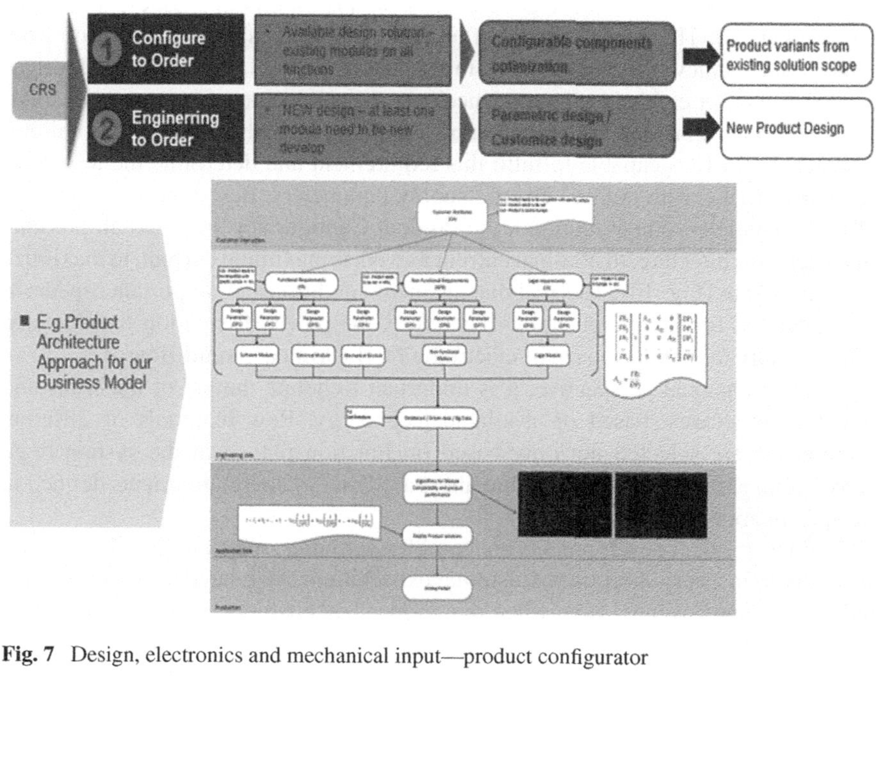


Fig. 7 Design, electronics and mechanical input—product configurator

(PVs) and the flexibility in the case of custom products or with reduced series of manufactures. In this way, the module interchangeability will be assured.

The three vertical levels of model are: customer interaction, engineering, application and production. The functional, non-functional and legal requirement CAs are further detailed in design parameters that are grouped in optical, electrical and mechanical modules and progressively refined to achieve the detail level necessary to materialize the product optimized solutions and further in developed product (production).

The main modules in designing this product, besides the collecting requirements module, are the optical module, electronic module, mechanical module, non-functional module, and legal module (Fig. 8). Choosing between modules of optical module means choosing the different optical design parameters, different design options and with individual values for independence and information content.

The **configuratordb—Lamp** open architecture product configure (Fig. 9) is developed on the platform-based structure, in which modules are connected, the configurator database contains information about different types of LEDs, electronics module, mechanical materials. The back-end of the application is based on Structured Query Language and connects to the database and process the information to achieve all technical options to fulfil the customer requirements. In the INPUT page, the customer requirements are introduced on the dedicated field (as presented in Fig. 8). After saving it, based on algorithms with constraints and compatibilities already defined, the Product Configurator can generate the result.

The result is all technical options with information from databases, to achieve the requirements for different stages during development process (optics, electronics and mechanics). The result consists of all technical options with information from databases, to achieve the requirements in the different stages of the product development process (optics, electronics, and mechanics requirements). First, based on lightning conditions related to the optical output requirement, in the Product Configurator is selected all LEDs capable to fulfil this requirement and determine the necessary number of LEDs with defined electric supply parameters. Based on the number of LEDs and supply voltage required, the Product Configurator proposed all possible electronics architecture, in nominal current as well as maximum current, to maximize the light output (Fig. 10). Thermal management conditions are important aspects to be considered for such applications; therefore, the power dissipation is calculated automatically to define the size of electronic PCB (Printed Circuit Board).

From mechanical perspective, it is important to define the list of materials and assembly processes, based on available technology. Raw materials of different components are selected not only based on function played in the system (e.g., Lens—transparent PMMA or PC) but also based on system requirements defined in the specification (Fig. 11).

The new product developments are complex activities, which required high level of creativity in a very short time. Customer expectations are related to short lead time and quick responsiveness even when requirement specification is often changed. In this context, an adaptable design with high degree of flexibility needs to be developed by each company. The new product development processes are complex activities,

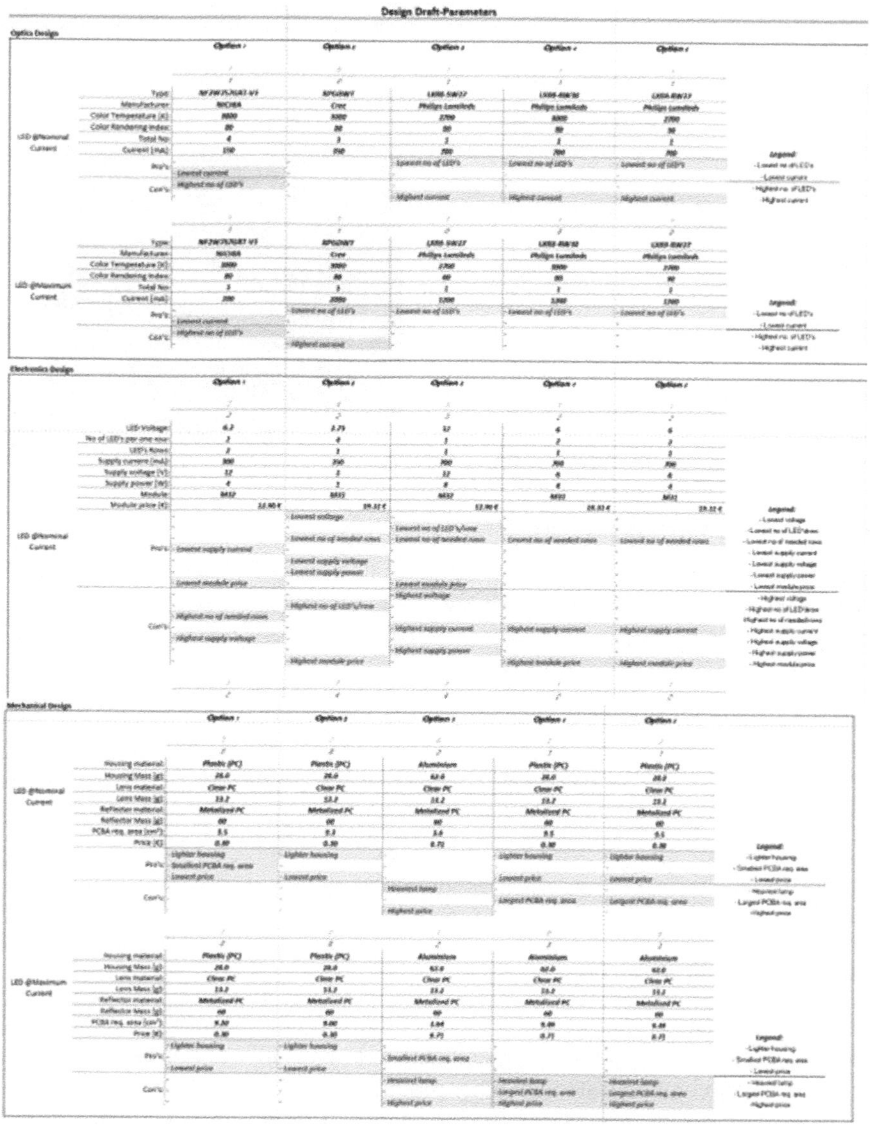

**Fig. 8** Design, electronics and mechanical input—product configurator

which required high level of creativity in a very short time. Customer expectations are related to short lead time and quick responsiveness even when requirement specification is often changed. As a response to this situation, an adaptable design process with high degree of flexibility needs to be developed by each company. In addition, because of the products' complexity constant increasing it becomes obviously that

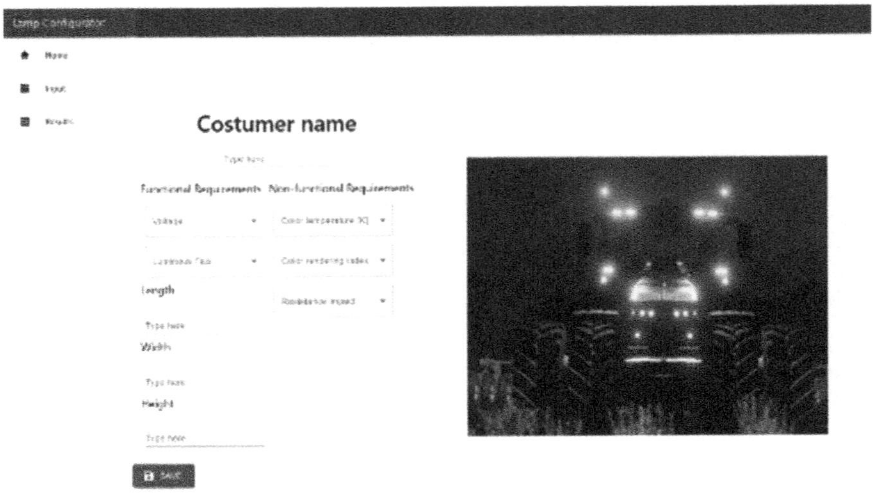


Fig. 9 Interface of lamp configurator

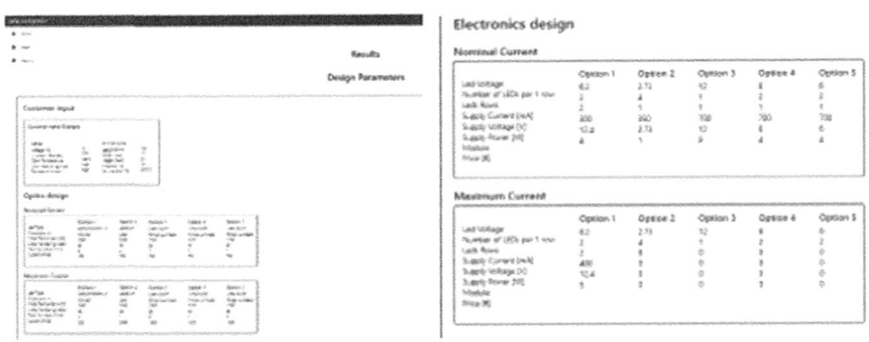


Fig. 10 Interface of design parameters

the use of intelligent systems incorporated to new innovative products can exchange information or communicate with other product as well.

In this context, the traditional design cannot satisfy the customer expectation in terms of time to market and cost efficiency. The iteration approach to define the optimum solution for individual customer simply takes too long time, and designers rarely can explore alternative solutions that depart significantly from their base-case assumptions, too often the final design is suboptimal.

By using cutting-edge technology and integrating digital simulation of a possible solution, the result can be available in a very short time, giving the chance to customer to accept or even change the requirements without additional cost. Moreover, customers can be a part of development process and contribute on product design, adapt specification and choose different features.

**Mechanical design**

**Nominal Current**

| | Option 1 | Option 2 | Option 3 | Option 4 | Option 5 |
|---|---|---|---|---|---|
| Housing Material: | Plastic (PC) | Plastic (PC) | Plastic (PC) | Plastic (PC) | Plastic (PC) |
| Housing Mass [g]: | 28.0 | 28.0 | 28.0 | 28.0 | 28.0 |
| Lens' Material: Clear PC | Clear PC | Clear PC | Clear PC | Clear PC | Clear PC |
| Lens' Mass [g]: | 13.2 | 13.2 | 13.2 | 13.2 | 13.2 |
| Reflector Material: | Metalize PC | Metalize PC | Metalize PC | Metalize PC | Metalize PC |
| Reflector Mass [g]: | 60 | 60 | 60 | 60 | 60 |
| PCBA requested area [cm^2]: | 3.5 | 3.5 | 3.5 | 3.5 | 3.5 |
| Price [€]: | | | | | |

**Maximum Current**

| | Option 1 | Option 2 | Option 3 | Option 4 | Option 5 |
|---|---|---|---|---|---|
| Housing Material: | Plastic (PC) | Plastic (PC) | Plastic (PC) | Plastic (PC) | Plastic (PC) |
| Housing Mass [g]: | 28.0 | 28.0 | 28.0 | 28.0 | 28.0 |
| Lens' Material: | Clear PC | Clear PC | Clear PC | Clear PC | Clear PC |
| Lens' Mass [g]: | 13.2 | 13.2 | 13.2 | 13.2 | 13.2 |
| Reflector Material: | Metalized PC | Metalized PC | Metalized PC | Metalized PC | Metalized PC |
| Reflector Mass [g]: | 60 | 60 | 60 | 60 | 60 |
| PCBA requested area [cm^2]: | 9.20 | 9.20 | 9.20 | 9.20 | 9.20 |
| Price [€]: | 0.3 | 0.3 | 0.3 | 0.3 | 0.3 |

**Fig. 11** Interface of mechanical design

Beside the possibilities to choose from the existing reliable set of solution (e.g., by introduce the algorithms and digital data simulation) it is needed to increase the processes to identify new product innovation opportunities or non-intuitive solution, which may never would be found with traditional processes. Design for open innovation is another advantage of product configuration system, when a pre-development set of solution is determined based on continuous customer feedback.

## 7 Conclusions

A review of the state of product development research in the context of the transition from Industry 4.0 to Industry 5.0 (preparation for the next development stage of industry) reveals the following limitations:

- Products are becoming increasingly complex, and the traditional process of developing new products is not being updated with new technologies;
- Intelligent product development (mechanical, electronic, software) is carried out with multi-disciplinary teams, but the risk of human error is not controlled in real time;

- Communication within distributed project teams is not always effective, partly due to internal organizational procedures, partly due to differences in culture, education, time zones etc.;
- Integration of customer specifications (i.e., customer involvement in the development process is partial).

The trend of personalized and individualized products with customer-centric product development required a new approach for process development to stay competitive in a very dynamic market.

Developing a model for integrating new technologies for the development of a product, optimizing production and investment costs, reducing time to market, for small and customized production runs, the secondary objectives that will ensure the achievement of the main objective are:

- Determining advanced technologies that can contribute and be integrated into the product development process.
- Definition of milestones in the design and development of new products with cost optimization, increasing product performance throughout the life cycle.

This chapter has presented a model of product development approach considering a new concept of product configurator consisting of a technical configurator (configure-to-order and engineering-to-order) and an e-service web application, which involve customer directly in the design process. In the first step, a theoretical configurator framework was defined and consisted of modularity approach on the technical side and web application on the customer side. Product development in open architecture approach provides the opportunity to integrate customer requirements, manufacturer expertise and supplier's technology in the same product platform. Second, we validated the theoretical framework by developing a product configurator architecture for a lamp product and prove that digitalization of product development process reduces the time-to-market, cost on one hand, and on the other hand encourage innovation.

Future research should support industry to achieve simultaneously two goals of reduction variation in production and promoting variation in their product configuration activities, by digitalized and integrate complete product lifecycle and supply chain.

## References

1. Abramovici M, Goebel JC, Savarino P (2017) Reconfiguration of smart products during their use phase based on virtual product twins. CIRP Ann 66(1):165–168. https://doi.org/10.1016/j.cirp.2017.04.042
2. Aheleroff S, Xu X, Zhong R (2020) A digital twin reference for mass personalization in industry 4.0. https://doi.org/10.1016/j.procir.2020.04.023
3. Aquilani B, Piccarozzi M Abbate T, Codini A (2020) The role of open innovation and value co-creation in the challenging transition from industry 4.0 to society 5.0: toward a theoretical framework. Sustainability 12(21):8943

4. Armengaud E, Sams C, von Falck G, List G, Kreiner C, Riel A (2017) Industry 4.0 as digitalization over the entire product lifecycle: opportunities in the automotive domain. In: Stolfa J, Stolfa S, O'Connor R, Messnarz R (eds) Systems, software and services process improvement. EuroSPI 2017. Communications in computer and information science, vol 748. Springer, pp 334–351. https://doi.org/10.1007/978-3-319-64218-5_28
5. Berger R (2015) Strategy consultants, commissioned by the Federation of German industries (BDI) (2015). The digital transformation of the industry. http://www.confindustria.ud.it/upload/pagine/Industria%2040/Il%20tema%20quarta%20rivoluzione%20industriale/Roland_Berger-Digital_transformation_of_industry%20-%202015.pdf
6. Casadesus-Masanell R, Almirall E (2010) Open versus closed innovation: a model of discovery and divergence. Acad Manag Rev 35:27–47
7. Chesbrough H, Crowther AK (2006) Beyond high tech: early adopters of open innovation in other industries. R&D Manag. 2006(36):229–236
8. Council for Science, Technology and Innovation Cabinet Office, Government of Japan (2015) Report on the 5th science and technology basic plan. https://www8.cao.go.jp/cstp/kihonkeikaku/5basicplan_en.pdf. Accessed 12 May 2020
9. Filieri R (2013) Consumer co-creation and new product development: a case study in the food industry. Mark Intell Plann 31
10. Hidayati R, Novani S (2015) A conceptual complaint model for value co-creation process. Procedia Manuf 4(2015):412–418
11. Holle M, Maisenbacher S, Lindemann U (2015) Design for open innovation individualization-oriented product architecture planning. In: 2015 annual IEEE systems conference (SysCon) proceedings, Vancouver, BC, pp 397–402. https://doi.org/10.1109/SYSCON.2015.7116783
12. Hu S (2013) Evolving paradigms of manufacturing: from mass production to mass customization and personalization. Procedia CIRP 7:3–8. https://doi.org/10.1016/j.procir.2013.05.002
13. Koren Y, Shpitalni M, Gu P, Hu SJ (2015) Product design for mass-individualization. In: CIRP 25th design conference innovative product creation. Procedia CIRP, vol 36, pp 64–71. https://doi.org/10.1016/j.procir.2015.03.050
14. Kuhl J, Krause D (2019) Strategies for customer satisfaction and customer requirement fulfillment within the trend of individualization. Procedia CIRP, vol 84, pp 130–135. ISSN 2212-8271
15. Levandowski CE, Jiao JR, Johannesson H (2015) A two-stage model of adaptable product platform for engineering-to-order configuration design. J Eng Des 26(7–9):220–235. https://doi.org/10.1080/09544828.2015.1021305
16. Mesa JA, Esparragoza I, Maury RH (2019) Trends and perspectives of sustainable product design for open architecture products: facing the circular economy model. Int J Precis Eng Manuf-Green Technol. https://doi.org/10.1007/s40684-019-00052-1
17. Nafisi M, et al (2016) Manufacturing involvement in new product development: an explorative case study in heavy automotive component assembly. In: 26th CIRP design conference, procedia CIRP, vol 50, pp 65–69
18. Pahl G, Beitz W, Feldhusen J, Grote K (2007) Engineering design, a systematic approach. Springer-Verlag London
19. Peng Q, Liu Y, Yun-Hui Zhang J, Gu P (2018) Personalization for massive product innovation using open architecture. Chinese J Mech Eng. https://doi.org/10.1186/s10033-018-0239-0
20. Reddy EJ, Sridhar CNV, Rangadu VP (2015) Knowledge based engineering: notion, approaches and future trends. Am J Intell Syst 5(1):1–17. https://doi.org/10.5923/j.ajis.20150501.01
21. Salvatore G (2018) Da Industria 4.0 a Società 5.0. https://www.academia.edu/35809134/Da_Industria_4_0_a_Societ%C3%A0_5_0_La_scelta_del_Giappone. Accessed 12 May 2020
22. Sanya IO, Shehab EM (2015) A framework for developing engineering design ontologies within the aerospace industry. Int J Prod Res 53(8):2383–2409
23. Schroll A, Mild A (2011) Open innovation modes and the role of internal R&D. Eur J Innov Manag 2011(14):475–495
24. Stauss B, Seidel W (2004) Complaint management: the heart of CRM Thompson, Phoenix, USA

25. Suh NP (1995) Designing-in of quality through axiomatic design. IEEE Trans Reliabil 44(2)
26. Trentin A, Perin E, Forza C (2012) Product configurator impact on product quality. Int J Prod Econ 135. https://doi.org/10.1016/j.ijpe.2011.10.023
27. Tseng M, Jiao R, Wang C (2010) Design for mass personalization. CIRP Ann Manuf Technol 59:175–178. https://doi.org/10.1016/j.cirp.2010.03.097
28. Voltolini R, Oliveira K, Borsato M, Peruzzini M (2018). Research and analysis of opportunities in product development cost estimation. https://doi.org/10.3233/978-1-61499-898-3-381
29. Zawadzki P, Żywicki K (2016) Smart product design and production control for effective mass customization in the industry 4.0 concept. Manag Prod Eng Rev 7(3):105–112. https://doi.org/10.1515/mper-2016-0030
30. Zheng P, Xu X, Yu S, Liu C (2017) Personalized product configuration framework in an adaptable open architecture product platform. J Manuf Syst 43:422–435. https://doi.org/10.1016/j.jmsy.2017.03.010

# Chapter 7
# Managing Complexity in Manufacturing Service Processes. The Case of Large Business Environments

**Frank Rennung, Caius-Tudor Luminosu, Mirabela Metzler, and Anca Draghici**

**Abstract** The chapter content will provide an overview of the present status of research on complexity management in manufacturing. Based on the findings, we will propose a model for managing complexity in large-scale manufacturing environments. The focus will be on the integrated approach of complexity, specific manufacturing-related complexity criteria, cyclical reporting, targeted measures to deal with the identified complexities, and specific measures. Finally, the developed Large Manufacturing Complexity Model will be illustrated in large-scale manufacturing environment and best practices will be presented.

**Keywords** Systems complexity · Management · Big project management · Stakeholders management · Services · Outsourcing

## 1 Introduction

Complexity Science is "the study of the phenomena which emerge from a collection of interacting objects" [20]. The ideas and theories of this science have permeated through other research fields, ranging from Mathematics to understanding organizations, but they have not always been applied in the same way. Although increasingly used for solving management problems, organizational contexts still translate these ideas into explanatory metaphors rather than theory and model testing [32]. Scholars

---

F. Rennung
Deutsche Telekom IT GmbH - Business Solution, Darmstadt, Hessen, Germany
e-mail: f.rennung@t-systems.com

C.-T. Luminosu · M. Metzler · A. Draghici (✉)
Faculty of Management in Production and Transportation, Politehnica University of Timisoara, Timisoara, Romania
e-mail: anca.draghici@upt.ro

C.-T. Luminosu
e-mail: caius.luminosu@upt.ro

M. Metzler
e-mail: mirabela.metzler@student.upt.ro

 
A. Draghici and L. Ivascu (eds.), *Sustainability and Innovation in Manufacturing Enterprises*, Advances in Sustainability Science and Technology,
https://doi.org/10.1007/978-981-16-7365-8_7

believe that complexity has two aspects: the structure of the complex, meaning that the system has many elements which have intensive interactions that form complex internal structures and the complex behavior of a system, revealing different or even conflicting sides depending on the perspective and the question posed to the system [12].

But complexity is more than working with mathematical models and has more to do with how we see the world. Rooted in knowledge management and complexity science, one such idea is the Cynefin Framework [23], developed as a tool for categorizing issues and devised to support decision-making in management. The framework shows five states ranging from the ordered to the unordered universe. The middle of the framework is occupied by the *Disorder* state, from which we must break out and move into one of the other states. The *Simple* and *Complicated* states are characterized by cause-and-effect relationships, based on facts. On the other side of the spectrum, *Complex* and *Chaotic* states don't have a clear relationship between cause and effect, which requires emerging patterns for finding solutions. Introducing chance in organizations causes "unpredictability and flux", making situations and decision-making a complex context. The Complex domain therefore is a place of novelty and evolution, meaning that "we can understand why things happen only in retrospect" [40]. By following the proposed action scheme "probe-sense-respond", we can discover developing practice, best suited for dealing with complex environments. Research scenarios that study complexity using the Cynefin model vary from managing cybersecurity risks [14], improving the useability of research data [28], or being used as a base for mapping risk models [42]. Cynefin application also has benefits for organizations, offering a modern, more holistic approach of process improvement [24] and an easy-to-use tool to illustrate and explore solutions to problems in complex situations [17].

Another approach in managing complexity is combining Agile[1] and Lean[2] methodologies for coping with complexity and improving project performance in the construction sector [41] or learning how to manage dynamic complexity from Shell's long tradition of scenario building as an attempt to "cooperate" with the future in an unpredictable and complex external world [21].

Due to the ongoing globalization affecting their longevity, companies are pressured into new markets, facing new competitors, and catering to individual customer needs. In addition to the market requirements, the company must manage ever-changing needs of internal and external customers, due to a variety of external factors like the latest information and telecommunications technologies. This context demands innovative companies to embrace "emergent complexity"[3] because

---

[1] Much of the Agile terminology that is used today has roots in the *Agile Manifesto* that includes four values and twelve principles explaining the key concepts of Agile development (full text at http://agilemanifesto.org).

[2] Lean methodology was developed for the manufacturing sector in the 1930s. For further reading refer to *Lean Thinking* by Jim Womack and Dan Jones.

[3] Emergent complexity refers to understanding how interactions between entities (even simple ones) can give rise to something completely unexpected, like "the swarming patterns of birds or the pattern of activity in an organization".

combining complex networks of ideas will create a solution with maximum value for the organization [36]. Within a brief time, the volume of business-related complexity has steadily increased, and a large part of the company has evolved from a complicated to a complex system. Complexity is believed to be an essential success factor for organizations [37] and offers immense opportunities for those companies that can properly address it, and which are able to offer robust processes and solutions to deal with it [10]. Moreover, market research showed that challenges of Industry 4.0 "lead to an increasing complexity and hence opacity" that must be addressed with proper know-how [19]. Complexity can become a desirable element in designing systems that are adaptable to changing environment, responding in "adequate and effective ways" [35].

In dealing with uncertainty of complex problems, critical-path methods and decision-trees are good for handling complexity in project management, but there must also be some methods to represent the potential influence of foreseen and unforeseen uncertainties [11]. Tools to measure complexity have been developed in different sectors and business environments but with a limited scope, for example, in public health to support only objectivity and reliability [33], an Agile hybrid model for process design [2], a control loop model of complexity management focused on the portfolio, product, processes, and resource level in smart services [39] or in product development as a method to improve decision-making by integrating the complexity perspective into decision-making [7]. In the era of Cyber-Physical Production Systems in manufacturing, the topic of complexity handling will receive much more emphasis as complexity issues will have much larger significance [30]. However, before adopting new tools and approaches following the latest management trends, organizations should calculate the costs and benefits of short cycles of change and implementation [9].

## 2 Concepts for Managing Complexity in Business Situations

The science of complexity management is still young in the field of research and a wide variety of approaches exist. Early research and studies on complexity management recommended following guidelines regarding the balance between internal and external complexity within an organization [15]. This was to be done by performing processes sequentially [13], following production-driven strategies for mastering complexity with organizational focus [5] or following specific steps based on system archetypes ([1], p. 59). The goal was to make complexity tangible and visible in the organizational network.

A first complexity model in large infrastructure projects was developed under the name "House of Project Complexity" that was able to graphically present complexity in the organization [25]. The core of this model supports the idea that complexity can be associated with the relationship between various project features and properties.

The result is two dimensions: institutional features and technical features that form the body of the "house". The authors expressed the belief that the concept can be generally extended to large manufacturing projects, or other industrial megaprojects.

Starting with 2010 researchers developed extensive criteria for complexity management. Notably, Wildemann proposed functional modules considered relevant for complexity management and developed three basic associated strategies: avoiding, mastering, and reducing complexity [46]. Schoeneberg recommends practical tools and methods of managing complexity comprised a holistic and interdisciplinary approach. He classifies complexity drivers in two categories internal and external and developed cluster with specific criteria for each category [37]. Based on the complexity dimensions of Schoeneberg [34], a holistic business complexity management model called the "House of Large Business Complexity Management" and a complementary methodology containing managerial methods and tools (including a template) for practical application was proposed. The model is created by mapping selected internal and external complexity criteria to the four dimensions of the "Balanced Scorecard Model"[4]—Learning & Growth, Business Process, Customer and Financial—adapted to large-scale business projects. By using the model in large-scale project situation, the involved parties are capable to predict complexity and prevent falling into the complexity trap. Furthermore, its use enables choosing the desired management strategy of either avoiding, managing, or reducing complexity. A literature review focusing specifically on the manufacturing industry identified 486 complexity drivers divided into three groups (external complexity, internal complexity, and general complexity) consisting of 22 main complexity driver categories [44]. Trying to find solutions to the increasing external and internal complexity of organizations, their environment and strategies, organizations must consider flexibility when elaborating and adapting strategies [31]. This flexibility should extend to managerial mindset as well, because organizational performance is related to the shift from individual capabilities as independent variables to the configurations of capabilities [47].

Existing approaches to address complexity in production can be grouped into three main categories: objective complexity (that neglects the workers and their role) and subjective or perceived complexity (omits the factors that are independent of people) and a joint analysis of both subjective and objective aspects of complexity [4]. A recent attempt to manage the risk in value creation chains posed by increasing complexity in production development processes was made by a group of researchers in Germany who presented a synergetic approach for mastering complexity using a model that merges a system model with business processes, but with a high up-front modeling effort [22].

Academic literature shows a positive trend and interest in complexity management in different disciplines, but not as much in manufacturing and supply chain management, although positive impacts on performance indicators are directly connected to best management of complexity in organizations [43]. Current research shows a

[4] The Balanced Scorecard (BSC) was developed in the 90s by Robert Kaplan and David Norton to measure strategic performance. Since then, BSC went through many transformations.

common thread in managing complexity by focusing on interdisciplinary models, instead of only on one or two capabilities and ignoring their interactions with each other. Our observation is that complexity has become a cardinal feature of future manufacturing companies and complexity management should be given a top priority.

## 3 Large Manufacturing Complexity Management

In this section, we describe a model for managing complexity in large manufacturing, which is the purpose of this research. First, we describe the concepts composing the model: manufacturing and management.

Economically, "manufacturing[5] is the transformation of materials into items of greater value by means of one or more processing and/or assembly operations", based on three inter-dependent building blocks of modern manufacturing (the materials, processes, and systems) [18]. Industry 4.0 allowed for automated, collaborative manufacturing models such as "Manufacturing as a Service" (MaaS), that harnesses the collective use of a manufacturing network to divide complex tasks necessary to produce goods by different companies in an interconnected environment, like a manufacturing ecosystem [16].

Large manufacturing companies are considered for their high revenue [8, 27], and are characterized by "various business units, a diversified business model portfolio, and complex IT landscapes" [3]. For the purpose of our research regarding large business environements, a large manufacturing company is profit-oriented, international, industry independent, with mass production capability and a large organization with over 1000 employees (sometimes distributed internationally), operating on a free, competitive market, having a long-term business strategy and a high percentage of production is done in-house (with purchased products and services), selling products wholesale and retail, with accompanying services.

Management allows to steer and control the company with different methods and instruments, one of these being the complexity model. The main task of management in a company is to advance the company and to ensure the company's long-term success, including the design of the company processes. Classic management, or management in an Agile environment, has a different structure and characteristics. In any case, it plays an essential role in the introduction and operation of processes and must be actively involved ([26], pp. 188–189). The classic management theory evaluates entrepreneurial requirements or challenges to be solved from the perspective of plannability and rationality. They can be controlled with the right expert knowledge, so that continuity and clarity is ensured. In the VUCA[6] world,

[5] The cited author's view is that production has a broader meaning than manufacturing, but it can be used interchangeably in some industries.

[6] VUCA is an acronym that stands for volatility, uncertainty, complexity, and ambiguity. It represents the challenges of a rapidly evolving business environment. Find more at https://www.vuca-world.org.

however, many individual influencing factors create a networked overall structure. The individual parts, as well as the interrelationships, are difficult to understand. Such a system is overly complex as its functional models can only be understood in the given context. Correct, i.e., successful, action can no longer be mastered with expert control and the principle of predictability. The greatest challenge is dealing with complexity and, above all, dealing with ambiguity, which requires an openness to all conceivable interpretations in relation to the external corporate environment ([6], p. 25). A complexity management tool can be used in organizations to show the strenght and number of inter-dependencies between units as evaluation criteria for determining the complexity [29]. One thing is certain: complexity management must be set up on a strategic level because of the critical impact it has on the entire business.

### *3.1 Large Manufacturing Complexity Management Model*

"House of Large Business Complexity Management" model developed by [34] was initially designed for the service sector. It features a model for managing large and complex projects based on the complexity dimensions of Schoeneberg.

In the next paragraph, the proposed holistic "Large Manufacturing Complexity Management" model is described. Figure 1 presents the proposed model that is the result of the critical overview, the analysis, and synthesis of the relevant references in the field of complexity management that were previously presented, adapted, and transformed. The model is the core of Manufacturing Complexity, and it indicates all relationships that are relevant in a large-scale manufacturing environment. The model should be used as a framework for the other elements that will be described in the following sections.

As a first step, a new set of large business complexity criteria are identified and developed, to accommodate the characteristics of manufacturing (section **A**). This area of the model forms the theoretical foundation, adapted to the type and character of the company or organization. Factors influencing the adaptation are, among other things, the company's industrial sector, size, and structure (e.g., degree of internationalization). The adaptation should be carried out very thoroughly for a company at the beginning of the implementation of the model. The result of the adaptation is the selection of the clusters and the definition of the associated complexity criteria. These principles must be made transparently accessible to all relevant stakeholders and users of the model and a uniform understanding must be achieved across all stakeholders. It makes sense to use the definitions as a user guide for ongoing reviews. These criteria are the basis for further instruments of complexity management. Next section will give a detailed description of individual criteria within each cluster, with definition and interpretation, as well as arguments for the evaluation of relevant characteristics of the company in the form of a Relevance factor.

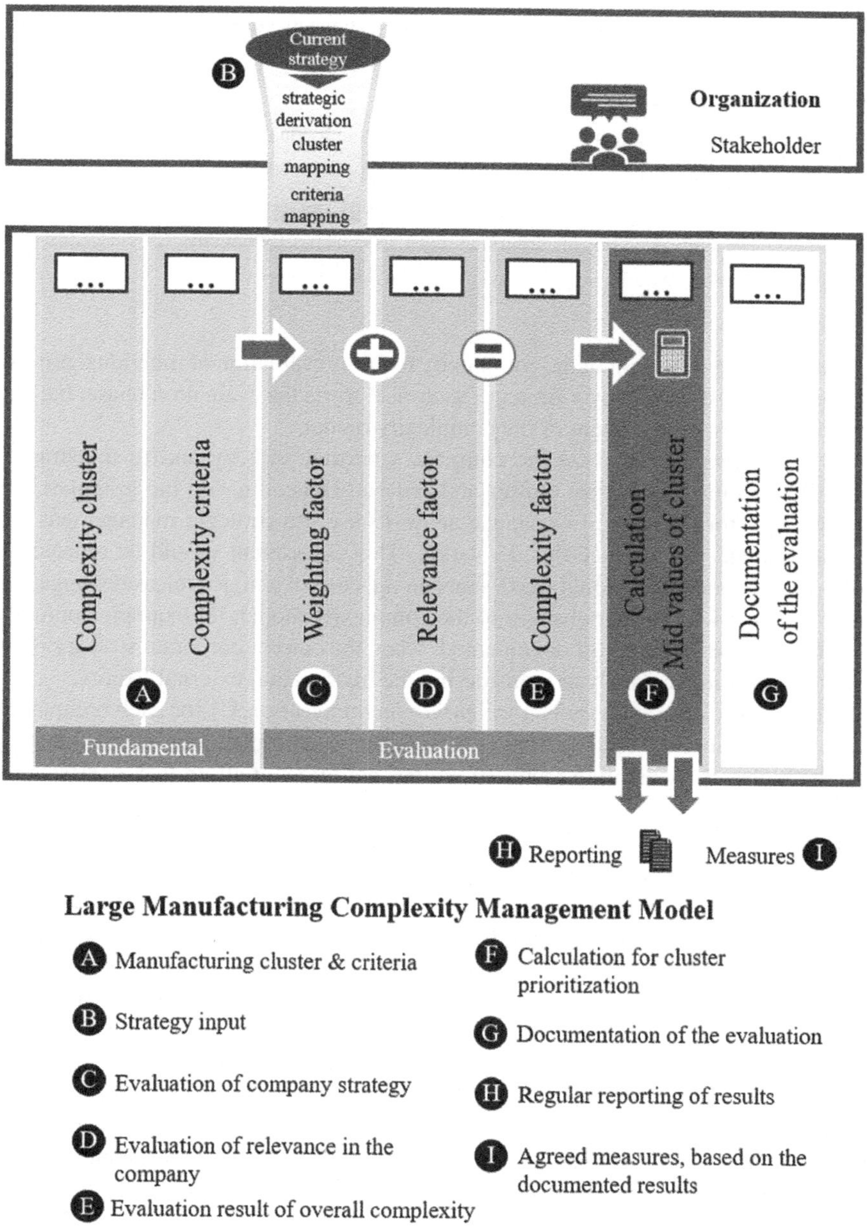


**Fig. 1** Large manufacturing complexity management model (own schema)

**Table 1** Values of the weighting factor for evaluating complexity criteria

| Value | Description |
|---|---|
| 0 | The strategy does not affect the criteria |
| 1 | The strategy affects the criteria only in a few individual aspects, rather indirectly |
| 2 | The strategy barely affects the criteria, mostly indirectly |
| 3 | The strategy affects the criteria in part, directly and indirect |
| 4 | The strategy affects the criteria in many aspects, mostly directly |
| 5 | The strategy affects the criteria directly to a large extent |

An overall assessment of the complexity requires evaluation of individual criteria, depending on the company's strategy. Section **B** starts the evaluation phase, but also serves to prepare the content of the complexity model.

It is important to analyze the company's current strategy and to transform it according to the complexity cluster and criteria. Depending on the granularity of the strategy formulation, it can make sense to use the strategic management tool "strategy map"[7] to produce this derivation. This assessment should be repeated in long-term periods; this would make sense in the case of a major release/change also in connection with further changes to the complexity model, for example, annually. All data will be gathered (or documented, when the current corporate strategy is not clear) from company analysis, considering the internal and external factors.

The result of the analysis is a *weighting factor* (section **C**) for each complexity criteria. The more the current strategy affects the criteria (in the sense of influencing it), the higher the weighting factor (0 = does not affect at all up to 5 = affects the criteria to a considerable extent, as described in Table 1. The weighting factor should reflect the company's strategic goals and should be the same in a mid and long-term perspective, to ensure comparability of ongoing evaluations, and only be adjusted in the event of significant strategy adjustments. The evaluation should be carried out by at least three people, with different perspectives and various backgrounds, to create more reliable results. If any changes occur, we recommend keeping a separate, parallel record, to achieve continuity and comparability. Often, a company must decide between working with actual and non-comparable data or not actual but comparable data; achieving the right balance is a strategic decision.

The *relevance factor* (**D**) is a dynamic value and reflects specific characteristics of the company, in accordance with the current business context. Collecting data for evaluation of the Relevance factor should be done in the same manner and at the same time across company units, throughout the monitored period. Here to, the evaluation should be carried out by at least three people, with different perspectives and various backgrounds, to create more reliable results. The relevance assessment must be justified; the more relevance the criteria have for the current situation in the company (in the sense of current characteristics in the company), the higher the

[7] More information on the topic can be found in *Strategy Maps: Converting Intangible Assets into Tangible Outcomes* by R. S. Kaplan, D. P. Norton.

**Table 2** Values of the relevance factor for evaluating complexity criteria

| Value | Description |
|---|---|
| 0 | The current characteristics in the company have no relevance to the complexity criteria |
| 1 | The current characteristics in the company are relevant to the complexity criteria only in a few aspects |
| 2 | The current characteristics in the company are directly and indirectly relevant to the complexity criteria |
| 3 | The current characteristics in the company are mainly directly relevant to the complexity criteria |
| 4 | The current characteristics in the company are directly relevant to the complexity criteria in many aspects |
| 5 | The current characteristics in the company have great relevance to the complexity criteria |

relevance factor (0 = currently no relevance up to 5 = high relevance of the criteria, Table 2).

When evaluating, it should be noted that there are visible or hidden complexities, or that some parameters can develop dynamically over time.

The weighting factor is strategy-derived and future-oriented, while the relevance factor is derived from the actual and current situation of the company. These are the two sides of the *complexity factor* (section **E**), calculated as the product of the weighting factor and the relevance factor. Complexity factors represent an evaluation of the overall complexity within the company.

Based on this result, each cluster will receive a value representing the mean of all individual complexity criteria within the cluster (section **F**), allowing the arrangements of items in order of importance relative to each other and leading to cluster prioritization. To achieve appropriate prioritization, it requires an analysis of the complexity driver (both quantitative values and a qualitative view), followed by a calculation of respective mean value of all complexity clusters, and determining a suitable size that will decide which complexity clusters should be considered further. The complexity clusters below the specified limit are no longer considered. We recommend to only consider a cluster that has reached a mean complexity factor value of 10 and higher. This way, company resources can focus on parts of the business that are truly dealing with elements of complexity.

An evaluation for prioritization at cluster level makes sense because the complexity criteria are not 100% selective. A consideration at cluster level compensates for this impossibility of sharpness in definition and evaluation. If necessary, a qualitative look at the values of the complexity criteria shows that there are a few outliers and then, these can be individually included in the prioritization. The result of this section is a list of complexity criteria which will be continuously evaluated in the future.

Section (**G**) serves to document the results of the discussion and to make them comprehensible, also as a reference for future evaluations of the respective criteria.

Cyclical reporting takes place in (section **H**). Depending on the requirements and practices of the stakeholders or the organization, the results can be presented graphically for better understanding. Regular reporting of the results is necessary for discussion in the management group and with stakeholders.

Targeted measures (section I) to deal with the identified complexities (and further development) are defined based on cyclical reporting and discussions with stakeholders. Here, too, it should be noted that the number of measures should be limited. When defining measures, the three main strategies of complexity management should be considered: avoiding, reducing, or mastering complexities. The measures themselves should be set up using Agile management[8] methods and procedures, for example, KANBAN and SCRUM frameworks, or other approaches via iterations and systemic loops. A survey evaluation of the current state and the setting up of suitable measures from the management should be carried out in a monthly cycle.

To develop complexity management in the company, we recommend defining (central) responsibility; this central function should be enhanced by further responsibilities in all locations.

## 3.2 *Stakeholder Analysis*

The stakeholders have an outstanding role in applying the Complexity management model. because, on the one hand, this group of employees are affected by the implementation of the complexity model and, on the other hand, they have legitimate claims based on their role in the organization. The dialogue between a process owner and the stakeholders, but also between various categories of stakeholders (on complexity management-related topics), should be permanent, mutual, and complementary. The main purpose is to consider the needs of the stakeholders, but also to strive for the participation of the stakeholders. The stakeholder dialogue can be established in different forms and can be supported by reporting ([38], pp. 9–11). The stakeholders can be divided into primary and secondary stakeholder groups. The primary stakeholders influence the process directly and immediately and can take on a process role in a process; secondary stakeholders influence the process indirectly and from a distance ([45], p. 265). This research focuses on the primary stakeholders that are listed in Table 3.

---

[8] Agile management reffers to the iterative, incremental method of managing, allowing flexibility and interaction.

**Table 3** Primary stakeholder analysis

| Stakeholder | Formulation of claims and requests |
|---|---|
| Process owner (with team) for the complexity management model | Suited framework to carry out the responsibility<br>Strategic anchoring in the organization<br>Space and prioritization to place complexity management in the organization |
| Supplying units of data for the complexity management model | Framework to support the process responsibility<br>Need of procedural and methodical support for the creation of the detailed data |
| Manager (responsible) of functional units /segments or similar of the organization[a] | Need of prepared data in the form of reports and presentations, to be able to evaluate appropriately and to consider detailed measures<br>Requirement for the process owner to participate in the interpretation of the data<br>Support with the adaptation (interpretation) of the key figures in the special department of the organization[b] |
| Management of the entire organization | Need of strategically prepared data in the form of reports and representations to be able to assess the overall complexity of the organization<br>Requirement for the process owner to participate in the interpretation of the data on the strategic level<br>High effectiveness and efficiency of the complexity management |

[a]Depends on the structure of the whole organization
[b]Because the key figures cannot be broken down into each individual area of the organizational structure, the possible contribution of individual areas must be interpreted

## *3.3 Recommendations for Implementation and Dealing with Changes*

In complexity management, there is no such thing as 100% correctness but only estimate. It is important to orientate oneself to the agile world, where 80% quality is sufficient as a minimum. Based on this philosophy, managers should try to get better and better in small iterative steps and to increase the level of maturity. Depending on the scope, requirements, possibilities, and sufficient time must be allowed for the implementation (in larger organizations it should be a few months).

The aim of implementing the Large Manufacturing Complexity Model is to map the current situation and foreseeable further developments. Even if everything works well, changes or adjustments will be required after a short period (e.g., just a few months), either because of changing influencing variables or framework conditions of the organization, or due to the internal handling of the complexity model. This section establishes guidelines recommended for addressing these changes.

The complexity model, with the criteria and the evaluation processes, should be kept unchanged to ensure comparability of values over the period of application, because only this way trends and developments can be analyzed. Changes to the model can be made, if necessary. However, this also leads to the disadvantage of the reduced comparability of the results when considering the entire cycle. This means that advantages and disadvantages of the change requirement must be assessed in advance. Suitable methods to compensate for these disadvantages can be

- Detailed weighing of the desired advantages of change (also considering the minimum 80% target achievement)
- Change (Release) cycle: instead of making numerous minor changes, it is better to make a major change to the model in a larger cycle (e.g., in 1 year). This ensures the comparability of the results and the evaluation of the effectiveness of the measures at least for a certain period.
- Parallelization of the preparation of results, with the previous logic and the changed model. In this way, comparability can be ensured, but the preparation of results is more complex. This must also be assessed.
- Type of implementation: Depending on the nature of the requirement, there are usually several ways of integrating the new requirements into the model. The way of integration can influence the stability of the model in the application or in the interpretation of the results.

In summary, during the implementation, as well as when considering a further development, the following guiding principles should be considered:

1. Mapping of the current and foreseeable complexities and uncertainties by the model
2. Following the 80% quality target for implementation and ongoing execution
3. Acceptance and applicability of the model (also for the units involved)
4. Feasibility and added value of the requirements for the model
5. Effectiveness and effort in applying the model.

## *3.4 Large Manufacturing Complexity Criteria*

Our working definition of complexity drivers is widely used in literature, meaning a “phenomenon, which actuate a system to increase their own complexity” [44]. The complexity drivers can be either internal or external to the company. They can be further classified according to complexity criteria adapted specifically to manufacturing, grouped in twelve clusters as shown in Fig. 2.

### 3.4.1 Society Complexity

Socio-political changes and developments can impact both the company and production, suddenly or develop slowly. Likewise, these developments can be visible, partially visible, or hidden and not predictable. Various dimensions can be relevant

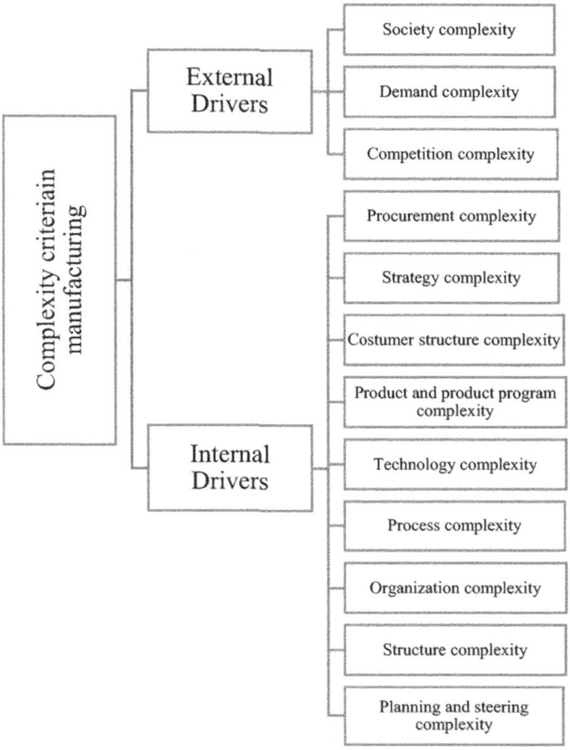


**Fig. 2** Complexity clusters for manufacturing environment. Adapted from [34] based on [37]

here, for example, ecological, economic, or political backgrounds. Another key question is to what extent society registers these developments or actively shapes them itself. In a production environment, this can affect all areas, from employees' attitude or ecological issues to purely political decisions that contain significant production guidelines (Table 4).

### 3.4.2 Demand Complexity

Almost all changes and developments (like globalization or increasing individualization of potential buyers) will ultimately lead to an unforeseeable increase of customer interface, in the form of customer demand characteristics and diversity of products. These requirements also affect the entire market in which the company is operating (including competitors and other market participants). Developments can be visible or invisible, develop slowly or very dynamic (Table 5).

**Table 4** Society complexity criteria

| Criteria | Definition and interpretation | Relevance factor evaluation |
|---|---|---|
| Changing values | Description of the social change of values that may affect the business | This can have an influence on the purchasing behavior of the manufactured products, production processes, but also on employee behavior and attitude to the company |
| Environmental awareness | Awareness level of the stakeholder to recognize trends and social environments and based on this will change their behavior | To what extent the stakeholders can recognize the changing trends and framework conditions of the market and can react to them |
| Ecological and economic influencing factors | Ecological and economic factors influence the behavior of internal and external stakeholder | This change behavior of the stakeholders can depend on economic framework conditions, like purchasing power or income trend of buyers, but also from ecological parameters of buyer's geographical location |
| Political framework | The political environment can have an impact on the overall business | The political framework can have an effect both inside and outside the company; they can be either short-term or long-term. The type of influence can be vastly different, among other things in the form of incentives for the buyer or as company guidelines |

**Table 5** Demand complexity criteria

| Criteria | Definition and interpretation | Relevance factor evaluation |
|---|---|---|
| Diversity of customer requirements | The diversity of customer requirements may have an impact on the complexity | The diversity of customer requirements can have an impact on different layers of the customer interface and the internal delivery of the factory. This is also derived from the market diversity, including the competition |
| Individuality of the demand | Individuality of customer requirements may have an impact on the complexity | The impact of the individuality of customer requirements can be reflected in design of the standard products and in the degree of acceptance of the standard products or separate requests besides the standards |
| Market dynamics | The market dynamic (as an external factor) can influence the customer general behavior, attitude, and ultimately also buying behavior | Market dynamics can also affect all relevant perspectives for a manufactory. These could be, for example, product life cycle, production process, investment behavior, and standard product design |
| Global requirements | Global and international circumstances can give rise to specific requirements | Country-specific requirements can lead to additional specific requirements, which can either have an impact on standard products or buyer behavior, and on production |

**Table 6** Competition complexity criteria

| Criteria | Definition and interpretation | Relevance factor evaluation |
|---|---|---|
| Degree of strength of competitors | The strength of the competitors can have an impact on the complexity of the market | The strength of competition in the market can also influence a lot of other complexity drivers. One of the main questions hereby is, if the competitors are so strong (potential and existing), that they will influence their own position in the market |
| Changing markets | Change and the rate of change in the market have an impact on complexity | Shifts in the market lead to its own re-positioning. These changes can be relevant for the products themselves, but also for value creation depths and chains |
| Competitive dynamics | The dynamics of the competitors in the market, in which the parties are acting, can have an impact on the complexity | This indirect influence is affecting the market dynamics. From this one a lot of perspectives for a manufactory could be affected; for example, product life cycle, production process, investment behavior, and standard product design |
| Number and structure of the competition | The number and the structure of potential and existing competitors are potential complexity drivers | Depending on the possibility of recording and the traceability of the competition and their market position, this can have an impact on the complexity to be managed |

### 3.4.3 Competition Complexity

The competition is an essential market participant of one's own company and thus has a potentially high impact on the production. The question here is to what extent existing or potential competitors are known, how stable the competitive situation is, or how dynamic the changes in competition are. It is also important how the strength of a company compares to competitors. Regarding manufacturing, production depths and widths, the company's own production and that of competitors, must also be considered. A list of relevant criteria is presented in Table 6.

### 3.4.4 Procurement Complexity

Depending on the degree of the company's own vertical range of manufacturing and production widths, external purchasing is relevant for production. Different dimensions and levels must be designed, for example, the contractual security, the development of a common production strategy, regulations on the exclusivity of the procurement of third-party parts or the supplier's manufacturing strategies (Table 7). The company must define its strategy and which supplier strategy is the most suitable (considering the company's current strategy).

Table 7 Procurement complexity criteria

| Criteria | Definition and interpretation | Relevance factor evaluation |
|---|---|---|
| Number of suppliers | The number of participating suppliers can have an influence on the complexity in manufactory | The number of suppliers has an impact on the complexity in different areas of the own production. Depending on integration level, interfaces and the own processes of the contractors must be considered into the evaluation |
| Procurement strategy and concept | The sales approach and strategy of the supplier can have an impact on the degree of complexity | Strategy and sales concept of the supplier have an impact on the complexity of (kind of) contract and B2B interface to the own company (agreements, processes, organizational aspects, etc.) |
| Uncertainty of the delivery or quality | Uncertainties and quality in the delivery of the service providers can have an impact on the complexity of the own production process and supply chain | The stability of delivery security and quality of the service provider effects the complexity regarding the necessity to manage these uncertainties to ensure the own production |

### 3.4.5 Strategy Complexity

The company's strategy and production must be considered when assessing the complexity. It is essential to what extent and in which areas (production-relevant) strategies must be adapted. By considering the dynamics of the adjustments, it is also necessary to evaluate the current degree of implementation of the strategy. Only two criteria are considered (Table 8).

### 3.4.6 Customer Structure Complexity

The characteristics of the existing and potential customers are important throughout the company, right up to production. The heterogeneity, the stability of the customer's characteristics, and the various customer structures (with your own further developments) must be considered (Table 9). Customers can take on a wide variety of roles, and interfaces between customers and their own company (including company areas such as production) are designed differently. The extent to which customers are integrated into their own company processes is significant.

### 3.4.7 Product and Product Program Complexity

The degree of predictability, planning options, and the stability of the further development of own product range contain potential complexities and uncertainties that affect the company directly or indirectly, depending on the company's vertical range and

**Table 8** Strategy complexity criteria

| Criteria | Definition and interpretation | Relevance factor evaluation |
|---|---|---|
| Dynamics of the strategy adjustment | The frequency and rate of change of the own strategy and targets may have an impact on the complexity | The potential complexity in the own company can be affected by the dynamics of the need to adjust the own strategy and strategic targets. These could be including, for example, financial, market, social, competition-relevant aspects. The complexity appears mainly in the implementation phase in the own organization |
| Maturity of current strategy achievement | The current state of fulfillment-maturity of the defined strategy can have an impact on managing the current state of complexity and the necessary measures | Depending on the ability to manage the existing complexity in the company and to what extent (appropriately) the company goals are met, this can influence the degree of complexity. This can lead to different assessments of complexities and be relevant for the implementation of measures derived from them |

**Table 9** Customer structure complexity criteria

| Criteria | Definition and interpretation | Relevance factor evaluation |
|---|---|---|
| Number and heterogeneity of customers and customer groups | The number, characteristics, and heterogeneity of customers can have a very pronounced effect on one's own complexity in the value creation process | The type of expression can be based on contractual framework conditions, functional and non-functional or technical requirements for the product or production, or also involve a variety of characteristics in the (customer-specific) value chain design |
| Level of participation | The degree of integration in the respective customer relationship between the customer and the own company may have an impact on the complexity | The type of design of the interface to the various customers, regarding customer integration, influences the complexity to be controlled. It is important to assess where the customers can directly or indirectly influence the production process, how similarly this integration occurs and how stable this integration is |

breadth of the product mix. Numerous areas of production are potentially affected, for example, investment behavior for the purchase of machines or the entire strategy for warehouse organization. Structure and dynamics are further considered (Table 10).

**Table 10** Product and product program complexity criteria

| Criteria | Definition and interpretation | Relevance factor evaluation |
|---|---|---|
| Structure and update period of products | The product structure and the corresponding product life cycle must be considered when assessing complexity | Depending on the product life cycle and the promised or expected serviceability of the products, this is usually ensured via (predictable) configurability. In certain constellations (for example, technical dependencies or availability bottlenecks) this can lead to complexities that have to be managed |
| Dynamics of the product changes | The dynamics and the predictability of release cycles of products n have a potential impact on the complexity | The dynamics of product changes may have a great impact on the complexity; in concrete terms, the degree of complexity depends on the degree of dynamics, the steering logic of product management, and the degree of self-determination in product development |

### 3.4.8 Technology Complexity

The technology used can be a driver for complexities in different areas of production. Among other things, the dynamics of the further development of the technology, the constellation of internal and external procurement, as well as the agreements with the suppliers in the case of external procurement must be considered (Table 11). It is also important to what extent the company invests in innovations in recent technologies

**Table 11** Technology complexity criteria

| Criteria | Definition and interpretation | Relevance factor evaluation |
|---|---|---|
| Technological change and lifecycle | The technology development and the characteristics of the technological change of the current product portfolio and the production may have impact to the company | Technology changes and lifecycle can, depending on the products and the manufacturing, have significant impact to the complexity. It is important to assess how often and to what extent technological changes take place, to what extent do they influence your own production and to what extent can this be planned in your own company, or what do these mean for buyer/market behavior and competitors |
| Availability of (innovative) technologies | The possibility and ability to use innovative technologies can influence the complexity | The question in the assessment is to what extent your own company has the opportunity and ability to develop innovative technologies itself, to access them in the case of third-party developments, and what opportunities does the competition have to use them |

and thereby influences the market itself. There are also cross-connections with other clusters, such as Demand Complexity.

### 3.4.9 Process Complexity

Every organization can design its internal processes for production which serve to fulfill the corporate strategy. As previously explained in the Demand Complexity criteria, the customer interface has distinctive characteristics and different degrees of integration. Depending on these, the customer will participate in internal processes. Process interfaces can also arise through the integration of partners or suppliers. This creates new dependencies and connections that can become relevant in terms of complexity. Specifically, this question is to be defined in the partner selection and supplier strategy. Three criteria are considered in Table 12.

Organization Complexity

In addition to the processes (operational organization), the organizational structure must also be assessed in terms of complexity. With increasing size and international character, the type of corporate governance and the degree of uniformity of the organizational structures become relevant (Table 13). A high degree of transparency

**Table 12** Process complexity criteria

| Criteria | Definition and interpretation | Relevance factor evaluation |
|---|---|---|
| Number and design of interfaces | The number of needed and established interfaces and design of in the company itself can have a high impact on the complexity | It is important to consider how many interfaces there are in the value chain, how often and how much they need to be adapted. Furthermore, procedural interfaces can also result in organizational (process) and IT-relevant changes |
| Degree of crosslinking of the processes | The extent and characteristic of linking processes between customer and the own production can have an impact on the complexity of the organization | Interaction with customers and other partners can lead to increased complexity of production. This applies specially to changes and the consideration of the ability to plan and influence these changes |
| Degree of standardization | The degree of standardization (as a target and the current state) can have an impact on the complexity in production | Standardized interfaces improve the ability to plan in the design and further development of interfaces. The standards can include aspects of processes, organization, and IT support. Individual (external and internal) increase dependencies and involve uncertainties and unpredictability |

**Table 13** Organization complexity criteria

| Criteria | Definition and interpretation | Relevance factor evaluation |
|---|---|---|
| Number and structure of hierarchical levels | The number and the structure of hierarchy levels can have an impact on the shaping of the complexity in the different areas of an organization | The structure and the number of hierarchy levels have an immediate and direct impact on the whole organization, and of the production process. This applies particularly to the creation of suitable framework conditions for production, for further developments and for the implementation of changes in which more than one production facility is involved. The importance grows with, for example, decentralized production structures and internationally distributed production sites |
| Degree of centralization and common governance | The degree of centralization (in an organizational perspective) or the existence of a common governance structure (for the process organization) can have an impact on the complexity | A centrally organized procedure for relevant topics in the company or a joint governance structure has an influence on the control and transparency in the company, this also affects the potential occurrence and extent of the uncertainties and the nature of the complexities to be managed. With the increasing size of the company, decentralization and internationalization, this aspect potentially increases |
| Uniformity and number of organizational units | The uniformity and the number of organizational units of the involved organizational units in any changes can have an impact in the business operation | The uniformity and the number of areas in a large company has an impact on controllability, the extent of the need for control and on the occurrence of organizational complexities. The relevance of this criteria increases especially in the case of changes in which numerous and different units are involved |

can support the identification of complexities and uncertainties, create a high impact when implementing measures, and avoid unpredictable situations and events.

### Structure Complexity

The complexity structure is a very far-reaching cluster when considering complexities in an organization. First, it must be considered where the organizational boundaries are and how the structural characteristics of the integration of external units are designed and implemented. There are some structural drivers within the organization, which usually run right through the entire organization, including production (Table 14). These are, for example, the communication structure, the degree of structural transparency of the organization, and the very far-reaching perspective: the cultural characteristics in the organization.

**Table 14** Structure complexity criteria

| Criteria | Definition and interpretation | Relevance factor evaluation |
|---|---|---|
| Cultural characteristics | The characteristics, knowledge, and the way of dealing with cultural characteristics can be important for the evaluation of complexities | Especially in international companies there can be significant complexities about corporate culture. Influencing factors are, for example, the characteristics of corporate governance, the company's history, the number and distribution of countries or regions of the world involved and the type of corporate management |
| Homogeneity and transparency of the corporate structure | The degree of consistency and transparency of the corporate structure and the interfaces to relevant partners have impact on potential complexity | These includes a lot of different dimensions in the organization: the roles of the employees, IT-infrastructure, governance aspects, leadership and process and organizational aspects. in the business operation. The involvement of the partners as an external element of the organization must also be structurally assessed |
| Communication structure and systems | The characteristics and availability of a communication system (including supporting IT) have an impact on the occurrence of complexities and uncertainties | Communication structures, processes and supporting IT can be structural elements in a company and thus influence the occurrence of complexities. This also applies to the exercise of complexity-relevant control mechanisms in the company. All directions of information are relevant here, for example, management information to the entire company, information from individual areas of the company to the company management, or the coordination of the areas in the company with one another |
| Vertical integration with partners | The organizational structure and vertical integration of the line organization (of own company and all partners) can have an impact and characteristics on the complexity in all dimensions (contractual, process, organizational, etc.) | The degree of the vertical integration of the organizations involved, is of immense importance regarding the complexity in all developments and changes, which are carried out. When the integration with the different partner constellations is very narrow, often the expression of the complexities to be managed increases |

### Planning and Steering Complexity

The "Planning and steering complexity" cluster has a prominent role. Here, it is necessary to reflect on and evaluate the extent to which the organization can recognize complexities, define strategies, and determine appropriate measures. It should also be

**Table 15** Planning and steering complexity criteria

| Criteria | Definition and interpretation | Relevance factor evaluation |
|---|---|---|
| Complexity infrastructure and supporting communication systems | The availability and characteristics of an infrastructure to practice complexity management in one's own company also influences complexity | The availability and characteristics of a suitable infrastructure is necessary to implement complexity management in the company. For this, instruments for communication and organizational requirements are necessary. Objectives are on the one hand to achieve transparency about the complexity in the company (complete and topic or area-specific) and to be able to implement measures to be derived from this in a targeted manner |
| Intensity and strategic importance of complexity management in the company | The intensity of the implementation and the strategic importance of complexity management in the company is relevant to whether the determination and implementation of measures themselves, uncertainties and complexities arise and must be controlled | Depending on the strategic necessity and importance of complexity management in the company, the intensity and characteristics should be established in the company. This affects the cyclical survey and the implementation of targeted measures. Strategies for dealing with complexities can include reducing, avoiding, and managing complexities. The company structure can intensify or reduce the criteria |

considered how the organizational structure is established in the whole organization, especially for manufactory challenges (Table 15).

## 3.5 *Large Manufacturing Complexity Management Process*

We suggest following a cyclic process consisting of six steps, to ensure that all crucial elements have been considered and there is a solid and repeatable process in managing complexity (Fig. 3), corresponding to the sections described in the model (Fig. 1).

In **step 1,** the entire preparation of the upcoming period should be carried out. If this is the first iteration of the complexity management cycle, this step must create the complete set up measures to transfer the model from theory into the organization. This includes the initial communication measure on the introduction of the complexity management system, establishing necessary roles, training the employees and stakeholders, as well as defining the process structure and the development of the supporting IT tools. The development of the adaption of the cluster and criteria are also part of this step. If this is not the first iteration, the changes from the previous step 6 (retrospective) must be embedded here into the new cycle. The result of step

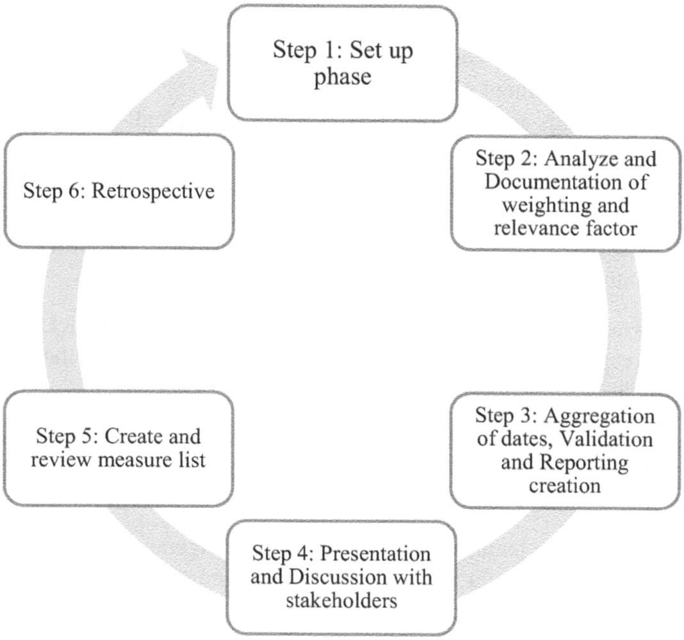


**Fig. 3** Large Manufacturing Complexity Management Process (own schema)

1 is a prepared model for the upcoming period (including Complexity cluster and criteria). This corresponds to segment A in the model. All stakeholder groups are involved in this step.

The cyclical evaluation takes place in **step 2**. This corresponds to the activities in sections B, C, and D. and partly G of the model. During the initial set-up of the process, the company's current strategy is derived, and the weighting factor is defined for the respective criteria. If a cycle has already been run through, a verification is made to determine whether the weighting factor is still valid. The determination of the relevance factor takes place with the help of the supporting units of the organization. This determination is the core of the periodic complexity survey. The result of this step is the respective complexity factors of each complexity criteria. The results are documented for later traceability (part of G). In this step the most involved stakeholders are

- Process owner (with team) for the complexity management model,
- Supplying units of data for the complexity management model,
- Manager (responsible) of functional units/segments or similar of the organization.

In **step 3**, the collected detailed data is aggregated. Furthermore, a quality assurance should take place and implausibility should be clarified. This is followed by an examination to determine whether the prioritization (Section F in the model) is still up-to-date, or whether it needs to be adjusted, or whether the prioritization is to be carried out the first time. The requested and defined reporting views are then created. In this step, the most involved stakeholders are the Process owners (with team) for the complexity management model.

In **step 4**, the results are presented to and discussed with all stakeholders, usually in various bodies of the organization. It makes sense to show this together with the accompanying measures. The discussion and feedback from the committees can form the basis for the next phase. Based on these discussions, target values can be defined and set, to effectively measure and review the measures in the process (regularly every 1–3 months). In this step the most involved stakeholders are

- Process owner (with team) for the complexity management model,
- Manager (responsible) of functional units/segments or similar of the organization,
- Management of the entire organization.

The list of measures is updated in **step 5**. This includes the potential inclusion of new measures, changes to measures, or the closure of existing measures. In this step, the most involved stakeholders are

- Process owner (with team) for the complexity management model,
- Supplying units of data for the complexity management model,
- Manager (responsible) of functional units/segments or similar of the organization.

**Step 6**: Retrospective is a method derived from the Agile methodology. In this phase, reflection is carried out between and with those involved from different perspectives. Basic questions to answer could be as follows:

- Are all roles performed as expected?
- Does the current (structural) status of the Complexity Management model reflect reality, or have basic framework conditions changed?
- Is the complexity management model perceived in the organization and the added value recognized?
- How stable (procedural, personnel, etc.) is the process, or are potential disruptions in the future already foreseeable today?
- Which improvements should be initiated in order to increase the degree of maturity of the process?

In this step, the most involved stakeholders are

- Process owner (with team) for the complexity management model,
- Supplying units of data for the complexity management model,
- Manager (responsible) of functional units/segments or similar of the organization,

### 3.6 *Use Case—A Detailed Description of the Illustrated Model*

To better understand the model of Large Manufacturing Complexity Management, an example will illustrate the full process. Our example organization is a manufacturing company that produces high-quality interior equipment for series cars, with an upscale line of equipment. We will further go through the steps listed in the model's instructions of use. The data used in this example are fictitious and exclusively used for the purpose of analysis and evaluation, to test and validate the proposed model.

Table 16 captures the results of the internal and external analysis conducted by the company. The core questions to answer are

- What does the result of the analysis mean, now and in the future, for the complexity that the company has to manage?
- In which areas does this complexity arise and must be continuously assessed to then derive suitable measures in strategic management?

According to the current strategy, the high-quality standard of the products should be increased. The company analysis of production has shown that it requires new, innovative investments that are currently not available. This leads to the strategic decision to outsource 80% of the high-quality individual parts. To ensure delivery stability, corresponding delivery contracts are to be concluded with three international companies.

In the evaluation phase described in Table 17, the weighting factor (Wf) and relevance factor (Rf) are allocated, and the resulting complexity factor (Cf) and mean value (Mid) are calculated. Summarizing the result, first we notice that the weighting and relevance factor have the same value for most criteria, meaning that the strategic orientation (reflected in the weighting factor) and the current assessment of the complexity (reflected in the relevance factor) are matching, meaning that there is a great need to permanently evaluate complexities in the transformation phase (in this example, outsourcing) and to derive suitable measures. Secondly, the Procurement complexity, followed by Process complexity was rated the highest; in view of the future strategy and challenges, this should be the focus. Also, the results have developed the focus areas of complexity: 18 criteria have reached the value of complexity factor of 10 or more; 17 values are less than 10.

Preparing the results for discussion with management and stakeholders implies selecting the focus criteria and setting a method for following their progress over a predetermined cycle, for example, a calendar year is shown in Table 18. An orientation target factor (Otf) was agreed for all selected criteria. Based on the evaluation for the month of January, there is a call to action for 7 evaluation criteria (where the difference is more than 4 points). The comparison of the 2 summations in the month of Jan (220 versus 316) shows that the managers in charge already see a clear need for action in dealing with complexity in the company. A list of measures was created, and an action plan was designed as shown in Table 19, having an expected start date and an allocation of a person responsible for the implementation of each measure.

**Table 16** Internal and external analysis

| Analysis factor | Component | Description |
|---|---|---|
| External (environment) | Market | B2B contracts are often concluded, either for a limited period (1–3 years) or based on the number of items (from 100,000 items per product). The market participants are known internationally. There is increasing cost pressure, also due to competitors from the Asian region. Increasing dynamic justified by new innovative replacement products (digitization of the car interior), or by new processes for the manufacture of competitor products. The product life cycle of automobiles is becoming increasingly shorter and the expectations of customers to get new features are increasing |
| | Competition | 10 direct competitors worldwide in the known area of the market. Half of the competitors are constantly developing their products; the other half of the competitors make significant changes to their own products every 2–3 years and invest significantly in new (manufacturing) technologies. The flexibility of the providers to offer customized adaptations to the design of the value chain is also increasing |
| | Political & Legal Framework | Minor changes every 2–3 years so that all providers can prepare for them with a reasonable lead time. This does not result in a market distortion; the costs are insignificant |
| | Customers | 110 customers, in majority European countries; outside of Europe in 30 countries. Customers are also medium-sized and large organizations (<10 000 employees). 60% of the customers are car manufacturers and 40% are accessories dealers and other individual customers. The worldwide customers are divided across the locations |
| Internal (company) | Company Structure | The company was founded 50 years ago in Germany and has 18 locations in Europe and 13 locations outside Europe. Production in 8 locations. Individual locations have their own profit responsibility. Headquarters have 600 employees including management functions. The current workforce is around 10,000, including around 6,000 in Germany |
| | Program Strategy | 80 basic products, with an average of around 12 different configurations (expansion stages) for each basic product; the average product life cycle is around 2–3 years; In around 50 of the products, a minor layout update takes place after 1.5 years |
| | Production | In most cases, product developments can be produced through configuration in existing production facilities and established production processes |
| | Diversity of demands | 30% of the products have minor country-specific requirements for material and Design in detail |

(continued)

**Table 16** (continued)

| Analysis factor | Component | Description |
|---|---|---|
| | Current strategy | Growing internationality, improved international customer care due to local organizations, as well as price stability with differentiation from the competition through higher quality. The entire range consists of 5000 individual parts. For around 50% there are tight tolerances and therefore very high-quality standards in production |

## 4 Conclusion

Complexity has always something to do with uncertainties, unclear conditions, and unforeseeable developments. The introduction of a comprehensive complexity management model in an organization is a very demanding and challenging project. The management not only has to make the strategic decision about the introduction of Complexity Management, but must also provide suitable framework conditions, for example, sufficient budget and resources; as well as actively accompanying and monitoring the implementation. This commitment is one of great importance. Once the decision has been made, the appropriate personnel capacities must be identified and built up. If these are not (yet) available, temporary use of external experts should be considered. The mere fact that the organization considers and discusses the existing uncertainties and complexities, is a clear progress in significantly reducing potential business risks. Managing complexity can have a positive impact and also create a unique selling proposition in the company.

The present research reflects a comprehensive strategic management tool to significantly reduce existing and potential future risks caused by complexities in organizations. Furthermore, this management instrument creates a clear competitive advantage in the market. Large Manufacturing Complexity Management model is a comprehensive, holistic, and specific; it might seem difficult to put together at first. A perfectly suitable way to start is the iterative, successive introduction of the model in the organization. This step-by-step introduction can either take place according to organizational segments or with a reduced subset of complexity criteria. Even when the entire organization is not yet involved, the positive effects of the model become visible.

**Table 17** Evaluation of complexity factors

| Cluster | Criteria | $W_f$ | $R_f$ | $C_f$ | Mid | Arguments for evaluation the relevance factor |
|---|---|---|---|---|---|---|
| Society complexity | Changing values | 1 | 1 | 1 | 12.5 | No particular importance in terms of complexity |
| | Environmental awareness | 1 | 1 | 1 | | No particular importance in terms of complexity |
| | Ecological and economic influencing factors | 1 | 1 | 1 | | No significant importance, but little higher level in ECONOMICS, in terms of complexity |
| | Political framework | 2 | 1 | 2 | | No particular importance in terms of complexity |
| Demand complexity | Diversity of customer requirements | 2 | 2 | 4 | 6.5 | No significant relevance to handle diversity of customer requirements |
| | Individuality of the demand | 2 | 2 | 4 | | No relevant development, in the framework of existing strategy and use case |
| | Market dynamics | 3 | 3 | 9 | | No direct impact under existing frame conditions (now), but in a long-term perspective increasing challenge |
| | Global requirements | 3 | 3 | 9 | | No direct impact under existing frame conditions (now), but in a long-term perspective increasing challenge |
| Competition complexity | Degree of strength of competitors | 3 | 3 | 9 | 8.25 | With increasing outsourcing, the risk increases that expertise (products, production, processes) becomes known and used in the market, specifically by competitors |
| | Changing markets | 3 | 3 | 9 | | With increasing outsourcing, the risk increases that expertise (products, production, processes) becomes known and used in the market, specifically by competitors |
| | Competitive dynamics | 3 | 3 | 9 | | With increasing outsourcing, the risk increases that expertise (products, production, processes) becomes known and used in the market, specifically by competitors |
| | Number and structure of competition | 3 | 2 | 6 | | Weak impact on foreseeable challenges |

(continued)

**Table 17** (continued)

| Cluster | Criteria | $W_f$ | $R_f$ | $C_f$ | Mid | Arguments for evaluation the relevance factor |
|---|---|---|---|---|---|---|
| Procurement complexity | Procurement strategy and concept | 5 | 5 | 25 | 21.7 | With the increasing outsourcing of relevant and highly qualified production parts, the risk increases that complexities at the supplier interface, or indirect complexities of the supplier, become relevant for your own company |
| | Uncertainty of the delivery or quality | 4 | 5 | 20 | | With the increasing outsourcing of relevant and highly qualified production parts, the risk increases that complexities at the supplier interface, or indirect complexities of the supplier, become relevant for your own company |
| Strategy complexity | Dynamics of the strategy adjustment | 4 | 4 | 16 | 16 | The dynamics of the strategy and the upcoming changes are extremely high. Depending on the development, this may have to be adapted to the situation and at short notice |
| | Maturity of current strategy achievement | 4 | 4 | 16 | | The maturity level of the implementation of the changed strategy (with outsourcing) is currently not exceedingly high, but relevant |
| Costumer structure complexity | Number and heterogeneity of customers and customer groups | 2 | 3 | 6 | 7.5 | The customer structure is not relevant affected by the strategy and the current project of outsourcing |
| | Level of participation | 3 | 3 | 9 | | Mainly because of the local-to-local agreement to the customer, this should be observed with respect to delivery and ongoing assessment |
| Product and product program complexity | Structure and update period of products | 3 | 4 | 12 | 13.5 | With increasing outsourcing, the uncertainties that arise from the suppliers themselves increase. This can then also affect the complexity of the product portfolio |
| | Dynamics of the product changes | 3 | 5 | 15 | | With increasing outsourcing, the uncertainties that arise from the suppliers themselves increase. This can then also affect the complexity of the product portfolio |

(continued)

**Table 17** (continued)

| Cluster | Criteria | $W_f$ | $R_f$ | $C_f$ | Mid | Arguments for evaluation the relevance factor |
|---|---|---|---|---|---|---|
| Technology complexity | Technological change and lifecycle | 2 | 2 | 4 | 4 | The technical-based further development initially has less impact on the complexity. However, this can increase if the availability of elements changes on the supplier side or if new, innovative technological innovations are to be used |
| | Availability (innovative) technologies | 2 | 2 | 4 | | The technical-based further development initially has less impact on the complexity. However, this can increase if the availability of elements changes on the supplier side or if new, innovative technological innovations are to be used |
| Process complexity | Number and design of interfaces | 3 | 4 | 12 | 20.7 | The process complexity is relevant due to the outsourcing strategy itself, the high and distributed proportion of third-party products and the individuality of the processes |
| | Degree of crosslinking of the processes | 5 | 5 | 25 | | The process complexity is relevant due to the outsourcing strategy itself, the high and distributed proportion of third-party products and the individuality of the processes |
| | Degree of standardization | 5 | 5 | 25 | | The process complexity is relevant due to the outsourcing strategy itself, the high and distributed proportion of third-party products and the individuality of the processes |
| Organization complexity | Number and structure of hierarchical levels | 2 | 2 | 4 | 18 | Regarding the hierarchy level, it is important in the event of a notable change that information is exchanged promptly and correctly in all directions |
| | Degree of centralization and common governance | 5 | 5 | 25 | | The organizational character in terms of uniformity, awareness and responsiveness is of particular importance in this strategy and the use case |
| | Uniformity and number of organizational units | 5 | 5 | 25 | | The organizational character in terms of uniformity, awareness and responsiveness is of particular importance in this strategy and the use case |
| Structure complexity | Cultural characteristics | 3 | 4 | 12 | 11.25 | The characteristics of the culture is relevant in a transformation phase or significant changes in the company, here we have an international organization with country/world region specific independencies and responsibilities |
| | Homogeneity and transparency of the corporate structure | 3 | 4 | 12 | | Because outsourcing affects many production sites, the relevance of organizationally relevant criteria is correspondingly high. The number of uncertainties and the occurrence of complexities is correspondingly high |

(continued)

**Table 17** (continued)

| Cluster | Criteria | $W_f$ | $R_f$ | $C_f$ | Mid | Arguments for evaluation the relevance factor |
|---|---|---|---|---|---|---|
| | Communication structure and systems | 3 | 4 | 12 | | Internal company communication and communication at the interface with partners and suppliers must be considered. This is meaningful in the current use case |
| | Vertical integration with partners | 3 | 4 | 12 | | The characteristics, stability, flexibility, and quality of how the external partners are connected is of vital importance for the strategic project |
| Planning and steering complexity | Complexity infrastructure and supporting communication systems | 4 | 3 | 12 | 14 | The cluster: "Planning and steering complexity" has a special status. The assessment here has a reflective character |
| | Intensity and strategic importance of complexity management in the company | 4 | 4 | 16 | | The cluster: "Planning and steering complexity" has a special status. The assessment here has a reflective character |

**Table 18** Selected and simulated course of complexity management over 1 year

| Selected criteria | | Otf | Values | |
|---|---|---|---|---|
| | | | Jan | Feb |
| Procurement complexity | Number of suppliers | 12 | 20 | ... |
| | Procurement strategy and concept | 15 | 25 | ... |
| | Uncertainty of the delivery or quality | 15 | 20 | ... |
| Strategy complexity | Dynamics of the strategy adjustment | 16 | 16 | ... |
| | Maturity of current strategy achievement | 16 | 16 | ... |
| Product and product program complexity | Structure and update period of products | 10 | 12 | ... |
| | Dynamics of the product changes | 8 | 15 | ... |
| Process complexity | Number and design of interfaces | 12 | 12 | ... |
| | Degree of crosslinking of the processes | 12 | 25 | ... |
| | Degree of standardization | 12 | 25 | ... |
| Organization complexity | Number and structure of hierarchical levels | 6 | 4 | ... |
| | Degree of centralization and common governance | 12 | 25 | ... |
| | Uniformity and number of organizational units | 12 | 25 | ... |
| Structure complexity | Cultural characteristics | 10 | 12 | ... |
| | Homogeneity and transparency of the corporate structure | 10 | 12 | ... |
| | Communication structure and systems | 10 | 12 | ... |
| | Vertical integration with partners | 10 | 12 | ... |
| Planning and steering complexity | Complexity infrastructure and supporting communication systems | 10 | 12 | ... |
| | Intensity and strategic importance of complexity management in the company | 12 | 16 | ... |
| Summary | | 220 | 316.0 | ... |

**Table 19** Action planning (status of Jan 2022)

| Complexity criteria | Description of measures | Start | Responsible |
|---|---|---|---|
| Dynamics of the product changes | Verification of which products have potential for reducing dynamics and carry out TOP 3 implementation | July | Mr. Meyer |
| Degree of crosslinking of the processes | Identification of where avoidable process cross-connections exist in production and implementation of the first 5 significant reductions in interfaces | July | Ms. Reder |
| Degree of standardization | Identification of significant reductions in individual processes and implementation of the first TOP 5 measures | Aug | Ms. Huber |
| Degree of centralization and common governance | Identification of significant potentials for increasing governance structures in an international context (leadership relationships, meeting structures, management information systems, …) and the first implementation steps take place | June | Mr. Bleder |
| Uniformity and number of organizational units | Identification of significant potentials for increasing the uniformity of international management structures and initial implementation in at least 5 countries | Sept | Ms. Remmle |
| Intensity and strategic importance of complexity management in the company | Conceptual detailing and implementation of complexity management in the company, including international locations, specifically: Establishing regular agreements for monthly evaluation | June | Mr. Schmidt |

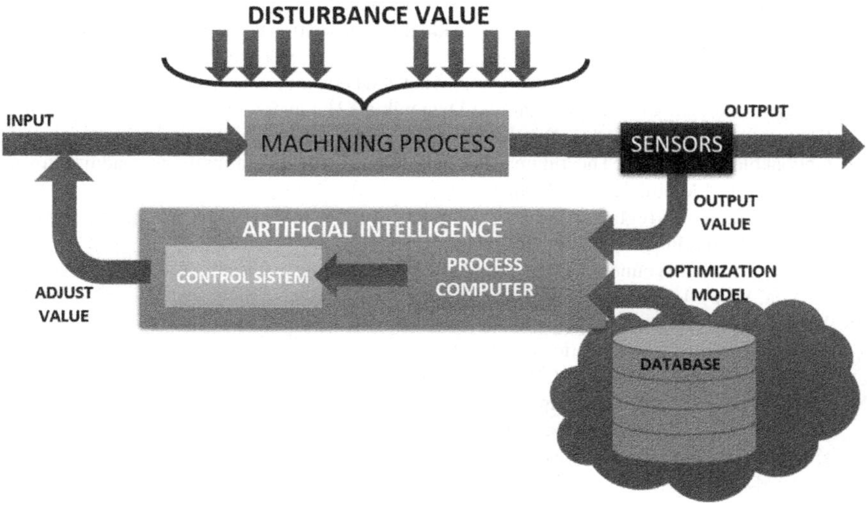

## References

1. Addor P (2010) Projektdynamik: Komplexität im Alltag. s.l., Reinhold Liebig
2. Albers A et al (2019) A systematic approach to situation-adequate mechatronic system development by ASD-Agile systems design. Procedia CIRP 84:1015–1022
3. Bilgeri D, Wortmann F, Fleisch E (2018) How digital transformation affects large manufacturing companies' organization. Seoul, South Korea, s.n., pp 1–9
4. Brinzer B, Schneider K (2020) Complexity assessment in production: linking complexity drivers and effects. Procedia CIRP 93:694–699
5. Brown A, Elser B, Messenböck R, Münnich F (2010) Mastering complexity. s.l., Boston consulting Group
6. Buchholz U, Knorre S (2019) Interne Kommunikation und Unternehmensführung. Springer, s.l.
7. Budde L, Nagler O, Friedli T (2015) A method to set up a complexity index to improve decision-making performance. In: CIRP 25th design conference innovative product creation, vol 36, pp 53–58
8. Cescon F, Costantini A, Grassetti L (2019) Strategic choices and strategic management accounting in large manufacturing firms. J Manag Gov 23(3):605–636
9. Chabane M (2021) Management tools: from complexity reduction to complexity creation. In: Encyclopedia of organizational knowledge, administration, and technology. s.l., IGI Global, pp 1842–1856
10. Czinki A, Hentschel C (2016) Solving complex problems and TRIZ. Procedia CIRP 39:27–32
11. De Meyer ACL, Loch CH, Pich MT (2002) Managing project uncertainty: from variation to chaos. MIT Sloan Manag Rev 43(2):60–67
12. Dittes FM (2012) Komplexität: Warum die Bahn nie pünktlich ist. Springer Publishing House, Berlin
13. Dörner D (2008) Umgang mit Komplexität. Industrial ecology. Vieweg+Teubner, Wiesbaden, pp 284–302
14. Dykstra JA, Orr SR (2016) Acting in the unknown: the cynefin framework for managing cybersecurity risk in dynamic decision making. In: international conference on cyber conflict (CyCon US), pp 1–6
15. Gerberich CW (2004) Managen der Komplexität und Dynamik-eine große Herausforderung für heutige Unternehmen. Komplexität und Dynamik als Herausforderung für das Management. Deutscher Universitätsverlag, Wiesbaden, pp 235–259
16. Ghobakhloo M (2018) The future of manufacturing industry: a strategic roadmap toward industry 4.0. J Manuf Technol Manag
17. Gorzeń-Mitka I, Okręglicka M (2014) Improving decision making in complexity environment. Procedia Econo Financ 16:402–409
18. Groover MP (2013) Fundamentals of modern manufacturing: materials, processes, and systems, 5th edn. Wiley, USA
19. Jäger J, Schöllhammer O, Lickefett M, Bauernhansl T (2016) Advanced complexity management strategic recommendations of handling the "Industrie 4.0" complexity for small and medium enterprises. Procedia CIRP 57:116–121
20. Johnson NF (2007) Two's company, three is complexity: a simple guide to the science of all sciences. Oneworld Pulications, Oxford
21. Knyazeva H (2020) Strategies of dynamic complexity management. Foresight STI Gov 14(4):34–45
22. Konrad C et al (2019) Enabling complexity management through merging business process modeling with MBSE. Procedia CIRP 84:451–456
23. Kurtz CF, Snowden DJ (2003) The new dynamics of strategy: sense-making in a complex and complicated world. IBM Syst J 42(3):462–483
24. Lepmets M, et al (2014) A Cynefin based approach to process model tailoring and goal alignment. s.l., IEEE Computer Society, pp 166–169

25. Lessard D, Sakhrani V, Miller R (2014) House of project complexity—Understanding complexity in large infrastructure projects. Eng Proj Organ J 4(4):170–192
26. Maximini D (2018) Scrum – Einführung in der Unternehmenspraxis. Springer, s.l.
27. Mcdermott CM (1999) Managing radical product development in large manufacturing firms: a longitudinal study. J Oper Manag 17:631–644
28. McLeod J, Child S (2013) The Cynefin framework: a tool for analyzing qualitative data in information science? Libr Inf Sci Res 35(4):299–309
29. Mikalsen M, Næsje M, Reime EA, Solem A (2019) Agile autonomous teams in complex organizations. Montréal, QC, Canada, s.n., pp 55–63
30. Monostori L, et al (2016) Cyber-physical systems in manufacturing. Cirp Ann 65(2): 621–6641
31. Negulescu OH (2019) Complexity and flexibility in strategic management. Rev Gener Manag 30(2):55–67
32. Pollack J, Adler D, Sankaran S (2014) Mapping the field of complexity theory: a computational approach. Emerg: Complex Organ 16(2):74–92
33. Ravishankar N, Mujja A, Lewis MG, Nair NS (2014) A tool to measure complexity in public health interventions. Clin Epidemiol Glob Health 2(2):80–86
34. Rennung FM (2016) Managing complexity in service processes. The case of large business organizations. Timisoara, Editura Politehnica
35. Salimi F, Salimi F (2018) A systems approach to managing the complexities of process industries. Elsevier, s.l.
36. Satell G (2016) The one big reason every business needs to embrace complexity. https://digitaltonto.com/2016/the-one-big-reason-to-every-business-needs-embrace-complexity/
37. Schoeneberg K-P (2014) Komplexitätsmanagement in Unternehmen: Herausforderungen im Umgang mit Dynamik, Unsicherheit und Komplexität Meistern. Springer Gabler, Wiesbaden
38. Schram B, Schmidpeter R (eds) (2016) CSR und Organisationsentwicklung: die Rolle des Qualitäts-und Changemanagers. Springer Gabler, Koln
39. Schuh G, Kuntz J, Stich V, Jussen P (2019) Managing complexity in product service systems and smart services. s.l., Elsevier, pp 410–414
40. Snowden DJ, Boone ME (2007) A leader's framework for decision making. Harv Bus Rev 85:68–76
41. Sohi, A. J., Hertogh, M., Bosch-Rekveldt, M. & Blom, R., 2016. Does lean & agile project management help coping with project complexity?. *Procedia-social and behavioral sciences*, pp. 252–259.
42. Stogsdill M, Ulfvengren P (2017) Mapping risk models/methods onto a complexity spectrum. Transp Res Procedia 28:133–140
43. Vidal GH, Jairo RCH (2021) Complexity in manufacturing systems: a literature review. Prod Eng 1–13
44. Vogel W, Lasch R (2016) Complexity drivers in manufacturing companies: a literature review. Logist Res 9(25)
45. Welge MK, Al-Laham A, Euler M (2017) Strategic management: basic process implementation. Springer Gabler, s.l
46. Wildemann H (2013) Komplexitätsmanagement: in Vertrieb, Beschaffung, Produkt, Entwicklung und Produktion. TCW Transfer-Centrum Verlag, München
47. Youngki P, Mithas S (2020) Organized complexity of digital business strategy: a configurational perspective. MIS Q 44(1):85–127

# Chapter 8
# From Industry 4.0 to Industry 5.0—An Overview of European Union Enterprises

**Lidia Alexa, Marius Pîslaru, and Silvia Avasilcăi**

**Abstract** Industry 4.0 refers to the transformation of industry through the adoption of techniques and processes based on information and communication technologies (ICT) to manage and optimize all aspects of the manufacturing processes and supply chain. Even if Industry 4.0 adoption has significant advantages for companies in terms of increased productivity and decreased costs, it also raises various challenges for companies. At the same time, more and more experts start to voice their concerns regarding Industry 4.0 and discuss the advent of Industry 5.0 with a focus on sustainability as opposed to productivity. During the discussions regarding Industry 5.0 and the necessity of an improved collaboration between humans and technological systems, it is relevant to evaluate the way companies from the manufacturing sector are adopting Industry 4.0, considering all the challenges and costs imposed by the process. The present study aims at providing an overview regarding the presence of the Industry 4.0 elements in European Union manufacturing companies using publicly available data from the European Statistical Office.

**Keywords** Industry 4.0 · Industry 5.0 · Digitalization · Manufacturing · Sustainability · European enterprises

L. Alexa · M. Pîslaru (✉) · S. Avasilcăi
"Gheorghe Asachi" Technical University of Iasi, Engineering and Management Department, Iasi, Romania
e-mail: marius.pislaru@academic.tuiasi.ro

L. Alexa
e-mail: lidia.alexa@tuiasi.ro

S. Avasilcăi
e-mail: silvia.avasilcai@academic.tuiasi.ro

 
A. Draghici and L. Ivascu (eds.), *Sustainability and Innovation in Manufacturing Enterprises*, Advances in Sustainability Science and Technology,
https://doi.org/10.1007/978-981-16-7365-8_8

# 1 Introduction

In an increasingly dynamic and complex world, manufacturing companies are constantly under pressure to innovate and become more competitive, and technological competitiveness has become a strategic element for organizations to differentiate themselves and improve their organizational performance both in terms of productivity and sustainability. We are currently living in an increased competitive environment that is constantly evolving into being customer-centric and demand-driven and so, in order to remain relevant, companies need to be able to make a precise, real-time decision while considering multiple ambiguous targets [29] and so, traditional approaches are no longer seen as an appropriate control tool [34].

So regardless if companies are ready or prepared to integrate new technologies and processes in their business models, the market forces them to adapt in order to remain relevant. Added to the general trend, the COVID-19 pandemic put extra pressure on companies that faced the need to rethink existing working methods and approaches and sped the adoption of digital technologies and practices.

# 2 Industry 4.0

Industry 4.0 refers to all the latest technological advances based on the Internet and supporting technologies that create the environment in which to integrate human actors, physical objects, smart machines, production lines, and processes throughout organizations in order to "create a new kind of intelligent, networked, and agile value chain" [35].

The term itself has been coined by the German Federal Government which considers Industry 4.0 to be the developing configuration through which logistics and manufacturing systems access the global information and communication network to automatically exchange information and match business and production processes [5]. Thus, Industry 4.0 has a significant impact over all industries, bringing disruptive changes in various business areas from supply chain management to business processes, business models, and market positioning [8, 33].

Among the new disruptive elements introduced by Industry 4.0 are "Big data and Analytics, Autonomous robots, Simulation, System Integration (Horizontal and Vertical), IoT (Internet of Things), Cybersecurity and Cyber Physical Systems (CPS), the Cloud, Additive Manufacturing, and Augmented Reality" [31].

**Big data and Analytics** refer to the practice of collecting and analyzing data from different sources as a means to provide support in the real-time decision-making processes [31]. According to Forrester's definition, there are four dimensions for big data, namely data volume, data variety, data velocity in terms of both generation and analysis, and data value Witkowski [39]. One of the main challenges with respect to big data refers to the fact that manufacturing companies need viable and up-to-date solutions that gather, analyze, and produce valuable and practical data from various sources and combine it in order to deliver around-the-clock real-time analytics.

**Autonomous robots** have a vital role in the modern manufacturing industry. According to a report published in 2014, the total number of multiuse industrial robots used in European Industry 4.0 companies has almost doubled from 2004 Roland Berger [1].

According to World Robotics 2020 Report, in 2019, almost 2.7 million industrial robots were functioning worldwide. Compared to 2018, there are 12% more industrial robots functioning in companies around the world and 68% more in comparison with 2016. In 2019, Europe reached an operational stock of industrial robots of 580,000 units (+7%). The European countries which leveraged most of the production advantages offered by industrial robots were Germany which had an operational stock of around 221,500 units, Italy with almost 74,400 units, France with 42,000 units, and the UK with 21,700 units [19].

These robots can carry out independent production methods and can finish specific tasks accurately and judiciously while also focusing on flexibility, safety, versatility, and collaboration [5]. With all these advantages, autonomous robots have a significant impact on the job market structure, as many jobs become obsolete, and companies need workers with a different set of competences and skills.

**Simulations** are using concurrent data to emulate the real world in a virtual model, which can incorporate machineries, goods, and persons [31]. Simulations' main advantage comes from the fact that it allows workers to assess and improve all the system's settings for future products inline with the virtual reality before the material switch, thus decreasing machine setup times and improving the end-product attributes. Thus, virtual engineering teams have been key elements of the successful application of these approaches and the decision-making processes of these types of teams should be an interdisciplinary approach [2].

**System Integration: Horizontal and Vertical System Integration.** "The two main mechanisms used in an industrial organization rely on integration and self-optimization" [34]. The entire Industry 4.0 theory is fundamentally outlined through three types of integration: (a) "horizontal integration across the complete value creation network", (b) "vertical integration and networked manufacturing systems", and (c) "end-to-end engineering throughout the entire product life cycle" Stock & Seliger [37].

**Internet of Things** refers to a "worldwide network of interconnected and uniform addressed objects that communicate via standard protocols" Hozdić [18]. Smart factories are all about intelligent planning, agility, and networking with all elements of the process interconnected and in constant real-time communication.

**Cybersecurity and Cyber Physical Systems (CPS)**. Considering the value of information in Industry 4.0, cybersecurity has become instrumental to "safeguard important industrial systems and manufacturing lines from cybersecurity risks" Vaidyaa et al. [38]. CPS refers to the systems in which the physical space consisting of natural and human-made systems is closely integrated with the cyberspace formed out of computation, communication, and control systems Bagheri et al. [35] and is characterized by the decentralization and autonomy of the production process.

**The Cloud**. Industry 4.0 implementation requires fast and reliable data sharing and so IT platforms are based on cloud function as a structure for communication and connection Landherr et al. [21].

**Additive Manufacturing** methods are also extensively used in Industry 4.0 to manufacture small lots of custom-made products, reducing time to market, and increasing customer satisfaction by proving products and services that better respond to the consumers' needs and expectations.

**Augmented Reality** systems support a variety of facilities meant to both increase productivity and decrease errors, such as providing employees real-time data to perform procedures and make decisions [31].

Regardless of the technological complexity of Industry 4.0, the main design principles of the transitional process from Industry 3.0 to Industry 4.0 refer to [17]: decentralization, interoperability, virtualization, real-time capability, modularity, and service orientation.

## 3 Industry 5.0

After more than 10 years of practice and research, many scholars have begun to vocally criticize Industry 4.0 due to the fact that its main goal focused solely on increased mass production and decreased costs which led to a perceived disregard toward environmental considerations [7] and human costs resulting from the optimization of processes [22].

In this context, there is a wide consensus in the literature that the next natural step in this evolution, namely Industry 5.0, should be focused on sustainability and waste prevention [30]. This vision is correlated with the fact that the bioeconomy vision of the European Commission is also centered around sustainability and bioeconomy is predicted to have a deep impact on businesses and industries [7, 32].

Considering that it is a relatively new term, there are a plethora of different visions for Industry 5.0: if Industry 4.0 was centered around technology and connecting devices together, Industry 5.0 brings back the human touch into the equation [20, 24] to leverage the synergy between humans and autonomous machines [22].

Østergaard also points that there is a significant shift in consumers behavior, and to meet consumers' increased demand for customized products, the next industrial revolution must rely on human involvement-human engagement Østergaard [25]. Furthermore, according to Atwell, Industry 5.0 focus is "increased collaboration between humans and intelligent systems" as more and more manufacturers are increasing the human component not only for customization purposes, but also for increased efficiency on the production line Atwell [3].

These discussions come to highlight that there is no clear demarcation line between Industry 4.0 and 5.0 and the two industrial revolutions are fundamentally connected and could actually be treated as one natural evolution. In Schwab's opinion [36], "to fully take advantage of the fourth industrial revolution, promising technologies should not be considered as simple tools that are entirely under human conscious control, or as external forces that cannot be controlled. The best solution is to try to understand how and where human values are embedded in new technologies, and how these technologies can be applied for the common good, environmental protection, and human rights".

### 3.1 Industry 4.0 Adoption, Opportunities, and Challenges

In this volatile context, manufacturing enterprises are currently confronted with considerable challenges in implementing the elements of Industry 4.0 and Industry 5.0 as it is becoming "more and more obvious that such a far-reaching vision will inevitably lead to an amplified complexity of all industrial processes, both on the micro and macro level" (Schuh et al. 2014).

Despite the clear advantages and opportunities provided, extensive research on companies' perspective over Industry 4.0 [10] has shown that enterprises face serious difficulties in grasping both the general idea and specific concepts related to Industry 4.0, among which we can enumerate the fact that:

- Most companies consider Industry 4.0 concepts to be extremely complex with limited to no strategic assistance provided;
- Most companies prove to be unable to relate Industry 4.0 to their specific area or their specific business strategy;
- Companies face serious difficulties in evaluating their Industry 4.0 development status and, therefore, are unable to establish a specific area for action or create plans and projects.

Another relevant aspect refers to the substantial costs implied by Industry 4.0 adoption process which significantly impact manufacturing SMEs (small and medium-sized enterprises). A study conducted in 2019 [27] highlights that, in terms of impediments for Industry 4.0 adoption, most companies are struggling with the lack of digital competences, lack of technology and infrastructure, and lack of skills and guidance for transformation. Ideas related to the needs for changes and transformations are being already announced by previous studies [28, 26].

## 4 Industry 4.0 Adoption Among EU Enterprises

The European Union understands the potential of advanced technologies to transform EU industries and so there are several European level policies aimed at encouraging and supporting the adoption of digital technologies.

The EU Commission began to implement a Digital Single Market Strategy in May 2015. This was one of the EU's top 10 political priorities for 2015–2019. This Digital Single Market Strategy consisted of sixteen plans aimed at covering three wide-ranging areas, namely "promoting improved online access for products and services across Europe, designing an optimal environment for digital networks and services, and providing support for companies to take full advantage of the digital economy as a potential driver for growth" [11].

In the same spirit, most EU governments have included Industry 4.0 among their priorities by adopting extensive strategies and programs to improve efficiency and competitiveness and improve the digital skills of their workforce.

Within this strategy, the European [11] has been working on measuring and characterizing the EU digital society by identifying and gathering relevant data, incorporating information connected with the digitization of the industrial processes. The EU has also constantly developed measures to provide support for cooperation between national research initiatives in the field of "Digital Manufacturing" and to provide various lines of funding within Horizon 2020 [6].

In February 2020, the Commission launched its vision for digital transformation: "Shaping Europe's digital future". Through this, the Commission plans to deliver a comprehensive use of technology "that works for people and respects EU's fundamental values" [13]. The first two pillars of the Commission's innovative digitalization strategy are the White Paper regarding Artificial Intelligence [14] and the European data strategy [15].

**Data covering Industry 4.0 Adoption in EU**

Most of the data used throughout this study are founded on the Digital Economy and Society Index (DESI) [12], which is a compound guide that condenses all significant indicators regarding Europe's digital development and evaluates the expansion of digitalization of Member States and Eurostat database which assesses periodically the status of ICT usage and e-commerce processes implementation in companies [9].

According to the DESI report [12], the most progressive digital economies in the European Union belong to Finland, Sweden, Denmark, and the Netherlands followed by Malta, Ireland, and Estonia. At the other end of the spectrum, Bulgaria, Greece, Romania, and Italy registered the lowest numbers according to the DESI index.

In terms of **Internet access**, according to Eurostat data, in 2019, 91% of EU companies employing at least 10 persons declared that they are using a fixed broadband connection for Internet access. This number went up 3% compared with 2011, suggesting that the adoption of this technology has reached saturation at the EU level. Considering that nearly all EU enterprises have an Internet connection, the policymakers' focus has shifted lately toward Internet speed, as this is a key aspect for both Industry 4.0 and Industry 5.0.

**Internet connection speed**. In Industry 4.0 context, speed is equally important to access, and so, between 2011 and 2019, the number of companies accessing fast Internet connections tripled. In 2018, 20% of EU companies declared that their Internet connection ranges between 2 Mb/s and 10 Mb/s. 24% of enterprises declared their Internet connection speed ranges between 10 and 30 Mb/s. 25% of enterprises stated that their connection ranges between 30 and 100 Mb/s, and 18% of EU companies have Internet connections of more than 100 Mb/s.

In 2019, 16% of companies included in the study declared that their Internet connection speed ranges between 2 and 10 Mb/s. 23% were having a connection ranging between 10 and 30 Mb/s. Over 27% stated that their Internet connection was ranging between 30 and 100 Mb/s, while 23% of companies were using an Internet connection of more than 100 Mb/s [9].

Considering the importance of data and information in Industry 4.0, cybersecurity is an important element to consider. According to the DESI report, in 2019, 34% of EU enterprises declared that they have a focus on cybersecurity and have ICT safety protocols establishing specific procedures that need to be followed by employees. 93% of companies stated that they have implemented at least one cyber safety protocol. The implementation of cybersecurity measures is extensive among large companies and small and medium enterprises alike: as almost 99% of large companies and 92% of small and medium enterprises declared that they employ several cybersecurity measures, but the security measures taken are very diverse. Most enterprises included in the study have set out simple cybersecurity measures such as constantly updating software (87%), device authentication using a strong password (77%), and constantly doing back-ups in separate locations including using cloud solutions (76%). Only a few companies declared that they have in place more sophisticated cybersecurity measures such as security tests (36%) or risk assessments (34%) and only 9.5% of companies implemented biometric methods for user identification and authentication.

In terms of **Cloud computing**, the Eurostat data showed that in 2018, 26% of EU companies declared that they have accessed cloud computing services. In 2020, the number went up by 12%, reaching 36% of EU enterprises using cloud computing. Most companies declared that they commonly use the cloud for e-email and storing files in electronic form.

At the same time, there are important variations that can still be observed across EU countries. In northern countries such as Finland (75%), Sweden (70%), and Denmark (67%), over 60% of enterprises declared that they constantly used cloud computing. At the other end of the spectrum, in Greece (17%), Romania (16%), and Bulgaria (11%), less than 20% of companies regularly employ cloud-based services.

Most companies employing cloud computing services are active in the information and communication sector (71%). For all the other economic sectors, the percentage of enterprises using cloud computing ranges from 27 to 43%. However, the manufacturing sector had the highest increase (+19%) in the use of cloud computing compared with 2018.

Another relevant element refers to the significant differences between large enterprises and small ones when cloud computing use is discussed: 65% of companies employing 250 persons or more use **cloud computing**, representing a rise of 12% compared with 2018 while only around 50% of SMEs declared that they used cloud computing in 2020. However, both small and medium-sized enterprises recorded an increase of 12% compared with 2018, reaching 33% and 46%, respectively.

**3D printing** or "additive layer manufacturing", refers to utilizing special printers either in-house, by the company itself, or through outsourcing, using 3D printing services offered by other companies to create three-dimensional physical objects employing digital technology.

In 2020, 5% of European companies employing at least 10 persons declared that they utilized 3D printing (in-house or outsourcing). This number represents an increase of 1% compared with 2018 [9]. The highest number of companies using 3D printing in 2020 was from Denmark (9%) and Malta (8%) while the smallest shares were reported by enterprises in Romania (2%).

There is a significant discrepancy in using 3D printing between large enterprises and SMEs: 13% of large enterprises use 3D printing as part of their manufacturing process and only 4% of SMEs access this technology. The technology was mostly employed by companies from the manufacturing sector (9%), followed by companies in activating in the professional, scientific, and technical activities (6%) and in information and communication (5%).

**Big data usage**. In the last years, due to the ICT advances, the quantity of digital data created, stored, and processed worldwide has been constantly growing exponentially, because every online activity leads to the generation of a series of digital marks which, due to their size, diversity, and speed, are referred to as big data.

According to the data gathered from EU companies, in 2018, 12% of companies employing at least 10 persons reported analyzing big data. The trend for big data analysis is like the other elements measured, being mostly done by large enterprises (33%) and medium-sized enterprises (19%). When evaluating the status of EU Member States, the highest number of companies employing big data analysis is based in Malta (24%), the Netherlands (22%), Belgium (20%), and Ireland (20%). Companies from Italy (7%), Bulgaria (7%), Hungary (6%), Austria (6%), and Cyprus (5%) declared in significantly lower numbers the use of big data analysis.

Enterprises that analyzed big data declared that they use a variety of data sources, the most popular ones used to be geolocation and social media data. Almost 50% of companies stated that they use for their big data analysis geolocation data from portable devices (49%) and data from social media networks (45%). Only 29% of companies analyzed big data from smart devices and 26% used data from other sources. For large enterprises, the analysis was done mostly by employees (90%) and 75% declared that they employ an external service provider for this activity. 42% of the SMEs included in the study declared that they rely on external service providers to analyze big data and 40% stated that the activity is performed by their own employees.

The DESI report (2020) also analyzes the **workforce** and its role in developing the digital economy in the European Union. According to the data, in 2018, a little over 9 million persons were working as ICT specialists all over the European Union. Companies from UK and Germany were employing each around 1.6 million ICT specialists, while companies from France had over 1.1 million ICT specialists. According to the data from 2019, almost 20% of companies declared that they hired experts to create, manage, or maintain ICT systems and applications. Here, there is also a significant difference between large enterprises (75%) and small and medium enterprises (19%). Another aspect related to the specifics of the ICT labor refers to the companies' abilities to recruit personnel, and in 2018, almost 57% of the companies which employed or attempted to employ ICT experts declared having problems in filling such positions. These difficulties were experienced by over 64% of large enterprises and 56% of small and medium enterprises. Companies from Romania and Czechia are reporting having even more difficulties when recruiting or trying to recruit ICT specialists, as 80% of the companies declared that they are unable to find staff to fill these specific vacancies. The aspect of the specialized workforce becomes even more relevant when discussing Industry 5.0, as these specific skillsets represent a central element in the successful evolution toward Industry 5.0.

## 5 Conclusions

Since the introduction of Industry 4.0 at the beginning of the 2010s, there have been constant discussions and predictions regarding its evolution and impact over all industrial sectors. In terms of efficiency and costs, research shows that by implementing Industry 4.0, companies have registered 10–30% in decreased production costs, 10–30% in decreased logistic costs, and 10–20% in decreased quality management [23].

However, wide-range Industry 4.0 adoption led to generating other types of societal costs and so, more and more voices begun to point out the inadequacies of Industry 4.0 and propose a natural evolution toward Industry 5.0, not in terms of chronology but of approach, because the shift from a machine-centric approach to a human-centric approach is not only desirable but necessary.

We agree with the scholars who consider that the two industrial revolutions are intrinsically connected and should actually be treated as one, because Industry 5.0 supplements and increases the characteristics and advantages of Industry 4.0.

The use of innovative digital technologies, such as IoT or big data analysis, can ensure companies an increase in productivity and efficiency and provide new development opportunities for EU enterprises from all areas of activity. The analysis of the presence of Industry 4.0 elements in European Union manufacturing companies revealed that in terms of Internet access and Internet connection speed, most European enterprises have managed to advance and harness the advantages, and this is correlated with the European and national efforts aimed at ensuring connectivity and Internet access.

Other elements of Industry 4.0, such as cloud computing or big data usage, are less adopted by EU enterprises, with significant differences between large enterprises and small and medium enterprises. This can be explained due to the costs and resources implicated in the process of technology adoption.

Another aspect that needs to be considered is the workforce skills and competences which need to be updated alongside the technology to ensure a successful transition to Industry 5.0.

## References

1. Berger R (2014) Industry 4.0–The New Industrial Revolution: How Europe Will Succeed. In: International Conference The Next Industrial Revolution Manufacturing and Society in the XXI Century. Turin
2. Aldea CC, Popescu AD, Draghici A, Draghici G (2012) ICT tools functionalities analysis for the decision making process of their implementation in virtual engineering teams. Procedia Technol 5:649–658
3. Atwell C (2017) Yes, Industry 5.0 is Already on the Horizon. Available at: https://www.machinedesign.com/industrial-automation/yes-industry-50-already-horizon
4. Bagheri B, Yang S, Kao H, Lee J (2015) Cyber-physical systems architecture for self-aware Mmchines in Industry 4.0 environment. IFAC Conference. pp 1622–1627
5. Bahrin M, Othman F, Azli N, Talib M (2016) Industry 4.0: a review on industrial automation and robotic. J Teknol (Sci Eng) (78):6–13
6. Buhr D, Stehnken T (2018) Industry 4.0 and European innovation policy. Big Plans Small Steps. https://library.fes.de/pdf-files/wiso/14455.pdf
7. Demir K, Cicibaş H (2019) The next industrial revolution: industry 5.0 and discussions on industry 4.0. In: Industry 4.0 from the MIS perspective. Peter Lang GmbH, pp 247–260
8. Dongback S (2017) Digital business convergence and emerging contested fields: a conceptual framework . J Assoc Inf Syst 687–702
9. EUROSTAT (2021) Eurostat. https://ec.europa.eu/eurostat/databrowser/view/isoc_eb_p3d/default/table?lang=en
10. Erol S, Jäger A, Hold P, Ott K (2016) Tangible industry 4.0: a scenario-based approach to learning for the future of production. Procedia CIRP 13–18
11. European Commission (2015) The digital single market strategy. https://eur-lex.europa.eu/legal-content/EN/TXT/?uri=celex%3A52015DC0192
12. European Commission (2020) Digital economy and society index (DESI). European Commission
13. European Commission (2020a) Shaping Europe's digital future. https://ec.europa.eu/info/sites/info/files/communication-shaping-europes-digital-future-feb2020_en_3.pdf
14. European Commission (2020b) White paper on artificial intelligence—A European approach to excellence and trust. https://ec.europa.eu/info/sites/info/files/commission-white-paper-artificial-intelligence-feb2020_en.pdf
15. European Commission (2020c) A European strategy for data. https://ec.europa.eu/info/sites/info/files/communication-european-strategy-data-19feb2020_en.pdf
16. Eurostat (2021) Digital economy and society statistics—Enterprises. Eurostat. https://ec.europa.eu/eurostat/statistics-explained/index.php?title=Digital_economy_and_society_statistics_-_enterprises
17. Hermann M, Pentek T, Otto B (2015) Design principles for industrie 4.0 scenarios: a literature review
18. Hozdić E (2015) Smart Factory for Industry 4.0: A Review. Int J Modern Manuf Technol 28–35
19. IFR IF (2020) Executive summary world robotics 2020 industrial robots. https://ifr.org/img/worldrobotics/Executive_Summary_WR_2020_Industrial_Robots_1.pdf
20. Johansson H (2017) Profinet industrial internet of things gateway for the smart factory. Master's thesis in embedded electronic system design. Chalmers University of Technology. https://publications.lib.chalmers.se/records/fulltext/249922/249922.pdf
21. Landherr M, Schneider U, Bauernhansl T (2016) The Application Centre Industrie 4.0–Industry-driven manufacturing, research and development. In: 49th CIRP Conference on Manufacturing Systems (CIRP-CMS 2016), Procedia CIRP, pp 26–31
22. Nahavandi S (2019) Industry 5.0—A human-centric solution. Sustainability 11(16):4371
23. Nayak N, Dürr F, Rothermel K (2015) Software-defined environment for reconfigurable manufacturing systems. In: Proceedings of the 2015 5th international conference on the internet of things (IOT). Seoul, Korea, pp 122–129

24. Østergaard EH (2016) Industry 5.0—Return of the human touch. https://blog.universal-robots.com/industry-50-return-of-the-human-touch
25. Østergaard EH (2018) Welcome to Industry 5.0. Available at: Int Soc Autom. https://www.isa.org/intech/20180403/
26. Paschek D, Rennung F, Trusculescu A, Draghici A (2016) Corporate development with agile business process modeling as a key success factor. Procedia Comput Sci 100:1168–1175
27. Paschek D, Mocan A, Draghici A (2019) Industry 5.0—The expected impact of next industrial revolution. In: Management knowledge learning international conference. Reinhausen, KMedu Hub, pp 125–132
28. Paschek D, Luminosu CT, Draghici A (2017) Automated business process management—In times of digital transformation using machine learning or artificial intelligence. In: MATEC web of conferences, vol 121. EDP Sciences, p 04007
29. Porter M, Heppelmann J (2015) How smart, connected products are transforming companies. Harv Bus Rev
30. Rada M (2018) Industry 5.0 definition. https://medium.com/@michael.rada/industry-5-0-definition-6a2f9922dc48
31. Rüßmann M, Lorenz M, Gerbert P, Waldner M (2015) Industry 4.0: the future of productivity and growth. Manuf Ind 1–14
32. Sachsenmeier P (2016) Industry 5.0—The relevance and implications of bionics and synthetic biology. Engineering 225–229
33. Schmidt R, Möhring M, Härting R, Reichstein C, Neumaier P, Jozinović P (2015) Industry 4.0—Potentials for creating smart products: empirical research results. In: Lecture notes in business information processing, pp 16–27
34. Schuh G, Scholz P, Nadicksbernd M (2020) Identification and characterization of challenges in the future of manufacturing for the application of machine learning. In: 61st international scientific conference on information technology and management
35. Schumacher A, Erol S, Sihn W (2016) A maturity model for assessing industry 4.0 readiness and maturity of manufacturing enterprises. In: Procedia CIRP 52 changeable, agile, reconfigurable & virtual production, pp 161–166
36. Schwab K (2017) The fourth industrial revolution. Currency Books, New York
37. Stock T, Seliger G (2016) Opportunities of Sustainable Manufacturing in Industry 4.0. In: 13th Global Conference on Sustainable Manufacturing - Decoupling Growth from Resource Use, Procedia CIRP, pp 536– 541
38. Vaidyaa S, Ambadb P, Bhoslec S (2018) Industry 4.0 – A Glimpse. In: 2nd International Conference on Materials Manufacturing and Design Engineering. Procedia Manufacturing, pp 233–238
39. Witkowski K (2017) Internet of Things, Big Data, Industry 4.0–Innovative Solutions in Logistics and Supply Chains Management. In: 7th International Conference on Engineering, Project and Production Management, Procedia Eng, pp 763–769

# Chapter 9
# Assessment of Economic Impact Generated by Industry 5.0, from a Readiness Index Approach Perspective. A Cross-Country Empirical Analysis

**Aura Domil, Valentin Burca, and Oana Bogdan**

**Abstract** Over time, industrial revolutions have marked human existence and the work environment, impacting efficiency, effectiveness, and productivity. In the new promoted concept, Industry 5.0, it is expected that humans interact with robots, known as cobots, in a collaborative working space using innovative technologies. This chapter brings some new insights on the way the components of Industry 5.0 readiness model influence value added reported by countries, for the period between 2017 and 2019, on four levels of analysis, namely: (i) country gross value added; (ii) environmental activities value added; (iii) ICT sector value added; (iv) labour productivity. Our results prove that selected components of the Industry 5.0 readiness model determine a positive marginal effect on the value added reported by countries. We also registered a significant influence on the development of the circular economy, in strong relation with the development of the knowledge economy. This highlights the fact that the results of the research are efficiently implemented on manufacturing operations by highly skilled employees, leading to increase firms' performance, improving at the same time the cooperation between humans and smart machines, as stated in the Industry 5.0 framework.

**Keywords** Human-centric approach · Sustainable development · Industry 5.0 · Economic assessment · Resilience · Factor analysis · Readiness Index · Environmental value added · Labour productivity

---

A. Domil (✉) · V. Burca · O. Bogdan
West University of Timisoara, Faculty of Economics and Business Administration, Timioara, Romania
e-mail: aura.domil@e-uvt.ro

O. Bogdan
e-mail: oana.bogdan@e-uvt.ro

 
A. Draghici and L. Ivascu (eds.), *Sustainability and Innovation in Manufacturing Enterprises*, Advances in Sustainability Science and Technology,
https://doi.org/10.1007/978-981-16-7365-8_9

## 1 Introduction

The world has developed and continues to evolve due to higher and higher targets set in terms of technology, professional evolution, knowledge and desire for self-transcendence, elements that characterizes the human species.

Over time, new technologies have been implemented, and industrial revolutions have left their mark on the development of modern society. The first industrial revolution—Industry 1.0 began in England and mechanized production. New methods of metal processing were used, and the first factories and industrial cities appeared. The second revolution, industry 2.0 can be characterized by the introduction of electricity, mass production and labour division. The third revolution, known as Industry 3.0, automated production. Industry 4.0, known as the fourth revolution, was built on the Industry 3.0 and the word that defines it is connectivity to interconnect machines, processes and systems to improve efficiency and productivity. Nowadays, we are facing new waves, in which humans, robots and smart machines collaborate and find ways to work together for maximum performance optimization, ensuring at the same time a more sustainable and resilient European industry. Industry 5.0 "recognizes the power of industry to achieve societal goals beyond jobs and growth to become a resilient provider of prosperity, by making production respect the boundaries of our planet and placing the wellbeing of the industry worker at the centre of the production process" [11]. Hence, Industry 5.0 (henceforth Industry 5.0) involves the use of smart machines to accelerate human performance which will give the employee the possibility to perform high value tasks, leaving the repetitive tasks to the cobots (i.e., collaborative robots). In this sense, robots are not replacing humans, rather complementing their capabilities, and relieving them of strenuous tasks. Are we able to embrace the benefits of new technologies? The founder of Alibaba stated that with the help of Artificial Intelligence humans will work less and will have more time to enjoy being human beings [22]. It is known that robots have been used for some time in medicine, but are we able to co-work with robots in other areas? Reality shows us that that it is possible. For instance, a timber construction firm implemented this collaboration between human's employees and robots in manufacturing process and stated that they complement each other ideally, although, at first, there were some hesitations [38]. What is certain is that over time, new technologies have changed people's lives and, automation has reduced the number of workers but increased productivity. What at first looked like a SF movie script turns into a normal thing over time. But what does the future hold for us? Specialists like [25] go further and talk about the sixth industrial revolution involving industrial emotions, namely establish a social and emotional link between humans and smart machines. But technological advancement should not be viewed with scepticism, because, like Noory Bechor, the Lawgeex's CEO said it "is similar to pilots, with the autopilot, it's not like we don't need pilots anymore" [37].

The aim of this chapter is to bring some insights on the relationship between innovation, economy, human capital and digitalization, the key elements of the Industry

5.0. To achieve our goal, we used econometric analysis, to investigate how the components of Industry 5.0 readiness model influence value added reported by countries on four levels of analysis, namely: (i) country gross value added; (ii) environmental activities value added; (iii) ICT sector value added; (iv) labour productivity. Our expectations are that the value added is positively impacted by the Industry 5.0 readiness index components.

This chapter is structured, as follows: introduction, the first part of our paper, presents the research context. The second part highlights the background and the relevant scientific literature. The methodology used to achieve our objectives is reflected in the third section. The next section, the fourth, presents and explains the results obtained and the last section, the fifth, summarizes the main findings and conclusions.

## 2 Background

Industrial revolutions have significantly influenced the way people live and the environment in which they work. Smart electronics, home appliances and lighting, online payments and orders, entertainment, Cloud platforms, Internet of Things, Big Data and Artificial Intelligence applications are the main drivers of the Industry 4.0 [21, 28] that showed us that new technologies have become a priority for all industries. Nevertheless, we believe their benefits are much more obvious due to the pandemic context caused by the novel coronavirus. Measures taken to limit the spread of COVID-19 required physical distancing and repeated lockdowns revealed the need to approach working methods from a new perspective. This pandemic context imposed, mainly, transposing the activity into the online environment. Thus, we were able to continue our activity, in remote, to live and stay connected in the virtual environment due to the benefits of the new technologies implemented. But this is just the beginning if we consider the significant advancements in Artificial Intelligence and robotics research that requires ongoing innovation, funding, and legislative support through government policies. Efforts have been made in this sense, so at European level, European Commission elaborated "the Digital Economy and Society Index", to motivate EU member states to develop and practice the latest technologies by presenting and following the evolution of the indicators regarding the performance of each country [43].

The growing trend of the technology industry can be also highlighted through figures. According to European Commission [11], in 2009, only Microsoft, made in the top 10 of traded entities by market capitalization. Ten years later, the first 5 companies from the top were all from the technology field. In 2020, in the Global Top 100 companies by market capitalization elaborated by PwC [34], 22 companies were from the technology field, out of which 5 were among top 10 companies. 6 companies activate in the oil & gas sector, 13 in consumer services, 15 in financials, and 18 in health care system. 12 entities operate in consumer goods domain, 5 in telecommunication, 6 in industrial sector, 1 in utilities and, 2 in basic materials sector.

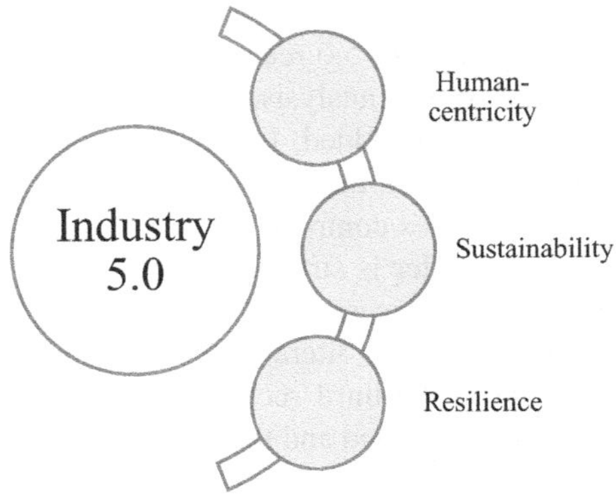


**Fig. 1** Industry 5.0 Main pillars (a synthesis after European Commission Report [11])

As we can see, the top of the most powerful companies is dominated by those from the tech field.

But society's evolution calls for more, and innovation in this domain shows no signs of slowing down. Based on the previous industry, namely 4.0, the new concept, Industry 5.0, represents a complementary approach to the digital industrial revolution that can be defined through three main pillars, outlined in the following Fig. 1.

As can be seen in this new approach, the workers well-being is put in the centre of the production process, in a sustainable environment focused on the circular economy and energy efficiency. As for resilience, the pandemic context generated by the novel coronavirus disease, known as COVID-19, highlighted the need to rethink the production process to mitigate the effect of disruptions, being able to ensure continuity within the supply chain even in pandemic times.

## 2.1 A Human-Centric Approach in the Context of Industry 5.0

Since the first industrial revolution, there have been concerns that the use of innovative technologies can lead to job losses, disrupting the workplace environment. In this context, will machines substitute human workers? We talk about "human vs. machines? or the discussion is actually about humans handling smart machines?

In the smart factory area, a new trend is appearing aiming to bring back the human touch in production to be able to fulfil the consumers' requirements that demand a high degree of individualization in the products they purchase. Hence, Industry 5.0 implements the idea of mutual completion between humans and robots, which thrive together, in a collaborative working process [41]. So, cobots take over the monotonous and strenuous tasks while human capital is valued and appreciated for the cognitive and critical thinking skills with which it is endowed to increase productivity [28]. But there are several concerns related to the cooperation between man and cobots [8]. One of the most important refers to the legal and regulatory issues arising from the integration of robots in the work environment. Authors [13, 14, 30] show interest for the aspects related to the safety of the collaboration between humans and smart machines, highlighting the need to implement a legal framework to handle the work with cobots. Demir et al. [8] depicted that the lack of regulations in this domain can cause serious problems for both companies and employees.

Roblek et al. [36] emphasized that in this new digital context, employees training is essential so that they can cope with the new demands of the workplace. In addition, it is in human nature to manifest preferences or aversions to something, so people may or may not like to collaborate with a robot. Some may be eager to begin this experience, while others, may feel that they can no longer benefit from peer interaction at work. In this sense, the transition to Industry 5.0 implies concrete and customized measures to ensure that both employees and companies are fulfilled by this collaboration.

## 2.2 *Sustainable Development Through Ongoing Innovations*

Sustainable development finds its applicability also in the case of industry 5.0 main pillars because economies around the world are constantly facing new challenges. Whether they react to external stimuli or global pandemics, to new technologies or anticipate new trends, their ability to change or to be reactive or proactive is tested every day. Hence, sustainability is a concept related to environmental, social, and economic issues. Setting sustainable development goals is a global consensus of the need to preserve the conditions that make life possible on this planet and focuses on the intersection of sustainability, environment and technology and their implications for corporations and corporate governance, government, institutions education, regions, and society, both now and soon.

Hence, sustainable development represents a difficult challenge for mankind. Population growth, amplification of climate change and the increased pressure created by food and water, combined with increased energy needs and environmental issues, require a leadership role with a strong commitment to the circular economic model. Information and the use of technology in cyber communications and the Internet of Things (IoT) lead to an innovative approach to improve linear business models and replace them with sustainable business models.

According to [5], industrial symbiosis seems to have found an innovative stimulus in the circular economy, a new approach of sustainable development, which has rapidly expanded worldwide with a particular focus on the potential role of symbiosis in achieving the goals set in the UN 2030 Agenda related to sustainable development. Baldassarre et al. [3] reiterated the idea that industrial symbiosis is a collective approach to competitive advantage in which separate industries create a cooperative network for the exchange of materials, energy, water and/or by-products. It plays a crucial role in the successful implementation of circular business models, addressing issues related to resource depletion, waste management and pollution. Nevertheless, the last Circularity Gap Report revealed that "the global economy was only 8.6%

circular, while two years ago it was 9.1%". These figures highlight that now the global economy is stuck in reverse, failing both people and the planet itself [6]. The transition to a circular economy needs to be accelerated, but this process can be slowed down by decisions based on misunderstanding of concepts, unclear indicators, and inadequate information (Dewick 2020).

Innovation can change this trend by smarter production planning with more energy efficient technologies. Hence, it is necessary to develop a framework that incorporates the dynamics of sustainability. In his paper, Jason [32] proposes a holistic quantitative framework for defining and assessing sustainability and unsustainability in terms of denoting the current geological era, seen as the period in which human activity was the dominant influence of climate and the environment [32]. This framework was applied to determine and analyse the fundamental dynamic relationships between the indicated levels of durability (S values) and different influencing factors. It is increasingly evident the configuration of a new research field based on understanding the methodology for evaluating the circular economy and sustainable development at the nano-level (individual paradigms and beliefs) micro-level (which includes products, services, or companies), meso-level (industrial symbiosis networks and reverse logistics chain) and macro-level (local community, national boundaries, and global approaches).

This new field contributes to the advancement of concepts and the determination of the methodological framework by facilitating the exchange of information in the cyber environment in which companies operate, monitoring progress, informing decision making and improving CEO decisions of circular business investments [20].

Braccini and Margherita [4] also investigated how the adoption of new technologies, namely, Industry 4.0 in manufacturing companies contributes to the economic, environmental, and social levels of sustainability. Their findings highlight those Industries 4.0 applications that support the main pillars of sustainable development process.

### 2.3 Resilience After Major Disruptions

According to European Union, resilience means "the ability of an individual, a household, a community, a country or a region to withstand, adapt and to quickly recover from stresses and shocks" [12]. The health crisis caused by the COVID-19 pandemic has showed us that the "normal" we know can change in a second. The measures taken to limit the spread of the novel coronavirus have generated major disruptions in the daily course of businesses worldwide for most industries and sectors and highlighted the weak points of our industries that need to be restructured to be more robust to face challenges [11].

The transition to "the new normal" after the pandemic period offers us the opportunity to reshape and renew the role of industry in society, with the right framework adapted to the new requirements of society. In 2017, Morrise and Prigge stated that "Industry 4.0 will change existing value chains on an international scale". We believe that Industry 5.0 will achieve much more to prevent another major disruption like the one caused by the COVID-19 pandemic.

## 3 Methodology

The design of our study is similar with Czvetkó et al. [7] who have developed an Industry 4.0 readiness index on a NUTS2 regional level of analysis. Their model incorporates multiple dimensions that describe countries' vision and strategy, Industry 4.0 enabling technology implementation, human capital characteristics, including education and digital skills. However, it does not incorporate additional dimensions of circular economy and sustainability development requirements which seem to become incorporate in the more recent Industry 5.0 framework. Instead [35], has developed a conceptual model of readiness assessment of the circular economy component of Industry 4.0 framework, showing that waste reduction can be achieved through processes automatization, highly skilled employees, and decision-making optimization, using Industry 4.0 enabling technologies.

Overall, there are few studies that analyse countries' readiness for Industry 4.0, as the focus is on a firm level analysis. However, as noted by [33], piloting seems to be essential on ensuring a flawless implementation of Industry 4.0, which means the state could involve on promoting and creating a general framework for Industry 4.0 initiative and offer grants as well. For this, there must be powerful institutions, flexible and future oriented to adapt to the extremely fast changes occurring in the private sector. That is why country assessment is an important element on the equation of implementing Industry 4.0, or nowadays, the Industry 5.0 evolved framework. Therefore, sustainable development goals agreed on a global level can be achieved through a more directed action of government towards the support of initiatives on Industry 5.0, as circular economy and sustainability represent core pillars of this evolved industrial revolution [23].

Our approach is to assess INDUSTRY 4.0 country readiness index starting from the dimensions identified by [17], who have made a literature review on the models developed for Industry 4.0 readiness assessment. They have underlined that technology, people, strategy, leadership, processes, and innovation represent key elements on assessing how ready are firms to implement Industry 4.0. The only approach which is not covered in our model is related to the process dimension, as it refers to the specific of the firms' process design and supporting infrastructure.

Instead, our readiness index is oriented on a country level analysis, that shows a systemic potential impact implementation of Industry 5.0 this time, which will be generated for each firm. For this purpose, we have first focused on countries' competitiveness, which provide signals already about the potential of firms' automatization, digitalization, and orientation towards circular economy. For instance, a first step on promoting on a national level the behaviour that support circular economy initiative is to have a proper national regulation on the area of renewable energy and waste management. Therefore, we add to those Industry 4.0 specific dimensions, additional dimensions related to Industry 5.0 framework, respectively the dimension of circular economy and sustainability.

Better view on the supportive stakeholders was made by Abonyi et al. [1], who have placed the government as a key responsible actor on the area of strategic view about evolving industrialization, highlighting states' role on financing such initiatives, especially on financing research and development projects addressing issues of sustainable development. If the state is aware of how important the efforts are to be made for economy digitalization, development and implementation of emerging manufacturing technologies, or improvement of people digital and analytical skills, they will finance related projects and will get benefits on a long-term, through economies of scale that firms will generate. Therefore, a systematic approach can be ensured on a country level on addressing the big challenges of Industry 5.0 implementation which are similar with the ones already noted in case of Industry 4.0, resumed to human capital development, research and development, digitalization, and sustainable economic development, including by increasing the weight of circular economy and knowledge economy of the national economy. On these circumstances, to check for the impact of the research and development initiatives, we will look as well for the quality of academic activity on this area, respectively searching for the amplitude of scientific research and the number of patents approved.

Additional to the country competitiveness characteristics, we have focused our attention to the specific of country governance quality, looking on the quality of national regulation and institutions' processes effectiveness. Otherwise, either the regulation can act as a constraint on Industry 5.0 initiatives or will bring no positive effect.

Table 1 summarizes the data included in our analysis. As a next step, we will proceed to a factor analysis to reduce the dimension of our model assessment. From the total of 42 variables included in our analysis, we keep for the construction of our Industry 5.0 Readiness index just the variables with a factor loading greater than 0.500, which means respective variables are faithfully represented by extracted factors. Additionally, those factor loadings show that those variables bring a significant contribution to the variation of the data analysed, meaning that distinction between countries through this index is significantly influenced by those variables [16].

The next step of our analysis is to understand how relevant our index is, through its components extracted. For this purpose, we will proceed to an econometric analysis, estimating several models that show how the components of Industry 5.0 readiness model influence the value added ($VA_{itk}$) reported on a country level per four levels, respectively: (i) country gross value added; (ii) environmental activities value added; (iii) ICT sector value added; (iv) labour productivity. Below we describe models that will be estimated:

$$\begin{aligned} \mathrm{VA}_{\mathrm{it}k} &= \alpha_0 + \alpha_1 \cdot \mathrm{HC}_{\mathrm{it}} + \alpha_2 \cdot \mathrm{R\&D}_{\mathrm{it}} + \alpha_3 \cdot \mathrm{ESG}_{\mathrm{it}} \\ &+ \alpha_4 \cdot \mathrm{DG}_{\mathrm{it}} + \sum_{\mathrm{l}=1}^{2} \alpha_{4+\mathrm{l}} \cdot \mathrm{ET}_{\mathrm{itl}} + \varepsilon_{\mathrm{it}} \end{aligned} \tag{1}$$

**Table 1** Data used to run factor analysis (authors projection)

| Dimension | Data | Source |
|---|---|---|
| Technology | ICT specialists, use of ERP, big data analytics, use of cloud solutions, connectivity, integration of digital technology | Eurostat |
| People | General skills, skills of current workforce, skills of future workforce, skillset of university graduates, individuals' digital skills, labour market | Global competitiveness report |
| Strategy | Institutions score, legal framework's adaptability to digital business models, government's responsiveness to change | Global competitiveness report |
| | Regulatory quality, government effectiveness, rule of law, corporate governance | World governance indicators |
| Leadership | Reliance on professional management, meritocracy and incentivization, future orientation of government | Global competitiveness report |
| Innovation | Growth of innovative companies, intellectual property protection, university-industry collaboration in R&D | Global competitiveness report |
| | Research and development expenditures, researchers in business sector | Eurostat database |
| | Published articles related to Industry 4.0 | VOS viewer |
| | Industry 4.0 patents | EPO database |
| Circular economy & sustainability | Gross Domestic Product (GDP), market capitalization | Eurostat |
| | Eco-innovation, generation of waste, circular material use rate, share renewable energy, patent related to circular economy, energy efficiency regulation, renewable energy regulation, commitment to sustainability, renewable electricity output, recycling rate of waste, natural resources depletion | Global competitiveness report |

where $k = \overline{1,5}$, $VA_{it}$ is the value added for country *i* reported on year *t*, $HC_{it}$ is the factor of human capital development, $R\&D_{it}$ is the factor of research & development, $ESG_{it}$ is the score that measure the perception on sustainability national regulation, $DG_{it}$ is the factor of digitalization, while the $ET_{itl}$ are the two Industry 5.0 enabling technologies, respectively the *artificial intelligence* and *the Internet of Things.*

All those data are collected from Eurostat database, except for the element of $ESG_{it}$ score, which are collected from the Global Competitiveness Report published by World Economic Forum. The period analyzed is 2017–2019.

As noted by [24], Industry 4.0 generate a positive effect on the labour market, education, changes in operational processes, or economic growth. Additionally, the changes on perception on how manufacturing process and entire supply chain should look like have changes and emerge towards more focus on sustainable development concerns, especially related to the circular economy component and the eco-innovation initiatives [10]. Therefore, our expectations are that the value added considered on different level of analysis is positively impacted by the Industry 5.0 readiness index components. The different levels of value added considered in our analysis are selected specially to show how those Industry 5.0 readiness index components influence each of the main pillars of a more sustainable development, respectively the economic dimension (*gross value added* analysed), the environmental dimension (*environmental value added* analysed) and the social dimension (*labour productivity* analysed).

# 4 Results and Discussions

## *4.1 Factor Analysis*

As a first step on our research, we reduce the dimensionality of the data, by resuming the 31 variables to a considerably smaller number of factors which describe the potential impact of Industry 5.0. Similar with Czvetkó et al. [7], we reflect the impact through the level of Industry 5.0 readiness, focusing not only on human capital and technology, but on environmental orientation and country institutional capabilities as well.

In Table 2 are provided the statistic for PCA results robustness. If the value of the statistic (Stat. = 0.634) is greater than the threshold of 0.60, the results obtained are acceptable [16].

**Table 2** KMO and Bartlett's Test (authors' own calculation using SPSS 22.0)

| | | |
|---|---|---|
| Kaiser–Meyer–Olkin measure of sampling adequacy | | 0.634 |
| Bartlett's test of sphericity | Approx. Chi-Square | 4566.9 |
| | df | 861 |
| | Sig | 0 |

**Table 3** Total variance explained (authors' own calculation using SPSS 22.0)

| Component | Initial eigenvalues | | | Extraction sums of squared loadings | | | Rotation sums of squared loadings[a] |
|---|---|---|---|---|---|---|---|
| | Total | % of variance | Cumulative % | Total | % of variance | Cumulative % | Total |
| 1 | 19.932 | 47.456 | 47.456 | 19.932 | 47.456 | 47.456 | 19.677 |
| 2 | 5.773 | 13.746 | 61.203 | 5.773 | 13.746 | 61.203 | 7.191 |
| 3 | 3.523 | 8.389 | 69.591 | 3.523 | 8.389 | 69.591 | 5.028 |
| 4 | 2.299 | 5.475 | 75.066 | 2.299 | 5.475 | 75.066 | 3.827 |
| 5 | 2.077 | 4.946 | 80.012 | 2.077 | 4.946 | 80.012 | 2.770 |
| 6 | 1.711 | 4.073 | 84.086 | 1.711 | 4.073 | 84.086 | 6.333 |
| 7 | 1.199 | 2.855 | 86.941 | 1.199 | 2.855 | 86.941 | 3.878 |
| 8 | 1.056 | 2.515 | 89.456 | 1.056 | 2.515 | 89.456 | 3.411 |

In Table 3 we show the contribution of each component extracted on the total variation of the data analysed. We observe our data is reduced to eight components, cumulating a total of variation explained over than 89.4%. However, the only components that generate a marginal contribution to the total variation over than significant level of 5% are the first two components, with a contribution of 47.46% and respectively a contribution of 13.75%. Considering the Kaizer criteria we retain for further analysis the first four components extracted, that sum-up more than 75% from the total variation on our sample data.

In Table 4 are summarized the composition of the main forth components extracted. We have proceeded to generic naming of resulted components. Henceforth, remaining components considered for construction of the composite index of Industry 5.0 country readiness reflect a human capital component, a component related to the financing of the research and development (R&D) sector and relation with the circular economy, a component describing countries' ESG regulation and respectively a component of digitalization.

Figure 2 has summarized the framework proposed for our further analysis.

The main component refers to the human capital development which seem to become increasingly a sustainable and essential resource for each country on the actual context of environmental, economic, and social constraints [39]. Moreover, in the light of the preparations for adoption of an Industry 5.0 framework on the future economies, this national resource becomes even more important, as human capital represent the central component [11, 28]. The increase role of Artificial Intelligence, Internet of Things and Big Data analytics technologies and the increase of cobots used starts from the human capital potential and people digital skills and programming abilities. The framework of Industry 5.0 is aimed to facilitate the coordination of all those elements and leads to a smooth and flawless launch of Society 5.0, considering sustainability requirements as well [44]. Afterall, Industry 5.0 differ significantly

**Table 4** Main factors extracted (authors' own calculation using SPSS 22.0)

| Factor | Variable | Factor loading | Comunalities ($R^2$) |
|---|---|---|---|
| Human capital | Skills workforce | 1.047 | 0.938 |
| | Regulatory quality | 1.022 | 0.963 |
| | Individual digital skills | 1.019 | 0.944 |
| | Government effectiveness | 0.990 | 0.949 |
| | Reliance on professional management | 0.979 | 0.969 |
| | Skills of current workforce | 0.977 | 0.769 |
| | Rule of law | 0.952 | 0.953 |
| | Skills of university graduates | 0.944 | 0.933 |
| | Skills of future workforce | 0.940 | 0.759 |
| | Institutions score | 0.914 | 0.980 |
| | Meritocracy and incentivization | 0.897 | 0.853 |
| | Growth of innovative companies | 0.877 | 0.845 |
| | Legal framework's adaptability to digital business model | 0.872 | 0.950 |
| | Research and development expenditures | 0.847 | 0.951 |
| | Labour market | 0.836 | 0.910 |
| | Intellectual property protection | 0.820 | 0.867 |
| | ICT specialists | 0.766 | 0.875 |
| | Government's responsiveness to change | 0.722 | 0.926 |
| | Corporate governance | 0.683 | 0.845 |
| | Government long-term vision | 0.641 | 0.852 |
| | Eco-innovation | 0.581 | 0.880 |
| | Future orientation of government | 0.532 | 0.841 |
| R&D financing and circular economy | Published articles in Industry 4.0 | 1.008 | 0.951 |

(continued)

**Table 4** (continued)

| Factor | Variable | Factor loading | Comunalities ($R^2$) |
|---|---|---|---|
| | Researchers in business sector | 0.943 | 0.968 |
| | Patents related to Industry 4.0 | 0.814 | 0.902 |
| | Generation of waste | -0.802 | 0.899 |
| | Gross domestic product | 0.742 | 0.959 |
| | Patents related to circular economy | 0.678 | 0.911 |
| | Market capitalization | 0.632 | 0.893 |
| ESG regulation and sustainability | Commitment to sustainability | 0.927 | 0.927 |
| | Energy efficiency regulation | 0.925 | 0.979 |
| | Renewable energy regulation | 0.897 | 0.946 |
| Digitalization | Use of ERP solution | 0.802 | 0.815 |

from Industry 4.0 especially through a higher orientation on sustainability development, a reorientation towards the objective of an efficient cooperation between human and machine, and an additional focus on circular economy requirements as [28] depicted.

The component of human capital contains mainly elements related to *people's skills* (e.g., skills current workforce, individual digital skills, reliance of professional management, skills of university graduates, skills of future workforce), *people motivation* (e.g., meritocracy and incentivization, intellectual property protection), with mostly the highest factors loadings.

However, this component incorporates *country institutional elements* as well (e.g., regulatory quality, government effectiveness, rule of law), elements that reflect *country strategic vision* towards sustainable development (e.g., government long-term vision, future orientation of government), country *institutional flexibility* (e.g., government's responsiveness to change), or even *private sector vision* towards sustainable development (e.g., growth of innovative companies; eco-innovation). Those elements are essential for a flawless implementation of Industry 5.0., as a macroeconomic strategic perspective of governments is essential on drawing-up the direction for the private sector as well [40]. Instead, all those elements are highly conditioned by the human factor, either we talk about public institution processes effectiveness, or the innovative spirit that should be highly promoted as this is the base for governments to ensure an optimal speed of alignment with the dynamic of the economic system.

The second components refer to countries' orientation to innovation through optimal financing of research and development projects, especially in the areas related

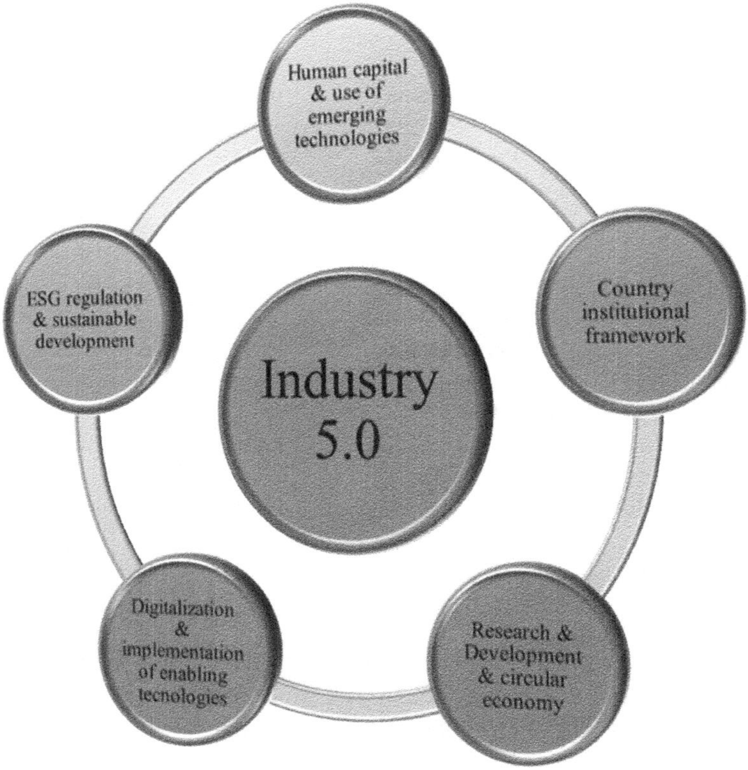


**Fig. 2** Industry 5.0 country profile readiness and enabling components (authors' projection)

to Industry 4.0 (e.g., published articles, researchers in business sector, patents related to Industry 4.0 area), or circular economy (e.g., patents related to circular economy, generation of waste). Human capital potential can be capitalized only if a high cooperation between academia, business sector and government exist on a country and regional level [26]. This Triple Helix model facilitate the development of a sustainable knowledge economy, that has on its centre the human capital, leading to an increase on the return of capital invested through R&D expenditures and direct positive effect on the acceleration coefficient in economy and human capital learning curve on a long term.

The third factor describe the perceived quality of ESG national regulation, which refer partially to the circular economy as well, through the energy national regulation. The component incorporates a score of commitment to sustainability as well, describing a measure of awareness of how important is that countries achieve the global SDG targets. Moreover, better results can be obtained on a country level, as long the legislative ensure a proper legal framework that set-up the main direction towards sustainable development. Expected benefits seem to be identified especially on the circular economy area, with direct impact on SDG 9 Industry and Innovation [23].

The last retained component in our analysis is related to the score of digitalization. Efforts on *digitalization* facilitate the adoption and implementation of Industry 5.0 framework, especially on the area of facilitating the use of Smart Manufacturing enabling technologies [23]. However, this factor must be analysed in relation with the other components, as only a mature national institutional and educational landscape can lead to benefits from digitalization, along with the evolution of human capital showing a rescaling to employees learning curve.

## 4.2 Descriptive Statistics

In Table 5 there have been provided the descriptive statistics of the variables included in our econometric analysis. Additionally, we provide statistics for our output variables as well.

Overall, we observe a heterogeneous sample when looking on the productivity of resources, as the standard deviation 1.197 is more than half of the mean. Instead, countries seem to be closer in terms of labour productivity, gross value added, or ICT sector value added, as their standard deviation is far from their mean value, because of the continuous EU cohesion policies that look for an increase of convergence of the national economies.

In Fig. 3 is provided a mapped representation of the score obtained per each country, considering the first four components extracted. There are no significant changes on a time-based analysis, as it can be seen on Fig. 4, where there have been calculated the mean values per each of the main three composite factors extracted, as the digitalization factor is unique in the composition of the fourth factor.

**Table 5** Descriptive statistics (authors' own calculation using SPSS 22.0)

| | Variable | Minimum | Maximum | Mean | Std. dev |
|---|---|---|---|---|---|
| Independent variable | Resources productivity | 0.366 | 4.957 | 2.061 | 1.197 |
| | Gross value added | 9.94 | 17.51 | 13.62 | 1.80 |
| | Environmental value added | 0.00 | 10.97 | 8.07 | 3.32 |
| | ICT value added | 5.71 | 7.92 | 6.95 | 0.67 |
| | Labor productivity | 66.10 | 129.7 | 97.35 | 19.26 |
| | Pay and productivity | 35.03 | 71.10 | 53.52 | 10.16 |
| Dependent variable | Human capital | −1.59 | 1.55 | 0.00 | 1.00 |
| | R&D finance | −2.56 | 1.44 | 0.00 | 1.00 |
| | ESG regulation | −3.00 | 1.56 | 0.00 | 1.00 |

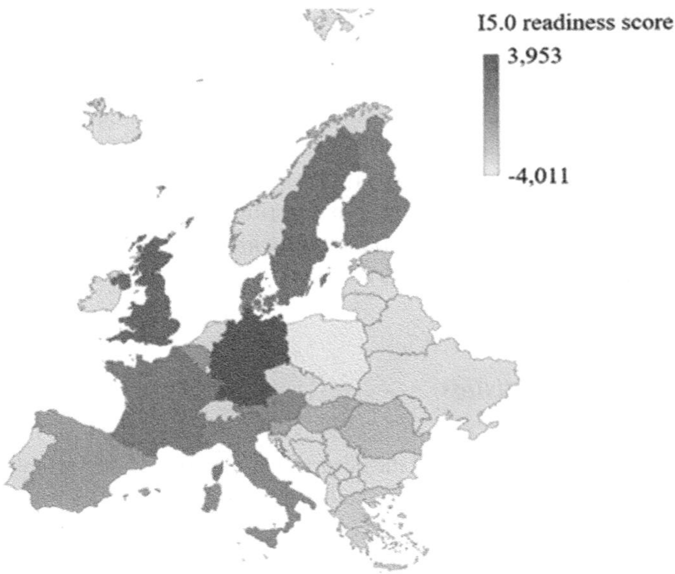


**Fig. 3** Industry 5.0 readiness score by country (authors' projection)

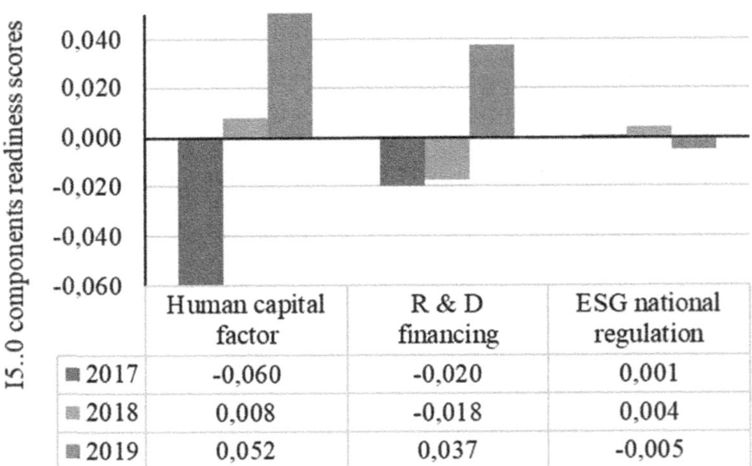


**Fig. 4** Time evolution of industry 5.0 readiness components authors' projection

Overall, the results show a continuous progress of human capital component score. R&D expenses have increase significantly in 2019, in the context of COVID-19 crisis and the need of research on the impact of COVID-19 crisis on socio-economic systems (OECD, 2021). Henceforth, human capital resources each country had become essential in providing viable solution in searching for a treatment, but

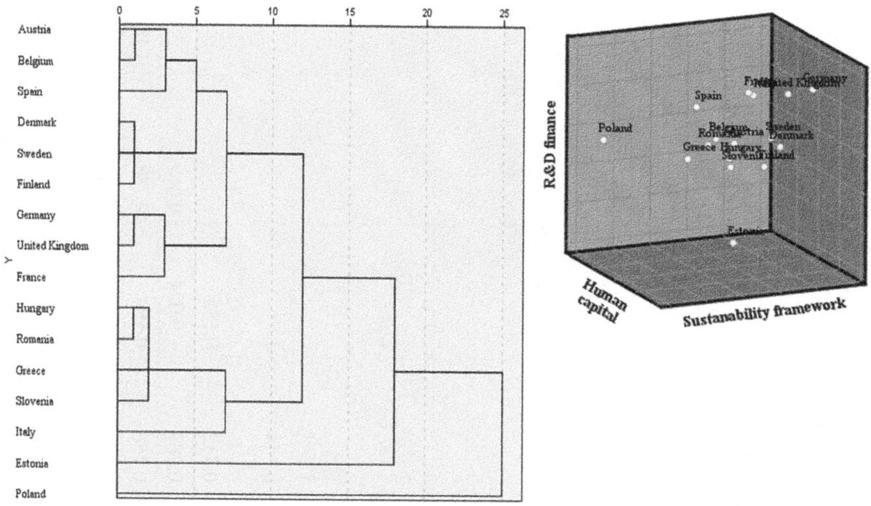


**Fig. 5** Representation of gaps on industry 5.0 potential (authors' projection)

also solutions for proper action plan on the economic and social public policies area. COVID-19 crisis has generated an instant increase as well on government uncertainty policy [31], with more evidence on the area of economic policy, with direct implications on the environmental and social components as well, which are the most affected.

Similar results are obtained also clustering countries analysed in our study against the components extracted. However, as can be seen in Fig. 5, countries are divided in two main clusters. In the first cluster we find EU developed economies, such as Austria, Germany, UK, or Spain. Within this cluster we observe the Nordic countries (Sweden, Denmark, Finland) which are well-known for their commitment on sustainability objectives [42]. On the other cluster there have been found less developed economies, such as Romania, Hungary, or Greece. Those results show how important is the potential of national economies economic growth and the economic capitalization achieved, as these represent the premises for the sources of finance used for financing the Industry 5.0 projects.

## 4.3 Correlation Analysis

In Table 6 are provided the Pearson correlation matrix. The main observation is that human capital is highly correlated with the ICT sector value added (0.906), the incentivization score (0.872), the digitalization score (0.897), and country competitiveness index (0.916). These strong correlations show once again that the current economic development is more related to a knowledge-based economy, for which the

**Table 6** Pearson correlation matrix (authors' own calculation using SPSS 22.0)

| Variable | Human capital | R&D finance | ESG regulation | Gross value added (GVA) | Environmental value added (EVA) | ICT value added (ICT VA) | Labour productivity (LP) | Pay and productivity (PP) | Resources productivity (RP) | Digitalization score | ESG score |
|---|---|---|---|---|---|---|---|---|---|---|---|
| Human capital | 1 | | | | | | | | | | |
| R&D finance | 0.300 | 1 | | | | | | | | | |
| ESG regulation | 0.298 | 0.278 | 1 | | | | | | | | |
| GVA | −0.053 | 0.543 | 0.218 | 1 | | | | | | | |
| EVA | 0.506 | 0.631 | 0.089 | 0.030 | 1 | | | | | | |
| ICT VA | 0.906 | 0.492 | 0.462 | 0.187 | 0.482 | 1 | | | | | |
| LP | 0.671 | 0.618 | 0.231 | 0.108 | 0.615 | 0.795 | 1 | | | | |
| PP | 0.872 | 0.153 | 0.150 | −0.099 | 0.405 | 0.699 | 0.445 | 1 | | | |
| Resources productivity | 0.306 | 0.764 | 0.432 | 0.218 | 0.475 | 0.580 | 0.631 | 0.144 | 1 | | |
| Digitalization | 0.897 | 0.110 | 0.169 | −0.090 | 0.399 | 0.805 | 0.572 | 0.712 | 0.185 | 1 | |
| ESG score | 0.611 | 0.065 | 0.232 | 0.061 | 0.178 | 0.607 | 0.525 | 0.289 | −0.024 | 0.669 | 1 |
| Competitiveness | 0.916 | 0.626 | 0.340 | 0.148 | 0.705 | 0.931 | 0.815 | 0.756 | 0.580 | 0.788 | 0.486 |

core element is the human capital, as set out in the report elaborated by the European Commission [11].

We observe a strong positive correlation between ICT sector value added and the incentivization score (0.699), which show that value creation is highly dependent on how employees are compensated for their work and their innovative and entrepreneurial mind set-up.

Labour productivity can be significantly increased through digitalization and automatization processes. However, this step on economic systems development requires skilled employees (0.897), which must be motivated with attractive remuneration packages (0.712). Such an example is the ICT sector that supposes a high-level digitalization process with impact on the value created (0.805) but constrained by higher competitiveness (0.931). Similar results are obtained by [36] who highlighted the need to ensure proper training of the employees to face the new technologies challenge.

### 4.4 Assessment of Industry 5.0 Components on Country Performance

In Table 7 are summarized the estimated econometric models that show a marginal impact on the value added of the basic pillars of Industry 5.0 framework, respectively the component of human capital, research, and development, and the ESG and circular economy regulation score. All models are relevant as their $R^2$ *adjusted* is between 0.549, in case of model 1, and respectively 0.929, for model 6. The models are statistically significant are they $\text{Sig.} < 0.1$ in case of all estimated models. We observe as well that models have a higher relevance in care the output variable is related to human capital. Those results show once again that Industry 5.0 does not just generate optimization on material use but provides relevant support for decision-making as well, which amplify prior employee's productivity, known as a basic principle of Industry 5.0 framework.

According to model 1 estimated coefficients, we see that human capital $(\text{Coef.} = 1.741, \text{Sig.} < 5\%)$ and the research & development financing $(\text{Coef.} = 0.816, \text{Sig.} < 5\%)$, determine a positive marginal effect on the gross value added reported by countries. The results are similar with Czvetkó et al. [7] who have shown also a significant positive association between their Industry 4.0 readiness index and country GDP.

Investing on human's capital development and in research and development sector generate value add which represent additional sources of public finance which could be directed to sectors with high multiplication coefficient in economy. If the state involves directly on the increase of employees' skills, by granting funds for trainings and research projects, the human capital potential can be capitalized, and benefits will come for firm because of generated economies of scale, created by qualified employees who do not prove just technical skills, but creativity and aligned with basic

**Table 7** Marginal effect of industry 5.0 components on country performance (authors' own calculation using SPSS 22.0)

| Dependent variable | Gross value added | Environmental value added | Resources productivity | Labour productivity |
|---|---|---|---|---|
| Models | (1) | (2) | (4) | (5) |
| Constant | 1,369 | 9,956 | 1,116 | −32,58 |
| | (3,942) | (6,439) | (1,817) | (28,287) |
| *Factors extracted* | | | | |
| Human capital | 1.741** | 1,742 | 0.911* | −3,665 |
| | (0,648) | (1,058) | (0,298) | (4,648) |
| R&D finance | 0.816** | 2.609* | 0.515* | 5.539** |
| | (0,321) | (0,524) | (0,148) | (2,301) |
| ESG regulation | 0,155 | −0,296 | 0,211 | −3,288 |
| | (0,229) | (0,375) | (0,106) | (1,646) |
| *Control variables* | | | | |
| Digitalization | 0,024 | 0,093 | 0.054** | 0,155 |
| | (0,049) | (0,079) | (0,022) | (0,349) |
| ESG composite index | 0,000 | 0,000 | 0,000 | 0,002 |
| | (0,000) | (0,000) | (0,000) | (0,001) |
| *Enabling technologies* | | | | |
| Artificial intelligence | 0,117 | 0,073 | 0.059** | 1.797** |
| | (0,062) | 0,101 | 0,028 | 0,443 |
| Internet of things | −0,043 | −0,064 | 0.042* | 0.196* |
| | (0,031) | (0,050) | (0,014) | (0,219) |
| *Model validation* | | | | |
| Sample size | 48 | 48 | 48 | 48 |
| $R^2$ adjusted | 0,549 | 0,648 | 0,784 | 0,798 |
| F stat | 5,926 | 8,969 | 17,74 | 19,28 |
| p (*F stat*) | 0,000 | 0,000 | 0,000 | 0,000 |

*Significant for 1% significance level; **Significant for 5% significance level

knowledge concerning the future Smart Manufacturing concept [2]. Great attention is to be made on the development of human capital and even more to the increase of awareness and opportunity of Industry 5.0 within the employees, as it represents one of the major challenges firms face with [33].

In the environmental area seem to be more need of research, as the area of research is quite recent. In model 2 we see a significant influence of the R&D component on the environmental value added (Coef. = 2.609, Sig. < 1%). Instead, there is no significant marginal effect of human capital component on the environmental area, as the focus of Industry 5.0 human capital component is to promote automatization of processes and increase labour productivity, while the orientation towards reduction of waste, or energy efficiency are placed on a second place. Overall, waste reduction is related to a more disciplined approach of economic activities, such the case of Lean manufacturing [19], or the more recent and directly Industry 4.0 related additive manufacturing [15], aimed to improve firms' sustainable performance, including the environmental cost burden firms support for waste management.

The positive impact of additive manufacturing on the efficiency of manufacturing operation and digital flexibility of operations planning that lead to better sustainable firm performance is a good example for the results described by our 3$^{rd}$ model, which show a significant impact on resources productivity generated by the human capital (Coef. = 0.911, Sig. < 1%) and respectively the financing component of the research and development sector (Coef. = 0.515, Sig. < 1%). On those circumstances, we see a significant impact of those two Industry 5.0 pillars on the development of the circular economy, in strong relation with the development of the knowledge economy as well. This way, the results of the research are efficiently implemented on manufacturing operations by highly skilled employees, leading to increase on firms' performance [33] and improved cooperation between machines and human, known as a basic principle of Industry 5.0 framework.

Sufficient funding of research lead to output that seem to determine a significant positive marginal effect on labour productivity (Coef. = 5.539, Sig. < 1%), as described on model 5. Industry 5.0 enabling technologies are the results of researchers' tremendous efforts which now have proven can bring notable benefits for firms that implement them on their operations. The benefits raise from the improvement of the decision-making, which is less constrained by economic uncertainty and more focused on automatic calculations and predictions meant to support decision makers, not only for strategic decisions but for operational decisions as well [29].

The Industry 5.0 readiness construct component of digitalization does not influence significantly our value-added macroeconomic measures. This association is expected, as digitalization leads to better operation planning and improved monitoring and control of manufacturing process [18], especially in case of horizontal and vertical integration of firm's systems and applications landscape [23].

Controlling for discretionary effects determined by two of the Industry 5.0 enabling technologies, respectively the Artificial Intelligence and IoT, we observe a positive influence of those elements on the measures of productivity, as shown on estimated model 4 and model 5. Because of higher volume of information processed by equipment used in operations, decision making process improves significantly on a long-term and short-term as well [18]. Important related to this observation is that the marginal effect of those enabling technologies is strictly conditioned by firms' capabilities to automize and digitalize processes and tasks that must be tracked using integrated ERP and MES systems [33].

## 5 Conclusions

While many manufacturers are still trying to implement Industry 4.0 in their sector, research and technological progress have created the premises for the development of the next phase of industrialization, known as Industry 5.0. Some researchers argue that this is not a new industrial revolution but only a new, improved version of the previous one, whose principles place first the well-being of humans. While in the other industrial revolutions automation was the key element, this new concept returns the human touch into industry, to create a collaborative working-space between humans and so called cobots, respectively collaborative robots. Environmental issues are not a priority in Industry 4.0 [28], so the main pillars of the Industry 5.0 are trying to mitigate these deficiencies in order to create a more sustainable environment, increasing prosperity and efficiency.

This paper aimed to investigate through econometric analysis, the way the components of Industry 5.0 readiness model influence value added reported by countries on four levels of analysis, respectively: (i) country gross value added; (ii) environmental activities value added; (iii) ICT sector value added; (iv) labour productivity. The results show a continuous progress of human capital component score that is highly correlated with the ICT sector value added score (0.906), the incentivization score (0,872), the digitalization score (0.897), and country competitiveness index (0.916). Our results also highlighted that human capital and the research & development financing determine a positive marginal effect on the gross value added reported by countries. We registered a significant influence of the R&D component on the environmental value added and no significant marginal effect of human capital component on the environmental area.

The positive impact of additive manufacturing on the efficiency of manufacturing operation and digital flexibility is outlined by the significant impact on resources productivity generated by the human capital and, respectively, the financing component of the research and development sector. In addition, sufficient funding of research lead to output that seem to determine a significant positive marginal effect on labour productivity.

We must be aware that human nature is directed towards knowledge. Technological progress no matter how high it will reach, it will still have to include the aspects related to the economic, social, and environmental issues to achieve sustainable development goals. But innovation must be supported through government policies to be effective and efficient and to bring value added for economies, people, and the environment itself.

## References

1. Abonyi J, Czvetkó T, Honti GM (2021) Are regions prepared for industry 4.0? In: The industry 4.0 + indicator system for assessment. Springer

2. Agolla JE (2018) Human capital in the smart manufacturing and industry 4.0 revolution. Dig Transform Smart Manuf
3. B Baldassarre M Schepers N Bocken E Cuppen G Korevaar G Calabretta 2019 Industrial symbiosis: towards a design process for eco-industrial clusters by integrating circular economy and industrial ecology perspectives J Clean Prod. 216 446 460
4. Braccini MA, Margherita EG (2019) Exploring organizational sustainability of industry 4.0 under the triple bottom line: the case of a manufacturing Company. Sustainability 11:36
5. Cecchin A, Salomone R, Deutz P, Raggi A, Cutaia L (2020) Relating industrial symbiosis and circular economy to the sustainable development debate. In: Industrial symbiosis for the circular economy. Strategies for sustainability. Springer, Cham
6. Circularity Gap Report (2021). https://www.circularity-gap.world/2021
7. Czvetkó T, Honti G, Abonyi J (2021) Regional development potentials of industry 4.0: open data indicators of the industry 4.0 + model. PLoS ONE 16(4):e0250247
8. KA Demir G Döven B Sezen 2019 Industry 5.0 and human-robot co-working Procedia Comput Sci 158 688 695
9. P Dewick M Bengtsson M Cohen J Sarkis P Schröder 2020 Circular economy finance: clear winner or risky proposition J Ind Ecol 24 1192 1200
10. Ejsmont K, Gladysz B, Kluczek A (2020) Impact of industry 4.0 on sustainability-bibliometric literature review. Sustainability 12:5650
11. European Commission (2021) Towards a sustainable, human-centric and resilient European industry. https://op.europa.eu/en/publication-detail/-/publication/468a892a-5097-11eb-b59f-01aa75ed71a1/
12. European Union (2016) Factsheet—Resilience—2016. https://ec.europa.eu/echo/files/aid/countries/factsheets/thematic/EU_building_resilience_en.pdf
13. Fosch-Villaronga E, Mahler T (2021) Cybersecurity, safety and robots: Strengthening the link between cybersecurity and safety in the context of care robots. Comput Law Secur Rev 41
14. Gleirscher M, Johnson N, Karachristou P, Calinescu R, Law J, Clark J (2020) Challenges in the safety-security co-assurance of collaborative industrial robots
15. Godina R, Ribeiro I, Matos F, Ferreira BT, Carvalho H, Peças P (2020) Impact assessment of additive manufacturing on sustainable business models in industry 4.0 context. Sustainability 12:7066
16. Hair JF Jr, Black WC, Babin BJ, Anderson RE (2019) Multivariate data analysis, 8th edn. Cengage
17. Hizam-Hanafiah M, Soomro MA, Abdullah NL (2020) Industry 4.0 readiness models: a systematic literature review of model dimensions. Information 2020 11:364
18. D Horvat H Kroll A Jäger 2019 Researching the effects of automation and digitalization on manufacturing companies' productivity in the early stage of industry 4.0 Procedia Manuf 39 886 893
19. M Iranmanesh S Zailani SS Hyun MH Ali K Kim 2019 Impact of lean manufacturing practices on firms' sustainable performance: lean culture as a moderator Sustainability 11 1112
20. E Lindgreen R Salomone T Reyes 2020 a critical review of academic approaches, methods and tools to assess circular economy at the micro level Sustainability 12 12 4973
21. Y Lu 2017 Industry 4.0: a survey on technologies, applications and open research issues J Ind Inf Integr 6 1 10
22. Ma J (2019) World artificial intelligence conference
23. Mabkhot M, Ferreira M, Maffei P, Podržaj A, Madziel P, Antonelli M, Lanzetta D, Barata M, Boffa J, Finžgar E (2021) Mapping industry 4.0 enabling technologies into united nations sustainability development goals. Sustainability 13:2560
24. Maresova P, Soukal I, Svobodova L, Hedvicakova M, Javanmardi E, Selamat A, Krejcar O (2018) Consequences of industry 4.0 in business and economics. Economies 6:46
25. Meyer R (2021) Leading edge technologies from the slide rule to the cloud
26. F Momeni AAM Yazdi SMS Najafi 2019 Changing economic systems and institutional dimensions of the triple helix model J Innov Entrep 8 1

27. Morrise M, Prigge C (2017) Design of a business resilience model for industry 4.0 manufacturers. In: Twenty-third Americas conference on information systems, Boston
28. Nahavandi S (2019) Industry 5.0—A human-centric solution. Sustainability 11:4371
29. J Para H Ser Del AJ Nebro U Zurutuza F Herrera 2019 Analyze, sense, preprocess, predict, implement, and deploy (ASPPID): an incremental methodology based on data analytics for cost-efficiently monitoring the industry 4.0 Eng Appl Artif Intell 82 30 43
30. Pauliková A, Gyurák Babel'ová Z, Ubárová M (2021) Analysis of the impact of human–cobot collaborative manufacturing implementation on the occupational health and safety and the quality requirements. Int J Environ Res Public Health 18:1927
31. G Pellegrino F Ravenna G Züllig 2020 The cost of coronavirus uncertainty: the high returns to clear policy plans Aust Econ Rev 53 3 397 401
32. Phillips J (2020) The sustainability dynamics framework—A holistic approach to define and evaluate sustainability and unsustainability in the Anthropocene. Environ Impact Assess Rev 84:106436
33. Pollak A, Hilarowicz A, Walczak M, Gasiorek D (2020) A framework of action for implementation of industry 4.0. an empirically based research. Sustainability 12:5789. https://doi.org/10.3390/su12145789
34. PwC (2020) Global top 100 companies by market capitalization. https://www.pwc.ro/en/press-room/press-release-2020/pwc-top-100-listed-companies-in-the-world---strong-recovery-in-m.html
35. Ramanathan K (2020) Industry 4.0 readiness with a circular economy focus: an integrated assessment framework, chapter in collection assessing the readiness for industry 4.0 and the circular economy. In: Economic research institute of ASEAN and East Asia, pp 56–107
36. Roblek V, Meško M, Krapež A (2016) A complex view of industry 4.0. SAGE Open 6
37. Rothenberg H (2019) How "Cobots" are transforming jobs in every industry, from fast food to law. https://www.linkedin.com/business/talent/blog/talent-strategy/hbo-vice-documentary-talks-cobotics-and-future-of-work
38. Schuster S (2020) Robotics and craftsmanship complement each other, KUKA blog
39. Šlaus I, Jacobs G (2011) Human capital and sustainability. Sustainability 3:97–154
40. TK Sung 2018 Industry 4.0: a Korea perspective Technol Forecast Soc Change 132 40 45
41. B Vanderborght 2019 Unlocking the potential of industrial human robot collaboration Publications Office of the European Union Brussels, Belgium
42. World Economic Forum (2019) Global competitiveness report. https://www.weforum.org/reports/how-to-end-a-decade-of-lost-productivity-growth
43. F Yang S Gu 2021 Industry 4.0, a revolution that requires technology and national strategies Complex Intell Syst 7 1311 1325
44. Zengin Y, Naktiyok S, Kaygın E, Kavak O, Topçuoglu E (2021) An investigation upon industry 4.0 and society 5.0 within the context of sustainable development goals. Sustainability 13(5):2682

# Chapter 10
# Trends in Teaching Artificial Intelligence for Industry 5.0

**Cristin Bigan**

**Abstract** As Industry 5.0 is about refining the collaborative interactions between humans and machines and so taking a step further, the efficiency and productivity of machines, processes, and systems together with humans, the Artificial Intelligence (AI) paradigm teaching at the university level for engineering has globally changed in last years, with local and regional exceptions, from a wide range of background of principles, tools, algorithms in the field of pattern recognition, clustering, intelligent decision-making education to a narrow band of AI basic principles and very specific, restricted useful tools/information such as: Machine Learning, especially Deep Leering, Convolutional Neural Networks, and Brain Computer Interface (BCI).

**Keywords** Artificial intelligence (AI) · Industry 5.0 · Skills · Competencies · Training · Courses · University curricula

## 1 Introduction—Public Policies and Leading Organizations Recommendations on AI Tuition for the Industry 5.0 Approach

According to Ref. [1], Industry 5.0 approach contributes to three of the European Union Commission's priorities, and the first element pertinent to Industry 5.0 is already part of its major policy initiatives as adopting a human-centric approach for digital technologies including artificial intelligence. The second one about up-skilling and re-skilling European workers, particularly digital skills, are not so open about specifically going as far as AI teaching (at least as basics) since no mention about Artificial Intelligence or Robotics could be found in none of Skills Agenda or Digital Education Action plan.

A priority of the European Commission on AI in context of Industry 5.0 is shown in the statements published in Ref. [2]. However, as produced by a policymaker legal body, the main part of this document is somehow centered on safety problems related to the use of such systems, and the main concern is that "changing functionality of

C. Bigan (✉)
Ecological University of Bucharest, Faculty of Managerial Engineering, Bucharest, Romania

 
A. Draghici and L. Ivascu (eds.), *Sustainability and Innovation in Manufacturing Enterprises*, Advances in Sustainability Science and Technology,
https://doi.org/10.1007/978-981-16-7365-8_10

AI systems," "can give rise to new risks that were not present when the system was placed on the market." Is nothing wrong on considering all aspects of the risk, but it seems that core issues were omitted by ignoring the huge effort of industrial companies, research units, together with universities to design, implement, continuously improve, produce, and deliver impressive high precision AI-based robots, autonomous systems, and other Internet of Things-AI-based tools.

All these, besides the high financial investment, rely on know-how, knowledge on AI theory and applications that is currently developing and spread to people mostly by teaching or by self-learning. As the fastest way to learn on specific technical issues such as robotics, AI implementations and maintenance are to subscribe and follow a course on the subject, it comes clear that education and all levels of organized tuition are great part of future Industry 5.0 approaches.

At the current stage, steps have been made toward covering all required knowledge for the educational purpose of all actors involved in appropriate building the collaborative interactions between humans and machines. Four main categories of actors in need for AI and its industrial applications tuition each at the appropriate level can be identified in decreasing order of required specific education level/amount:

- Engineers, researchers, developers, advanced users, and maintenance teams members of practical implementations such as robotics operating in Industry 5.0 framework.
- Decision-makers either coming from the State policy Regulatory bodies either from the top management/stake holders of industrial companies to provide realistic laws/recommendations/industrial business decisions about the future of Industry 5.0.
- Human Workers together with AI learning machines, processes, and systems increase the productivity and quality up to the Industry 5.0 level.
- General Public/Consumers requires personalized products or somehow benefits from the newly efficient production ways, resulting from Industry 5.0 level manufacturing.

In Ref. [3], the things are going furthermore on safety precautions, but in the same way of not considering the help of the information provided as training, practical skills, and work experience of users with such AI, IOT and robotic systems and mainly suggesting just the need of "Explicit obligations for producers could be considered also in respect of mental safety risks of users when appropriate (ex. collaboration with humanoid robots)."

The [4] document is the one that raises the problems of the Skills, up-skilling, and re-skilling in the context of evolution of technologies especially digital skills and artificial intelligence knowledge and understanding. It is perfectly true that some digital skills may not have been on the curriculum at the time workers concluded their education and training, but the idea that the educational and training institutions are unable to respond to skills shortage is far from being true. There is a gap in training workers at the re-conversion level, in the field of digital skills at the higher education level (e.g., universities), including lifelong learning courses there is an advanced training even for AI techniques that were not yet put into practice for industry use.

The recommendation for basic understanding of how AI works and to know the potential benefits and limitations of this technology in Ref. [4] fits only for the fourth category of actors identified before: General Public/Consumers.

As shown in recent document [5] on Industry 5.0, the AI role, perception, and its main suite of core technologies are somehow minimized by the statement that AI "often still referring to advanced correlation analysis technologies," which is not entirely true as a lot of existing techniques include causality based. All 10 recommendations of further AI development were already scientifically discovered and reported and can be again implemented to any appropriate specific industrial field if the problem requires so, although there are very well and reliable pure correlation or other statistical based technologies that fit much better for a specific problem than deterministic-based AI, as the focus is to do the job properly and increase efficiency.

The true expectations from the Industry 5.0 regarding **increasing collaboration and interconnectedness between human workers and AI-based machines such as robots** were clearly defined by a second type of identified actor (the decision-making) in the person of Founder and CTO of Universal Robots in Ref. [6].

According to Ref. [7], Industry 5.0 applications in the medical field that relies on artificial intelligence is the combination of big data, artificial intelligence-specialized software (including machine learning), and the Internet resulting in collaborative robots to efficient perform precise and complicated surgery and increasing accuracy of detecting diseases, so better treating patients. A specific hint is on future development of sensors with artificial intelligence that analyses the data quickly.

Based on the keyword analyses and the systematic literature review in Industry 4.0 made in [8] the only term related to Artificial Intelligence in their search that is Machine Learning was placed on the 17th position, however, with a simple search now on Industry 5.0; there are few references that can miss the (key) word artificial intelligence or AI.

As discussed in Ref. [9], it is expected that Industry 5.0 will create a new manufacturing role: Chief Robotics Officer (CRO)—an individual with expertise in understanding robots and their interactions with Humans, adding or removing them from the factory as needed. Such person needs to be trained in robotics, artificial intelligence, human factors modeling, and human–machine interaction.

Based on this new role of CRO and on great industrial experience in 2017, North America's expectations, as described in Ref. [10], were great as *realizing that education must occur to build the bridge from Industry 4.0 to 5.0 and hoping that* North America is uniquely positioned to lead the next, through industry 5.0. A study on future global competitiveness, by Deloitte Global and the U.S. Council on Competitiveness, predicted that the USA will dislodge China as the most competitive manufacturing nation in the world in 2020. This fact not just that did not happen but one of world's leaders USA president Joe Biden stated recently that a long-term strategy competition with China must be prepared together with Europe and others, so also the industry competitiveness is forced to do steps forward.

Meanwhile, Japan has defined Industry 5.0 as "Society 5.0" a "human touch" revolution: "a human-centered society that balances economic advancement with the resolution of social problems by a system that highly integrates cyberspace and physical space."

The debate on the research problem presented in this chapter started with the analysis of specific courses covered by AI curriculum offered as part of BSc or MSc by top universities around the world. This inventory is presented in Appendix and grouped three categories:

A. Specific AI courses and programs from top universities in the USA—**Looking at first five Best Online Artificial Intelligence Degrees for 2021** at top 10 universities in the USA from Ref. [11] and joining the course titles from their AI curriculum; it shows that there is a shift from AI to ML as term usage in the last decade. Here, it is stated an equivalence ("Artificial Intelligence, also known as Machine Learning") but realistic, it remains an inclusion that Machine Learning is included in Artificial Intelligence wider range of topics. Technologies behind AI include **machine learning, deep learning, neural networks, and algorithms. Thus, five universities were included in the analysis.**
B. Specific AI courses from European universities (as filtered from word ranking)—examples from top first five universities in the university ranking.
C. Examples of AI courses and programs from two top Asian universities.

The selection of the universities was piloted by the Academic Ranking of world universities—Shanghai Ranking's Global Ranking of Academic Subjects 2020—Computer Science & Engineering as from Ref. [12]. Table 1 includes the top 25 universities, considering that AI is included in Computer Science and does not have a separate ranking itself in this classification.

The word cloud be generated with the selected specific courses covered by AI curriculum offered as part of BSc or MSc by top universities around the world. Figure 1 shows the most often domains or topics presence in the curricula. The actual "obsession" of teaching staff is very well expressed by a metaphor "Learning—Vison—Computer" or may be better considered as Computer Learning Vision, but in the actual vision together with "Robotics—Deep—Data" and accompanying by "Data," "Theory," and "Human." These key words discovered by the basic semantic analysis summarize the key topics and approaches that have been introduced into the academic curricula in the last years to better align the training programs to the Industry 5.0 new context and requirements (also known as the twenty-first century skills).

## 2 Hi-Tech Job Requirements for the Industrial Areas that Are Meeting Conditions to Become 5.0 Ready

As the current worldwide job offers and application announcements are mainly from large companies that exhibit often transnational even transcontinental borders, the requirements concerning the knowledge acquired by the candidates during their degree tuition do not seem to include visible geographical or cultural differences.

**Table 1** Computer science and engineering University ranking from Ref. [12]

| | |
|---|---|
| 1 | Massachusetts institute of technology (MIT) |
| 2 | Stanford University |
| 3 | University of California, Berkeley |
| 4 | Carnegie Mellon University |
| 5 | ETH Zurich |
| 6 | Harvard University |
| 7 | Tsinghua University |
| 8 | Nanyang Technological University |
| 9 | University of Toronto |
| 10 | University of Oxford |
| 11 | University of California, Los Angeles |
| 12 | Princeton University |
| 13 | University of Technology Sydney |
| 14 | Cornell University |
| 15 | University of Southern California |
| 16 | The University of Texas at Austin |
| 17 | University of North Carolina at Chapel Hill |
| 18 | National University of Singapore |
| 19 | Shanghai Jiao Tong University |
| 20 | Zhejiang University |
| 21 | Harbin Institute of Technology |
| 22 | The Chinese University of Hong Kong |
| 23 | University College London |
| 24 | Peking University |
| 25 | University of Science and Technology of China |

This does not apply to work experience requirements, which is considered second after degree or specific courses or knowledge specified as compulsory, or as a plus.

To analyze the matching between the real job needs as to be realistic into the position of transitioning to the Industry 5.0 era for those who are pioneering and then for all those who will be tracked by the competition to evolve, according to this study, were identified four large types of mentioning the degree/specific courses/knowledge in the job posts/announcement having no visible hierarchy or index as a matter of better or faster recruitment. The search was done to focus on how well the initial undergraduate or postgraduate degree earned by candidates from the universities is fulfilling the needs described by the recruitment. To describe the four main types, letters will be used to reveal that no style proved yet to rank over another.

***A type*** *of mentioning the job required general degree and some of AI-studied subjects as knowledge.*

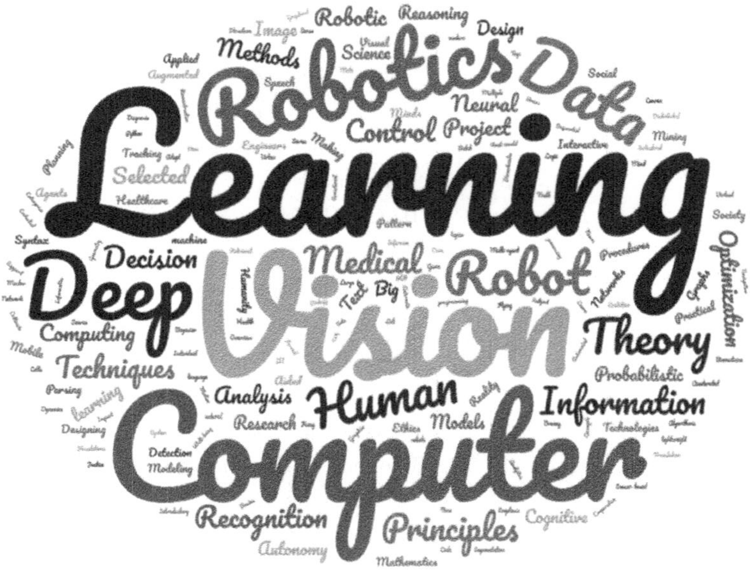


**Fig. 1** Word cloud be generated with the selected courses related to AI field

Surprisingly, some top companies that might be the core of Industry 5.0 transitioning as the first in Robotics (Universal Robots) are providing just the degree with poor detailing or none about the specific subjects or knowledge required from the field of AI, but about fundamentals as mathematics, physics, and programming. By example in Refs. [13, 14],respectively, one can find.

"A degree in a technical major such as computer science, robotics," "disciplines required to create user-facing software for a collaborative robot."

"**Can prove your talent during your studies.**"

- Expert in microcontroller programming.
- Writing high-quality, efficient, clean, maintainable code.
- Fluent in written and spoken English.

And "**It is a plus if you have knowledge and experience with any of following**":

- Robot engineering
- Motor control
- Mathematics and physics
- Python
- C++
- Functional safety/safety-related software
- Analog and digital electronics
- RTOS/real-time systems
- FPGA, VHDL
- Continuous integration and revision control systems (GIT).

***B type** of mentioning the AI job required related degree and none of AI studied subjects as knowledge but just as work experience*

For this, we are taking the automotive industry as example from Ref. [15].

"B.S. or higher in Computer Science or Engineering"

On an example, healthcare job announcement [16] requiring to "Be an ML/AI thought leader" as qualifications it is only specified.

"Master's degree in a Science, Engineering, Mathematics, Statistics, Data Science, Analytics, Computer Science, Health Science or related fields; Ph.D. preferred."

"University degree in computer science, automation, electronics and telecommunications, informatics, mathematics or comparable" is mentioned in example from Ref. [17] also.

***C type** of not mentioning the job required university degree but some of AI studied subjects as knowledge and others as work experience*

As in example from Ref. [18], only the following AI domain applies:

"Publications or experience in machine learning, AI, computer vision, optimization, computer science, statistics, applied mathematics, or data science."

In the Lockheed Martin example from Ref. [19], there are only possibly studied subject as:

"Familiarity with AI/ML domain programming languages (e.g. Python, C++, Java, GO, MatLab)."

"Familiarity integrating and deploying AI/ML models, with a knowledge of machine learning frameworks and deep learning toolsets."

***D type** of not mentioning the job required university degree but some of AI subjects only as work experience.*

In job announcements as in Ref. [20], only work experience is mentioned as: "You have a good understanding of stats and ML. You have solid knowledge of text classification (spaCy, Huggingface, Scikit-Learn). You have some experience with conversational AI (familiarity with at least a open-source chatbot framework, e.g. RASA, Snaps, DeepPavlov), You have knowledge of exploratory data analysis (e.g. clustering, visualizations), Hands-on experience with deep learning libraries like PyTorch, Tensorflow or equivalent., Hands-on experience implementing production-grade NLP architectures."

**At work courses/tuition/learning by experience**

In general, the "at work courses" or practical experience is very good but that does not exceed the few days limit or the practical information delivered does not lead to innovative conduct as the one required by the Industry 5.0 behavior in fulfilling its specifications as collaborative between human and robots. This could be only achieved by a very strong initial study performance during the degree expected to be hold by the employer at academic level completed by innovative ideas in most cases. By example in Ref. [21], the outcome result of the training at work is summarized as follows:

"After attending this 2-day factory-authorized course, you will have the hands-on experience with a UR robot that you need in order to feel comfortable with key aspects of operating one."

**AI jobs and tuition matching comments as prospective evolution of the domain**

According to [22], many of the company states, in the job description, that they want somebody with AI or ML knowledge and the know-how to apply AI and ML to solve current problems, but once one gets hired, it is a completely different reality as in the following:

- Do just data analysis in excel or R or Python because no AI/ML currently exists.
- Data are sparse and that the company does not have the right features to do what it wants.
- Real-world problems show a high degree of non-linearity and the AI/ML solutions that you learned in university studies do not really apply.

By example: A typical real-world scenario is, you run a AI model and the model comes back with a 45% accuracy, which is not much better than a 50% random guess.

## 3 Conclusions on AI-Related Studied as Universities Offer and Real-World Needs for the Transition to 5.0 in Industry

It has become clear that AI domains are wide, and its applications field gets wider everyday. Also, the AI as science, including Machine Learning and Deep learning, also Robotics as science were positively interfering in a lot of manufacturing devices, leading to any new level of work optimization due collaborative human to robot stage. All main advances as theory and practice were launched by academic level of universities, colleges, research institutions, either of them belonging to a top national or worldwide education system or coming as a part of research team of a company. From this large and interesting offer of study offered as bachelor (undergraduate) or master level (graduate) or even PhD/Post Doc research studies by the universities around the world, only a targeted part is specifically required at now stage by the must of job announcements in the field. It is obviously that there remains not specified a lot of subjects' knowledge that are probably assumed to be known for each candidate that claims to have studied advanced level of it.

However, the worry specified in [4] by EU regarding a possible gap between the tuition and employers' expectations from the employees in the AI for main Industry 5.0 achievements does not exist as general rule in that sense. At contrary, there are many AI job applicants willing and up to being enthusiastic to use a lot of theirs learned knowledge in those new Machine Learning algorithms for the new job. For each of such new job advertised as the ones like [20], until the deadline of application closure there are tens or hundreds of people to submit request, also proofing there is a huge human resources potential for the work on transition to Industry 5.0.

Possible evolutions on the field of AI study integration for the benefit of Industry 5.0 concepts as part of every company's strategy to move forward according to its own resources, strategy planning and also competition, which is determinant factor as always are summarized in the next issues:

– The troubleshooting/non-optimistic way that things might move around.

One of the greatest mistakes any low experienced computer programmer can do is to release first instance produced code at the level that in works, this level including inside a lot of unexpected bugs and poor overall quality as regard to the produced software and its re-usability. Similarly, any good intention that was not punished yet as the one to build AI implementations for the human to robot collaboration as by the book can have all good intentions but starving by quality coding of software implementation if the AI designer is not as talented for programming. This really applies when at the firm/company level the high-tech personnel are not well integrated to the software development team.

Another possible, not wanted outcome from all these AI tuition and jobs is that for the now stage of transitional impact of novel implementations in collaborative work we expect that the usage of let's say recurrent neural networks or convolutional architectures are going to be useful for a long time, but in fact, the field will have moved away from that onto something else. That else without university-level tuition, we do not yet understand. One can say that this is a proof that changings in AI field take it closer to engineering and far from math's but remember that it is basic principles were not changed and those initials considered big at the time delivered useful software as regard of available memory and computational capabilities and data amount to process for that time. As in many other fields, there is no obvious path for modeling the needed AI for the collaborative human to robots as there is none from perceiving the task and to building a complex/effective model of the world of 5.0 ready.

There is a high risk that at this stage, even with all the good intentions and wiling of the world, the costs of personalized items produced by the newly concept of implementing strategies defined in Industry 5.0 by means of a lot of AI and robotics will increase as much so, as well as the self-driving cars could be too expensive for worldwide use soon.

Following any of these possible not wanted paths, it is possible that investors who can make more money shortly will pass the now discussed focusing idea of more use of AI knowledge for advance towards 5.0.

***An optimistic way***: The invention of a super-intelligent system as from Ref. [23]: "It has tremendous potential to benefit the world and its peoples, with immense value in growing a smart economy, enhancing global sustainability and national securities and opening new frontiers in science, technology, education, medicine, communication and deep space exploration."

This could be the positive and optimistic way of advances for the Industry 5.0 strategy, also leaving to the near future as expected by 2025 the benefits that we can only imagine now. Of course that the step by step way of building the AI tools required to perform optimal and safe the collaborative interaction between humans and machines (intelligent robots in our case) seem the most realistic but we have to take into account that the solution is obviously not unique and one can find better such a system configuration than the others and so a newly super-intelligence delivering technique could benefit to the unification of scientific discoveries efforts toward the same goal.

According to the Ref. [23], "the man-AI symbiosis is any type of a close and long-term social-cybernetic interaction between two different species, where each termed a symbiont. There are five main **symbiotic relationships**: mutualism, commensalism, predation, parasitism, and competition."

As a final conclusive remark is that at his present stage, there is no gap in Artificial Intelligence areas tuition (especially machine learning) and the most required and challenging jobs offer in the field of further manufacturing process automation in the sense of the Industry 5.0 applications. There are many situations when advanced AI knowledge from enthusiastic machine learning experts or scientists is stopped of putting into practice their ideas, proposals, and knowledge by management or bureaucratic or even lack of IT support and enough open minded to accept major changes due to such intelligent applications. This does not apply always for the work experience way of gathering new knowledge as in this study only qualifications were considered. However, as changes and progress is very high in this area, a most benefit way of update remains to take a course/tutorial that is offered at required standard manly by the universities and very often online due to the recent pandemic conditions.

## Annex 1

A. **Specific AI courses and programs from top universities in the USA Specific AI courses and programs from top universities in the USA**

1. Carnegie Mellon University as from http://coursecatalog.web.cmu.edu/schools-colleges/schoolofcomputerscience/artificialintelligence/#curriculumtextcontainer

    - Concepts in Artificial Intelligence
    - Artificial Intelligence: Representation and Problem Solving
    - Introduction to Machine Learning (SCS Majors)
    - Computer Vision
    - Natural Language Processing
    - Neural Computation
    - Autonomous Agents
    - Truth, Justice, and Algorithms
    - Cognitive Robotics: The Future of Robot Toys
    - Planning Techniques for Robotics
    - Mobile Robot Algorithms Laboratory
    - Robot Kinematics and Dynamics
    - Deep Reinforcement Learning & Control
    - Intermediate Deep Learning

- Machine Learning for Structured Data
- Machine Learning for Text Mining
- Introduction to Deep Learning
- Advanced Methods for Data Analysis
- Speech Processing
- Computational Perception
- Computational Photography
- Vision Sensors
- Design of Artificial Intelligence Products
- Designing Human Centered Software
- Human Robot Interaction
- Artificial Intelligence and Humanity
- AI, Society, and Humanity
- Human Information Processing and Artificial Intelligence
- Perception
- Human Memory
- Visual Cognition
- Language and Thought

2. Stanford University as from https://ai.stanford.edu/courses/

   - Applied Machine Learning
   - Computational Logic
   - Continuous Mathematical Methods with an Emphasis on Machine Learning
   - Artificial Intelligence: Principles and Techniques
   - Introduction to Robotics
   - Natural Language Processing with Deep Learning
   - Natural Language Understanding
   - Machine Learning with Graphs
   - Social and Information Network Analysis
   - Experimental Robotics
   - Machine Learning
   - Machine Learning Theory
   - Deep Learning
   - Computer Vision: From 3D Reconstruction to Recognition
   - Convolutional Neural Networks for Visual Recognition
   - Reinforcement Learning
   - Principles of Robotic Autonomy
   - Principles of Robot Autonomy II
   - Decision Making Under Uncertainty
   - The Human Genome Source Code
   - Computational Biology: Structure and Organization of Biomolecules and Cells
   - Topics in Advanced Robotic Manipulation
   - Advanced Robotic Manipulation

- Machine Learning Systems Design
- Deep Multi-Task and Meta Learning
- Representation Learning in Computer Vision
- Machine Learning Methods for Neural Data Analysis
- Artificial Intelligence for Disease Diagnosis and Information Recommendations
- Designing AI to Cultivate Human Well-being
- Interactive and Embodied Learning
- Seminar in Artificial Intelligence in Healthcare

3. Massachusetts Institute of Technology as from https://ocw.mit.edu/courses/find-bytopic/#cat=engineering&subcat=computerscience&spec=artificialintelligence

   - Identification, Estimation, and Learning
   - Artificial Intelligence
   - Introduction to Machine Learning
   - Mobile Autonomous Systems Laboratory
   - Autonomous Robot Design Competition
   - Robocraft Programming Competition
   - The Battlecode Programming Competition
   - Machine Vision
   - The Human Intelligence Enterprise
   - Techniques in Artificial Intelligence
   - Under actuated Robotics
   - Natural Language and the Computer Representation of Knowledge
   - Advanced Natural Language Processing
   - Machine Learning
   - The Society of Mind
   - Knowledge-Based Applications Systems
   - Computational Models of Discourse
   - Adventures in Advanced Symbolic Programming
   - Introduction to Deep Learning
   - Machine Learning for Healthcare
   - Statistical Learning Theory and Applications
   - Pattern Recognition for Machine Vision
   - Ethics for Engineers: Artificial Intelligence
   - Prediction: Machine Learning and Statistics
   - Principles of Autonomy and Decision Making
   - Cognitive Robotics
   - Statistical Learning Theory
   - Mathematics of Machine Learning
   - Minds and Machines
   - Medical Artificial Intelligence
   - Medical Decision
   - Medical Decision Support

- Affective Computing
- Ambient Intelligence
- A Course on Computer Systems That Adapt To, and Learn From, Context
- Common Sense Reasoning for Interactive Applications
- Relational Machines
- Special Topics in Media Technology: Cooperative Brave New Planet
- Brains, Minds and Machines
- Exploring Fairness in Machine Learning for International Development

4. University of California, Berkeley as from http://ai.berkeley.edu/more_courses_berkeley.html

   - Introduction to Artificial Intelligence
   - Machine Learning
   - Intro to Data Science
   - Probability
   - Optimization
   - Cognitive Modeling
   - Machine Learning Theory
   - Vision
   - Robotics
   - Natural Language Processing

5. Harvard University as from https://www.seas.harvard.edu/computer-science/courses

   - Artificial Intelligence
   - Advanced Topics in Programming Languages
   - Decision Theory
   - Machine Learning
   - Introduction to Computational Linguistics and Natural-language Processing
   - Advanced Topics in the Theory of Machine Learning
   - Tiny Machine Learning
   - Research Topics in Human-Computer Interaction
   - Topics in Machine Learning: Interpretability and Explainability
   - Multi-Robot Systems: Control, Communication, and Security
   - AI for Social Impact
   - Advanced Machine Learning
   - Topics in Machine Learning: Batch Reinforcement Learning
   - Computer Vision
   - Optimization Algorithms for Robotics
   - Multi-Agent Systems
   - Machine Learning for Natural Language
   - Advanced Topics in Computer Vision
   - Biologically-inspired Multi-agent Systems

B. **Specific AI courses from European universities (as filtered from word ranking)**

1. University of Oxford as from https://www.ox.ac.uk/admissions/graduate/courses

   - Artificial Intelligence
   - Machine Learning
   - Advanced Topics in Machine Learning
   - Advanced Machine Learning

2. University of Cambridge as from https://www.cl.cam.ac.uk/teaching/1819/

   - Machine learning for programming
   - Further Human–Computer Interaction
   - Interaction with machine learning
   - Introduction to Natural Language Syntax and Parsing
   - Machine Learning for Language Processing
   - Advanced topics in machine learning and natural language processing
   - Introduction to Natural Language Syntax and Parsing
   - Natural Language Processing
   - Overview of Natural Language Processing
   - Machine Learning and Real-world Data
   - Formal Models of Language
   - Further Human–Computer Interaction
   - Computer Vision
   - Artificial Intelligence
   - Machine Learning and Bayesian Inference
   - Machine Learning
   - Probabilistic Machine Learning

3. ETH Zuric has from http://www.vvz.ethz.ch/Vorlesungsverzeichnis/sucheDozierendePre.view?lang=en

   - Probabilistic Artificial Intelligence
   - Introduction to Machine Learning
   - Machine Learning for Health Care
   - Information Retrieval
   - Big Data
   - Deep Learning
   - Computational Intelligence Lab
   - Machine Perception
   - Data Mining: Learning from Large Data Sets Information
   - Computational Biomedicine
   - Information Theory
   - Natural Language Processing
   - Big Data for Engineers

4. University of Edinburgh as from https://course.inf.ed.ac.uk/previous_session_index.shtml

   - Accelerated Natural Language Processing
   - Algorithmic Game Theory and its Applications
   - Automated Reasoning
   - Automatic Speech Recognition
   - Bioinformatics 1
   - Doing Research in Natural Language Processing
   - Human–Computer Interaction
   - Image and Vision Computing
   - Individual Project in Advanced Natural Language Processing
   - Informatics 2D—Reasoning and Agents
   - Introduction to Vision and Robotics
   - Introductory Applied Machine Learning
   - Machine Learning Practical
   - Machine Learning and Pattern Recognition
   - Natural Computing
   - Natural Language Understanding, Generation, and Machine Translation
   - Reinforcement Learning
   - Robotics: Science and Systems
   - Text Technologies for Data Science

5. Technical University of Munich as from https://www.in.tum.de/en/current-students/modules-and-courses/module-catalog/

   - Image Understanding II: Robot Vision
   - Augmented Reality
   - Computer Aided Medical Procedures
   - Computer Aided Medical Procedures II
   - Image Understanding I: Machine Vision Algorithms
   - Virtual Machines
   - Techniques in Artificial Intelligence
   - Machine Learning
   - Robotics
   - Sensor-based Robotic Manipulation and Locomotion
   - Knowledge-based Systems for Industrial Applications
   - Principles of Computer Vision
   - Robot Motion Planning
   - Tracking and Detection in Computer Vision
   - Computer Vision II: Multiple View Geometry
   - Combinatorial Optimization in Computer Vision
   - Computer Vision I: Variational Methods
   - Introduction to Surgical Robotics
   - Medical Augmented Reality

- Practical Research Project in Bioinformatics
- Advanced Deep Learning for Physics
- Robot Programming and Control for Human Interaction
- Autonomous Navigation for Flying Robots
- Machine Learning for Graphs and Sequential Data
- Probabilistic Graphical Models in Computer Vision
- Convex Optimization for Computer Vision
- Statistical Modeling and Machine Learning
- Interdisciplinary Project in an Application Subject
- Introduction to Deep Learning
- Advanced Deep Learning for Robotics
- Autonomous Driving
- Machine Learning for Computer Vision
- Natural Language Processing
- Advanced Deep Learning for Computer Vision
- Fundamentals of Human-Centered Robotics
- Control of modern lightweight robots
- Computer Vision III: Detection, Segmentation, and Tracking
- Advanced Robot Control and Learning
- Statistical Foundations of Learning
- Selected Topics in Artificial Intelligence and Robotics
- Selected Topics in Computer Graphics and Vision
- Selected Topics in Machine Learning and Analytics
- Selected Topics in Robotics

C. **Examples of AI courses and programs from two top Asian universities**

1. Tsinghua University as from https://ac.cs.tsinghua.edu.cn/curriculum.html

    - Machine Learning
    - Distributed Systems
    - Human Computer Interaction Technologies
    - Introduction to Big Data Systems
    - Natural Language Processing
    - Deep Learning

2. Nanyang Technological University as from http://scse.ntu.edu.sg/Programmes/ProspectiveStudents/Graduate/msc-AI/Pages/CourseContent.aspx

    - Introduction to AI and AI Ethics
    - Machine Learning: Methodologies and Applications
    - Deep Learning and Applications
    - Mathematics for AI
    - Python Programming
    - Computer Vision
    - Text Data Management and Processing

- Time Series Analysis
- Neuro Evolution and Fuzzy Intelligence
- Multi-Agent System
- Advanced Computer Vision
- Deep Neural Networks for Natural Language Processing
- Urban Computing
- AI Master Project

## References

1. https://ec.europa.eu/info/research-and-innovation/research-area/industrial-research-and-innovation/industry-50_en
2. European Commission (2020) White paper: on artificial intelligence—A European approach to excellence and trust. https://ec.europa.eu/info/sites/default/files/commission-white-paper-artificial-intelligence-feb2020_en.pdf. Accessed 12Mar2021
3. Report on the safety and liability implications of Artificial Intelligence, the Internet of Things and robotics
4. Research and Innovation Industry 5.0 Towards a sustainable, human-centric and resilient European industry DF (ISBN 978-92-76-25308-2). https://doi.org/10.2777/308407KI-BD-20-021-EN-N
5. In Enabling Technologies for Industry 5.0 Results of a workshop with Europe's technology leaders
6. Welcome to industry 5.0: The "human touch" revolution is now under way May 23rd, 2019, Published in Articles: EngineerIT, Featured: EngineerIT by EsbenØstergaard, Universal Robots
7. Haleem A, Javaid M, Industry 5.0 and its expected applications in medical field. Curr Med Res Pract. https://doi.org/10.1016/j.cmrp.2019.07.002
8. Li D, Landstrom A, Fast-Berglund A, Almstrom P (2019) Human-centred dissemination of data, information and knowledge in industry 4.0. ScienceDirect CIRP 84:380–386; 29th CIRP Design 2019 (CIRP Design 2019) Published by Elsevier B.V
9. Nahavandi S (2019) Concept paper industry 5.0—A human-centric solution. Sustainability
10. Rendall M, North America takes lead in industry 5.0 by necessity 06/15/17, 08:37 AM Robotics tomorrow Online Robotics Trade Magazine Industrial Automation. Robots and Unmanned Vehicles. https://www.roboticstomorrow.com/article/2017/06/north-america-takes-lead-in-industry-50-by-necessity/10198
11. https://successfulstudent.org/best-artificial-intelligence-colleges/
12. http://www.shanghairanking.com/Shanghairanking-Subject-Rankings/computer-science-engineering.html
13. https://jobs.teradyne.com/UR/job/Boston-Software-Director%2C-Universal-Robots-MA-02108/714768800/
14. https://jobs.teradyne.com/UR/job/Odense-S-Firmware-Engineer-Sout/722190300/
15. https://jobs.teradyne.com/AutoGuide/job/Chelmsford-Software-Engineer-MA-01824/692966200/
16. https://www.indeed.com/q-Artificial-Intelligence-Machine-Learning-jobs.html?vjk=8a00aefbf3879db2
17. https://www.linkedin.com/jobs/search/?currentJobId=2398702904&keywords=machine%20learning%20engineer&originalSubdomain=ro
18. https://www.indeed.com/q-Artificial-Intelligence-Machine-Learning-jobs.html?vjk=83bfc2e14cdf1193

19. https://www.indeed.com/q-Artificial-Intelligence-Machine-Learning-jobs.html?vjk=9f0a0d3425e7f2a9
20. https://www.linkedin.com/jobs/search/?currentJobId=2388903694&keywords=machine%20learning%20engineer&originalSubdomain=ro
21. https://academy.universal-robots.com/fpe-automation/our-trainings/authorized-core-training-march-17th-18th-2021/
22. https://www.quora.com/
23. Abdoullaev A, Engineering a symbiotic superintelligence by 2025: meeting Musk's concerns for $100 billion, Engineering Superintelligence by 2025: AI for Everything and Everyone (AI4EE),White Paper | January 17, 2021 | EIS Encyclopedic Intelligence Systems LTDhttps://www.linkedin.com/pulse/engineering-superintelligence-2025-meeting-musks-azamat-abdoullaev/?published=t&fbclid=IwAR0x0OP3jYamcqbujspVu1zuSfqyxRmtZ8BGvrpGJy3MeEFa5_JJqJWBGhA

# Chapter 11
# Trends for Manufacturing Industry: A Strategic Roadmap Toward Industry 5.0

**Florin Dragan, Larisa Ivascu, and Ben-Oni Ardelean**

**Abstract** Innovation has led to the rapid development of the manufacturing industry. The future directions of the manufacturing industry will highlight new competencies that employees will have to have and requirements that companies will have to have. The evaluation of these directions and the proposal of some organizational strategies represent necessary elements for the industry, experts and those who want to get started in the production activity. This chapter evaluates the implications of Industry 4.0 and makes an inventory of the technologies used in this industrial era. To consolidate the importance of this revolution for the manufacturing industry, a presentation of the situation in Romania is made from the perspective of various important indicators for Industry 4.0. At the end of the chapter is presented a market research and the strategic roadmap for the future of manufacturing. The case study highlights the fact that companies have a major interest in automation and digitization, but there are also several barriers that must be overcome. These barriers can be overcome with the help of facilitators that are presented at the end of the chapter.

**Keywords** Innovation · Industry 4.0 · Industry 5.0 · Production · Manufacturing industry · Competitiveness

## 1 Introduction

Industry plays an important role in local, national, and global development. The current modern industry has gone through several stages of development. The current

F. Dragan
Politehnica University of Timisoara, Department of Automation and Applied Informatics, Faculty of Automation and Computers, Timisoara, Romania
e-mail: florin.dragan@upt.ro

L. Ivascu
Politehnica University of Timisoara, Faculty of Management in Production and Transportation, Timisoara, Romania
e-mail: larisa.ivascu@upt.ro

B.-O. Ardelean (✉)
University of Bucharest, Faculty of Baptist Theology, Bucharest, Romania

 
A. Draghici and L. Ivascu (eds.), *Sustainability and Innovation in Manufacturing Enterprises*, Advances in Sustainability Science and Technology,
https://doi.org/10.1007/978-981-16-7365-8_11

Industry 4.0 appeared in 2011 because of a highly technological project in Germany. The emergence of this concept was based on initiatives to contribute to increasing the competitiveness of German industry. This project promotes computerized production. This advancement of technology has contributed to the improvement of production capacity, efficiency, and loss reduction and contributes to the development of sustainable production systems. Industry 4.0 is also called I4.0 or I4. It is also identified with the Fourth Industrial Revolution. This period can be identified with the digital transformation of production and the creation of the value of processes. The means of production and the different products can be connected, forming networks within which they communicate efficiently, and optimizations are made in real time. The basis of Industry 4.0 is cybernetic systems. Industry 4.0's main goal is customized mass production using process automation and computerization, with cost-effective costs and shorter life cycles [6, 15, 44].

These systems contribute to the development of intelligent factories. Now, production models can be customized, service services can be adapted to customer needs, production system inefficiencies are reduced, the costs of intermediaries in supply chains can be reduced, and irrelevant costs are eliminated. Industry 4.0 focuses on the customer and appreciates the speed, efficiency, and development of innovative services demanded by the market [24, 30, 45, 52].

Most manufacturers should embrace this industrial revolution to maintain their competitive capacity in an extremely competitive market. Fully digital production is the basis of a well-developed strategy that must be based on the principles of action of I4.0 and anticipate the following market and technology requirements [21, 30, 43, 45].

This chapter aims to make an inventory of the evolution of Industry 4.0, then continues with key technologies and applications I4.0. The next part of this chapter covers the market and operational challenges. Because, at present, steps are being taken for Industry 5.0, this chapter presents the possible characteristics of Industry 5.0. At the end of the chapter, it presents a case study referring to the situation of Romanian companies regarding Industry 4.0. Finally, the strategic roadmap for the future of manufacturing is presented.

## 2 The Evolution of Industry 4.0

The Industrial Revolution 4.0 presents a series of stages of development starting with the 1780s. For over 240 years, the first three industrial revolutions took place which were marked by important stages for the development of production capacity. In 1780, production was mechanized. This stage was an important one for production, representing the basis of the evolution of today's revolution so important for the business environment. At this stage, the manual work of our ancestors begins to be mechanized, being a first step in developing the capacity for globalization. The next stage is that of electrification, starting with the period 1870–1880. It can be seen that the second industrial revolution appeared after about 90–100 years [2, 14, 25, 53].

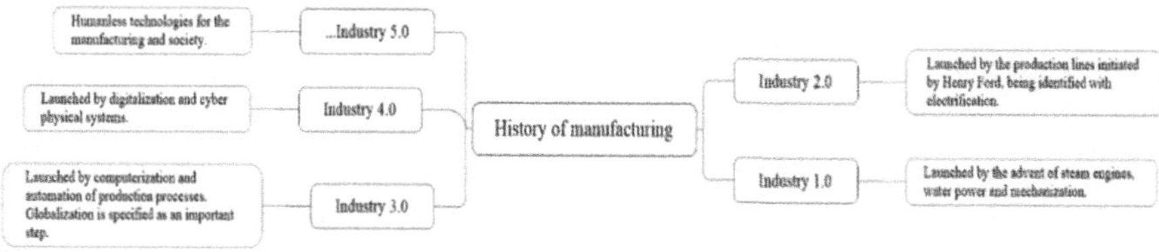


**Fig. 1** History of manufacturing (from Industry 1.0 to Industry 5.0)

The second revolution was the one that led to the realization of the first assembly line, the mass production is realized and, in the production, the electric current is used. It is an important step for today's digitalization that consumes a significant amount of electricity. The third revolution of industrialization knows the third stage, that of computerization and automation of production processes. This revolution occurs in the period 1969–1970. It is an important step in consolidating the basis for digitization and the use of various technologies. In this era of transformation, robots began to be used to perform various man-made activities. The next industrial age, industry 4.0 is called the era of cyber-physical systems (CPS) [5, 29, 40, 50]. CPS includes systems that are capable of autonomously changing the amount of information, launching activities, monitoring, and controlling. The entire exchange of information is based on the Industrial Internet of Things (IIoT). IIoT connects thousands of sensors and transfers data to a local or cloud server to develop predictive models based on this available data [14, 33]. This complex analysis contributes to the decisional industrial system that contributes to the increase of the industry's competitiveness, to the improvement of the industrial processes, of the logistic chain and to the management of the life cycle of the products. Evaluating the elements presented above, it can be seen that there is an important difference between Industry 3.0 and Industry 4.0 [11, 12, 25]. In simple terms, Industry 4.0 is based on machines that operate autonomously without human intervention or presence. The machines used in industry 3.0 can be used in Industry 4.0 if they are updated and are able to exchange data using IIoT and related sensors. Figure 1 shows an evolution of the history of manufacturing, starting from industry 1.0 to Industry 5.0. Industry 5.0 represents the next, anticipated stage of the production industry, based entirely on human absence and automation and connection of machines [17, 27].

## 3 Key Technologies and Application for Industry 4.0

The technologies used for the development of some applications used in industry contribute to the improvement of the production activity and support the decisional process of the enterprise. Today, more and more complex technologies are being implemented. These technologies connect people with systems, communicate, and generate very good results. Production activities are improved, and organizational activities are streamlined. These systems can exchange data, communicate, and detect errors independently and with minimal human input. Industry 4.0 presents a series of key principles that contribute to its intense and appropriate use [4, 36, 41, 49].

- Interoperability—is a very important feature for a production system. This feature refers to the ability of a system to exchange data with other systems and to have the ability to use that data. The interoperability feature offered by the present industrial revolution represents a solid support for the present needs.
- Information transparency—this feature provides operators with useful information for the decision-making process. These data can be used in the activities of the production process and in the decision-making system of the enterprise.
- Technical assistance—the human factor in Industry 4.0 no longer has the role of operator, but to solve certain problems that may arise and to make appropriate decisions.
- Flexibility—Industry 4.0 has increased flexibility and thus can improve the company's ability to meet customer requirements and improve the productivity of production systems. Artificial intelligence and cloud computing offer increased flexibility for production systems.

There are a number of key technologies that characterize Industry 4.0. Each technology develops or contributes to the efficiency of production processes and thus a series of benefits for the enterprise are registered. Table 1 systematizes a selection of technologies used in the fourth industrial revolution, that of automation and computerization [28, 32, 38, 46, 54].

## 4 The Market and Operational Challenges

In the conditions of digitalization of the manufacturing industry, an evaluation of the market requirements is required. These changes not only have several implications regarding raw materials, labor regulations, organizational risk management, global competition, global competition, and globalization but also changes related to transport, market volatility, and price reduction according to demand market. All these changes are shown in the map in Fig. 2 [20, 38, 39, 47].

The new industrial revolution requires operational and market changes. These operational changes are influenced by digitization, automation, and globalization. From the perspective of operational changes, the main changes aim at the involvement of many organizational departments; one of the operational changes refers to the increase in productivity. Each organization wants an increase in production capacity, but this increase must be proportional to the response to market demand, the efficiency of the customer control system and improved communication with suppliers [39]. What is clearly defined is that manufacturers in the industry will invest in digital technologies and advanced manufacturing systems. The acquisition of new capital equipment requires an analysis of the cost–benefit ratio. An efficient acquisition will contribute to the improvement of the production capacity. Among the priorities

**Table 1** Key technologies for industry 4.0

| Technology | Industry 4.0 |
|---|---|
| Augmented reality (AR) | It is a central part of Industry 4.0. Allows access to digital information. This information can overlap with the physical world |
| Automation and industrial robots | Automation and industrial robots are the defining elements for Industry 4.0. This digital revolution is characterized by automation and computerization, i.e., these aspects |
| Additive manufacturing (AM) | This industrial stage is characterized by flexibility, efficiency, and automation. Independent production systems are implemented that require minimal human intervention. This led to the emergence of smart factories and AM. AM is characterized by the fact that the final product can accept the addition of additional layers of material. This reduces the amount of waste and scrap, the production time is reduced, costs are reduced, and the digitalization of business is accentuated |
| Simulation and modeling | Existing modeling and simulation techniques are important for Industry 4.0. Through these activities, an efficiency of the production systems and an improvement of the costs are achieved. SE speaks in the literature of over 15 modeling techniques used mainly |
| Cyber-physical system (CPS) | CPSs are physical entities activated on the Internet, in-corporate with various computers, sensors, control components, and other elements. This unit can generate useful information for the production process, to communicate and to carry out the self-monitoring process |
| Semantic technologies | Interoperability between automated systems can be deficient and that is why semantic technologies are used |
| Internet of things (IoT) | It is one of the technologies that underlies Industry 4.0. This technology refers to the connections between different physical objects, such as production machines, sensors, or the Internet |
| Internet of service (IoS) | It plays an important role for Industry 4.0 along with IoT and CPS |
| Internet of people (IoP) | It involves the digitalization of employee relations, forming a collective intelligent network that stimulates communication using digital devices, the Internet, and data sharing |
| Internet of data (IoD) | A complex network concept composed of IoT entities |
| Cloud computing | Contributes to improving the flexibility of production systems |

(continued)

**Table 1** (continued)

| Technology | Industry 4.0 |
|---|---|
| Big Data analytics | Big data analytics play an important role and are used in smart factories and other production systems. For example, production machine sensor data is used to identify maintenance and upkeep periods. This will make the production process more efficient |
| Blockchain | Blockchain can provide information about the product, distribution, components, subcomponents, and its assembly. In this way, an improvement of the quality is achieved |
| Cybersecurity | It is a key element for Industry 4.0 because all businesses are exposed to risk of attack. The multitude of things connected at this stage requires security and safety, reliability, and a stable decision-making process |
| Cognitive computer | Using cognitive computers simulates different computer models based on human thinking to obtain answers to ambiguous and uncertain situations |
| Enterprise resource planning (ERP) | It is a business process management tool that is used to manage enterprise information |

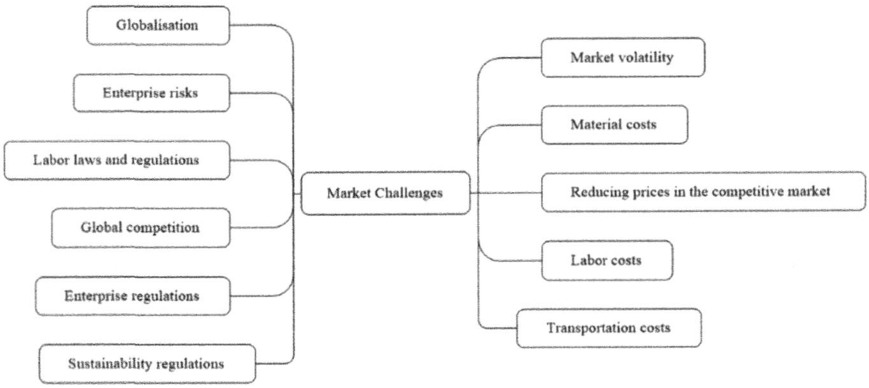


**Fig. 2** The market challenges

of production managers are investments in improving collaboration, more efficient communication, a non-polluting and efficient supply system, operational visibility, competitive strategic execution, flexibility, and several dimensions regarding the marketing function [3, 6, 16, 18, 44, 45, 52]. All these technological improvements are presented in Fig. 3.

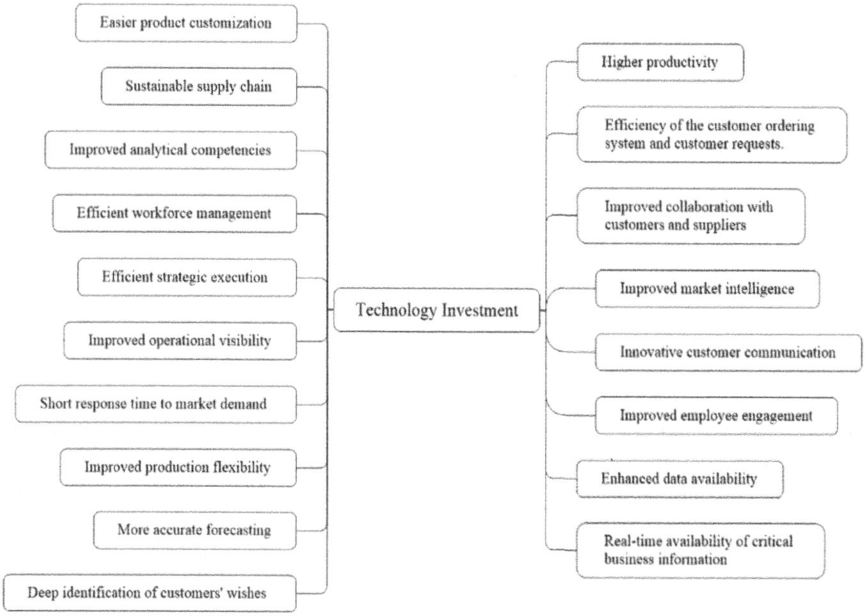


**Fig. 3** Operational changes and investments in technology

## 5 From Industry 4.0 to Industry 5.0 in the Sustainable Development Context

We are currently in the era of digitalization and automation in which we want mass production to be customized, and the resources used to be efficient. In the dynamics of information technology, we try to identify the stage of future development. Industry 4.0 combines the real world with "virtual twins". Industry 5.0 is defined by: human-centric, sustainable, and resilient. If we are currently talking about digital manufacturing, in the context of Industry 5.0, we are talking about digital society [13, 18–20, 34, 42].

Industry 5.0 will aim to eliminate commercial workers from the manufacturing industry. In addition to this central element, new technologies will be developed that will contribute to the substantiation of the concept. These new technologies must be adopted and implemented in accordance with the principles of sustainable development. The 17 objectives of sustainable development must be addressed and respected. The question is often asked whether information technology contributes sustainably to sustainable development or consumes organizational resources and does not contribute to sustainable development. The answer to this dilemma is complex and depends on many factors. So far there are a number of visions and interpretations. The sensors used in the network will be able to contribute to reducing the amount of

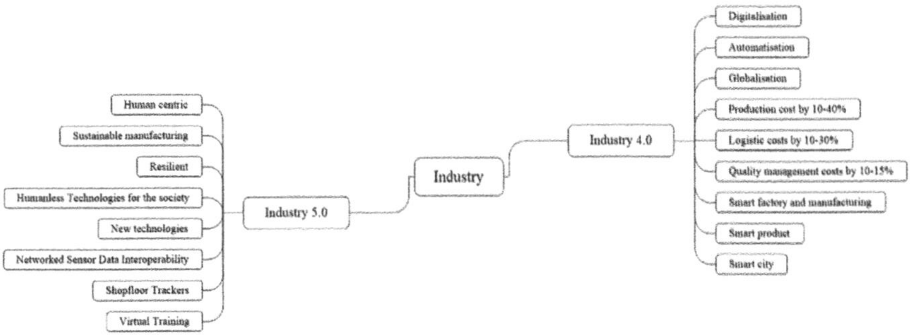


**Fig. 4** Industry 4.0 and industry 5.0

data transferred and communicated. Shopfloor Trackers will help track production and associate customer orders [4, 13, 23, 26, 37, 47, 48, 51].

Virtual Training is an opportunity for training and reduces the costs associated with these activities. Figure 4 shows the map of the elements that define the two industrial revolutions Industry 4.0 and Industry 5.0 (future requirements) [7–9, 48].

## 6 Industry 4.0 in Romania

To emphasize Romania's ability to adapt to the principles of Industry 4.0, a series of data series are presented below. The data series were obtained from the National Institute of Statistics in Romania [22]. To emphasize this capacity, certain dimensions were used, selected according to their importance, as follows: High speed Internet, companies in the industry (investments, financial indicators, number of employees, net investments), European strategy on production customization, employee competencies and higher education in the field, and grants programs. Romania has the capacity to absorb support programs, but the absorption rate can be improved. The analyzed period is 2015–2020. From Industry 4.0 perspectives, Romania has several advantages [6, 35, 43, 55].

- *High-speed Internet*

Romania is among the countries with good broadband internet connection. In a ranking (xxx), Singapore ranks first with about 200 Mbps, Hong Kong ranks second with 180 Mbps, and Thailand ranks third with 160 Mbps. Romania ranks fifth with 152 Mbps, thus having the potential for connections and data transfer in industry.

- *Manufacturing industries*

Table 2 presents the situation of selected indicators for manufacturing enterprises. These indicators are presented for the period 2015–2019. It can be observed that the number of employees, the number of enterprises, and most of the selected indicators

**Table 2** The situation of selected indicators for manufacturing enterprises [22]

| Dimension | Unit | 2015 | 2016 | 2017 | 2018 | 2019 |
|---|---|---|---|---|---|---|
| Number of enterprises | No | 48,405 | 48,349 | 49,837 | 52,449 | 53,123 |
| Average number of employees | No | 1,192,380 | 1,196,306 | 1,207,044 | 1,196,882 | 1,200,011 |
| Average number of people employed | No | 1,203,212 | 1,209,753 | 1,215,909 | 1,206,653 | 1,212,132 |
| Staff costs | Euro | 8,233,609,149 | 9,279,351,952 | 10,916,174,841 | 11,571,845,289 | 1,157,206,555 |
| Salary expenses | euro | 6,718,222,850 | 7,562,620,674 | 8,895,971,590 | 11,190,064,600 | 11,202,309,089 |
| Fiscal value | Euro | 66,551,836,735 | 68,885,918,367 | 78,184,897,959 | 85,613,061,224 | 85,722,448,980 |
| Exercise production | Euro | 60,900,612,245 | 63,310,408,163 | 72,291,632,653 | 79,407,959,184 | 79,594,489,796 |
| Gross value added at factor cost | Euro | 13,932,684,717 | 15,527,444,211 | 17,475,309,804 | 4,148,269,000 | 18,997,324,340 |
| Gross operating surplus | Euro | 5,699,075,568 | 6,248,092,259 | 6,559,134,963 | 37,447,959,184 | 7,527,643,743 |
| The gross result of the exercise | Euro | 3,154,970,238 | 3,550,572,915 | 3,345,648,685 | 4,148,269,000 | 4,150,226,946 |
| Direct exports | Euro | 3,1063,265,306 | 31,557,551,020 | 36,027,959,184 | 37,447,959,184 | 37,650,000,000 |
| Net investments made | Euro | 3,388,992,050 | 3,405,861,544 | 4,017,730,107 | 4,329,630,302 | 4,330,448,873 |
| Gross investment in tangible assets | Euro | 5,225,714,650 | 5,247,141,537 | 6,594,000,484 | 6,734,141,622 | 6,736,182,063 |

**Table 3** Net investments made in Romania and in the manufacturing industry [22]

| | Year (million Euro) | | | | |
|---|---|---|---|---|---|
| Dimension | 2015 | 2016 | 2017 | 2018 | 2019 |
| Total | 20,181.23 | 19,625.09 | 18,580.72 | 20,755.61 | 24,610.49 |
| Manufacturing industry | 3490.67 | 3454.84 | 4030.55 | 4482.76 | 4629.04 |

show a favorable evolution for the analyzed period. This underscores the ability of the manufacturing industry to adapt to the requirements of Industry 4.0 and future industrial developments if there is government support to innovate.

Net investments in the manufacturing industry are increasing during the analyzed period. It can be seen in Table 3 that investments increased in the analyzed period 2015–2019. In the manufacturing industry, the increase of investments takes place annually and presents a favorable evolution for the new technologies that can be used and for the increase of the innovation level [10, 31, 43].

- *European strategy on production customization*

The European strategy that also applies to Romania provides regulations and procedures for increasing the national capacity to customize production. Given the situations presented above, it can be stated that the country has the capacity for innovative development, graduates with higher education are growing, the number of enterprises in the manufacturing industry is growing, and financial indicators are favorable for this approach.

- *Employee competencies and higher education in the field*

Table 4 shows the evolution of the number of graduates in the period 2015–2019. In 2015, it is the highest level of graduates. In the period 2016–2017, there is a decrease in the number of employees, following the period 2018–2019 to bring an increase in the number of graduates on the assessed levels. Therefore, at the Romanian level, there is a potential for the development of adequate Industry 4.0 competencies and

**Table 4** Graduates at different levels in the period 2015–2019 [22]

| Dimensions | Year | | | | |
|---|---|---|---|---|---|
| | 2015 | 2016 | 2017 | 2018 | 2019 |
| Total graduates on all levels | 557,418 | 498,889 | 497,632 | 501,802 | 503,086 |
| University education—graduates with a bachelor's degree | 85,028 | 80,815 | 80,035 | 82,848 | 83,210 |
| University education—graduates with a master's degree and postgraduate education | 44,458 | 38,713 | 39,327 | 41,580 | 39,629 |
| University education—graduates with a diploma—doctorate and advanced research postdoctoral programs | 3,992 | 2,260 | 1,888 | 1,843 | 1,920 |

future industrial revolutions because of the graduates' interest to pursue forms of studies in the country.

The number of employees in Romania experienced a significant decrease during the analyzed period 2015–2019. In 2015, the number of employees is 1248.1 thousand of employees and, in 2019, it is 12198.3 thousand of employees. The number of employees in Romania are presented in Fig. 5.

The employment rate is calculated as the ratio between the civilian employed population and the labor resource, being expressed as a percentage. In 2015, the employment rate of human resources is 66.8 and, in 2019, the employment rate of human resources is 69.6 (Fig. 6).

The number of employees in Romania is presented in Table 5. In the analyzed period 2015–2019, the number of employees increased from one year to another. In 2015, there were 5,041,186 employees, of which 1,433,610 are employed in industry. It is observed that a percentage of 28.5 works in industry. The year 2019 registers an

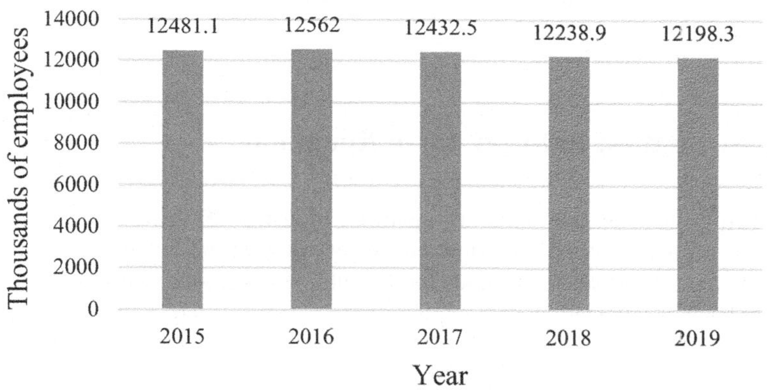


**Fig. 5** Labor resource in Romania [22]

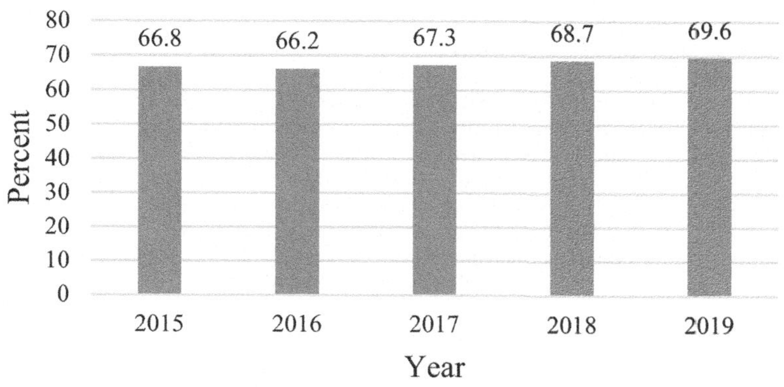


**Fig. 6** Human resources employment rate [22]

**Table 5** The situation of the number of employees in Romania [22]

| | | Year | | | | |
|---|---|---|---|---|---|---|
| | | 2015 | 2016 | 2017 | 2018 | 2019 |
| Total | | 5,041,186 | 5,223,767 | 5,362,346 | 5,426,272 | 5,481,143 |
| From which | Male | 2,653,507 | 2,744,962 | 2,804,590 | 2,837,523 | 2,841,193 |
| From which | Female | 2,387,679 | 2,478,805 | 2,557,756 | 2,588,749 | 2,639,950 |
| Industry | | 1,433,610 | 1,459,873 | 1,498,425 | 1,486,020 | 1,455,690 |
| From which | Male | 802,209 | 816,402 | 837,404 | 828,245 | 812,504 |
| From which | Female | 631,401 | 643,471 | 661,021 | 657,775 | 643,186 |

increase of 9% of the number of employees compared to 2015. At the level of the same year, the number of employees in the industry is higher by 1.5%. This supports the idea that Romania has the capacity to adapt to the new technological approaches that are being implemented in companies because the number of employability requests has increased.

## 7 Case Studies: Industry 4.0 in Romania

This section presents the situation regarding Industry 4.0 in Romania. An online questionnaire was applied for companies from different fields. The 30 most important companies were selected (based on the public image, financial results, competitiveness level, digitalization, and employee competencies). The research was applied online in 2020. Below is a selection of the results obtained. Figure 7 presents present the technologies used by the responding companies. Each company was able to select at least three technologies used. There is a distribution of technologies used, among

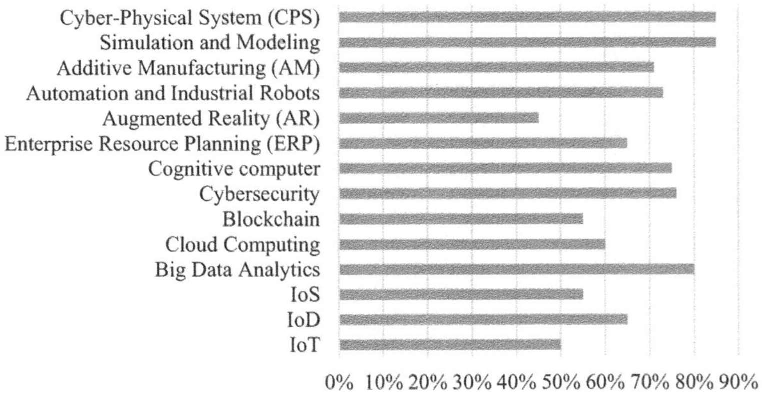


**Fig. 7** The three most important practices in industry 4.0 [22]

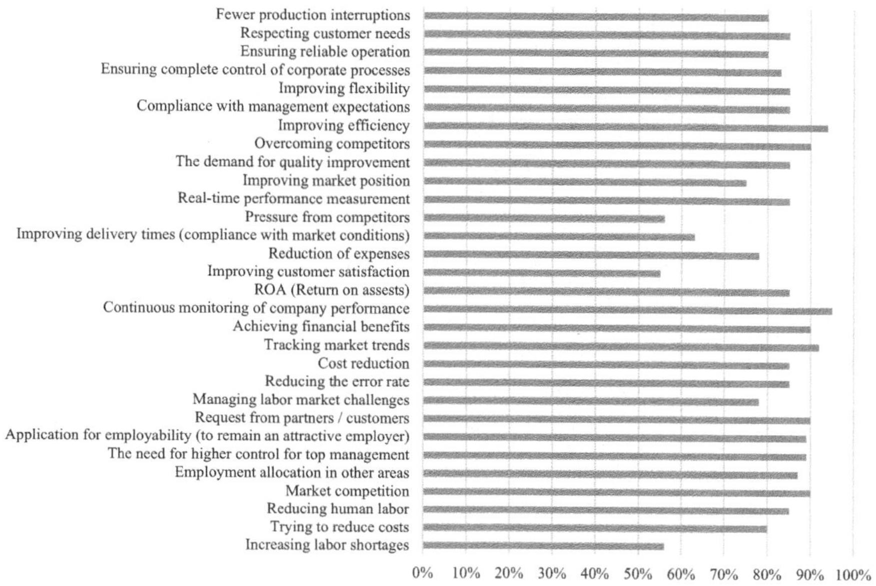


**Fig. 8** The facilitators of industry 4.0 [22]

the most used being CPS, simulating and modeling, big data analytics, cognitive computer, cybersecurity, AM, robots, and ERP.

In order to investigate the opinions of the respondents regarding the facilitators for Industry 4.0, Fig. 8, the following attributes were used which were evaluated: increasing labor shortages, trying to reduce costs, reducing human labor, market competition, employment allocation in other areas, the need for higher control for top management, application for employability (to remain an attractive employer), request from partners/customers, managing labor market challenges, reducing the error rate, cost reduction, tracking market trends, achieving financial benefits, continuous monitoring of company performance, return on assets, improving customer satisfaction, reduction of expenses, improving delivery times (compliance with market conditions), pressure from competitors, real-time performance measurement, improving market position, the demand for quality improvement, overcoming competitors, improving efficiency, compliance with management expectations, improving flexibility, ensuring complete control of corporate processes, ensuring reliable operation, respecting customer needs, and fewer production interruptions. Respondents were able to select one or more attributes [1]. Among the most appreciated attributes are the attributes related to the labor force, the improvement of the image, and the productivity.

To identify implementation barriers, Fig. 9 the following barriers were used: lack of necessary skills within the company, lack of financial resources, lack of skilled labor, lack of managers with adequate skills, longer learning periods (employee training), lack of willingness to cooperate (at the level of the supply chain), lack

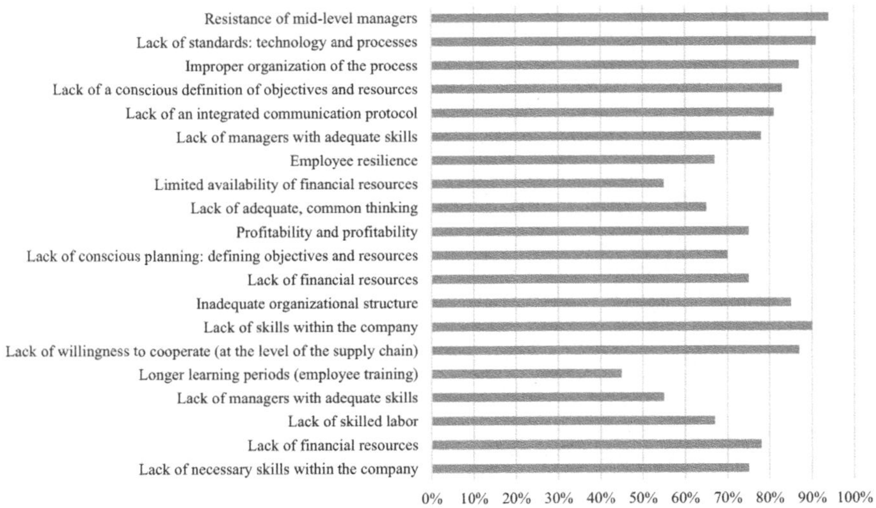


**Fig. 9** The barriers of industry 4.0 [22]

of skills within the company, inadequate organizational structure, lack of financial resources, lack of conscious planning: defining objectives and resources, profitability and profitability, lack of adequate, common thinking, limited availability of financial resources, employee resilience, lack of managers with adequate skills, lack of an integrated communication protocol, lack of a conscious definition of objectives and resources, improper organization of the process, lack of standards: technology and processes, and resistance of mid-level managers.

# 8 The Strategic Roadmap for Future of Manufacturing

The decision-making process for intelligent manufacturing must be efficient and resilient. It must consider all the principles of sustainable development and be based on a series of principles presented in the figure below. It must also be based on the future directions of Industry 5.0 which are presented on the left side of the map. All the strategic aspects of the map are presented on the right side, including the return of personalized production, fast and cheap customization of products, investments in productivity by prioritizing activities and smart decisions. Figure 10 represents a strategic roadmap for future manufacturing.

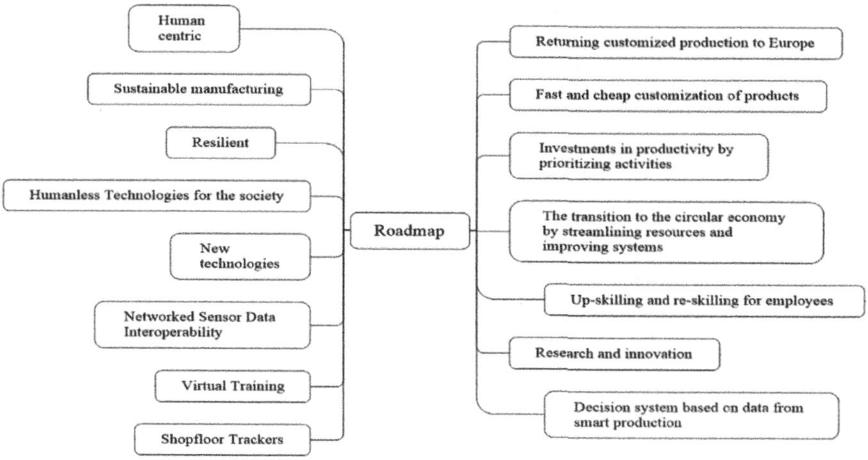


**Fig. 10** The strategic roadmap for future of manufacturing

# 9 Conclusions

It can be concluded that Industry 4.0 is a challenge for many companies. This industrial revolution brings a series of benefits to organizations. In the implementation of this approach, a series of barriers can be identified that pose from competences to the barrier to change. This I4.0 can be characterized by digitization, automation, and globalization. Steps are being taken for the next industrial revolution which will be characterized by independent processes in which human intervention is very low.

Adopting these approaches is important for some industries, but many organizations do not have the resources to fully digitize or automate. These aspects depend on several factors: organizational resources, company size, and much more.

# References

1. Albulescu CT, Draghici A, Fistis GM, Trusculescu A (2016) Does ISO 9001 quality certification influence labor productivity in EU-27? Procedia Soc Behav Sci 221:278–286. https://doi.org/10.1016/j.sbspro.2016.05.116
2. Asadollahi-Yazdi E, Couzon P, Nguyen NQ, Ouazene Y, Yalaoui F (2020) Industry 4.0: revolution or evolution? Am J Oper Res. https://doi.org/10.4236/ajor.2020.106014
3. Bai C, Dallasega P, Orzes G, Sarkis J (2020) Industry 4.0 technologies assessment: a sustainability perspective. Int J Prod Econ. https://doi.org/10.1016/j.ijpe.2020.107776
4. Bal HÇ, Erkan Ç (2019) Industry 5.0 and a critique of industry 4.0. Procedia Comput Sci
5. Bonilla SH, Silva HRO, da Silva MT, Gonçalves RF, Sacomano JB (2018) Industry 4.0 and sustainability implications: a scenario-based analysis of the impacts and challenges. Sustainability (Switzerland). https://doi.org/10.3390/su10103740
6. Büchi G, Cugno M, Castagnoli R (2020) Smart factory performance and industry 4.0. Technol Forecast Soc Change. https://doi.org/10.1016/j.techfore.2019.119790

7. Demir KA, Döven G, Sezen B, Bryndin E, Faruqi UAl, Martynov VV, Shavaleeva DN, Zaytseva AA, Han J, Wu S, Zhang X, Ozkeser B, Koluman OE, Carayannis, EG, Draper J, Bhaneja B (2019) Artificial intelligence and industry 5.0. J Knowl Econ
8. Draghici A, Mocan M, Draghici G (2011) On-line training and certification solution for business process managers. In: International conference on enterprise information systems. Springer, Berlin, Heidelberg, pp 380–389. https://doi.org/10.1007/978-3-642-24358-5_38
9. Draghici A, Olariu C, Rozman T (2014) Certified business process manager: a transfer of innovation experience. Int J Innov Learn 16(1):12–33. https://doi.org/10.1504/IJIL.2014.063371
10. Draghici A, Baban CF, Ivascu LV, Sarca I (2015) Key success factors for university-industry collaboration in open innovation. In: 8th annual international conference of education, research and innovation, Spain. ISBN: 978-84-608-2657-6
11. Ebrahimi M, Baboli A, Rother E (2019) The evolution of world class manufacturing toward industry 4.0: a case study in the automotive industry. IFAC-PapersOnLine. https://doi.org/10.1016/j.ifacol.2019.10.021
12. Erboz G (2017) How to define industry 40: the main pillars of industry 4.0. Managerial trends in the development of enterprises in globalization Era
13. Ferreira CM, Serpa S (2018) Society 5.0 and social development. Preprints
14. Gaiardelli P, Pezzotta G, Rondini A, Romero D, Jarrahi F, Bertoni M, Wiesner S, Wuest T, Larsson T, Zaki M, Jussen P, Boucher X, Bigdeli AZ, Cavalieri S (2021) Product-service systems evolution in the era of Industry 4.0. Serv Bus. https://doi.org/10.1007/s11628-021-00438-9
15. Ghobakhloo M (2018) The future of manufacturing industry: a strategic roadmap toward industry 4.0. J Manuf Technol Manag. https://doi.org/10.1108/JMTM-02-2018-0057
16. Ghobakhloo M (2020) Industry 4.0, digitization, and opportunities for sustainability. J Clean Prod. https://doi.org/10.1016/j.jclepro.2019.119869
17. Gisli T, & TOM, P. (2018). THE EVOLUTION OF 3D PRINTING AND INDUSTRY 4.0. *I-Manager's Journal on Future Engineering and Technology*. https://doi.org/10.26634/jfet.14.1.14600
18. Gogan LM, Rennung F, Fistis G, Draghici A (2014) A proposed tool for managing intellectual capital in small and medium size enterprises. Procedia Technol 16:728–736
19. Granrath L (2019) Large scale optimization is needed for industry 4.0 and society 5.0. Springer Optim Appl. https://doi.org/10.1007/978-3-030-28565-4_1
20. Haleem A, Javaid M (2019) Industry 5.0 and its applications in orthopaedics. J Clin Orthop Trauma. https://doi.org/10.1016/j.jcot.2018.12.010
21. Hariharasudan A, Kot S (2018) A scoping review on digital English and education 4.0 for industry 4.0. Soc Sci. https://doi.org/10.3390/socsci7110227
22. Institutul Național de Statistică (2021) INSSE—Baze de date statistice—TEMPO-Online serii de timp. Website. National Institute of Statistic (NIS)
23. Javaid M, Haleem A, Singh RP, Ul Haq MI, Raina A, Suman R (2020a) Industry 5.0: potential applications in COVID-19. J Ind Integr Manag. https://doi.org/10.1142/S2424862220500220
24. Javaid M, Haleem A, Vaishya R, Bahl S, Suman R, Vaish A (2020b) Industry 4.0 technologies and their applications in fighting COVID-19 pandemic. Diabetes Metab Syndr: Clin Res Rev. https://doi.org/10.1016/j.dsx.2020.04.032
25. Jeevitha T, Ramya L (2018) Industry 1.0–4.0: the evolution of smart factories. APICS Mag
26. Lau WY, Lam WKJ, Yeung CK (2020) From industrial revolution (Industry 1.0) to Surgery 4.0. Chin J Dig Surg. https://doi.org/10.3760/cma.j.cn115610-20200821-00573
27. Liu Z, Xie K, Li L, Chen Y (2020) A paradigm of safety management in industry 4.0. Syst Res Behav Sci. https://doi.org/10.1002/sres.2706
28. Longo F, Padovano A, Umbrello S (2020) Value-oriented and ethical technology engineering in industry 5.0: a human-centric perspective for the design of the factory of the future. Appl Sci (Switzerland). https://doi.org/10.3390/APP10124182
29. Maskuriy R, Selamat A, Ali KN, Maresova P, Krejcar O (2019a) Applied sciences industry 4.0 for the construction industry—How ready is the industry ? Appl Sci

30. Maskuriy R, Selamat A, Maresova P, Krejcar O, David OO (2019b) Industry 4.0 for the construction industry: review of management perspective. Economies. https://doi.org/10.3390/economies7030068
31. Meißner S (2016) Industry 4.0. Chernye Metall. https://doi.org/10.4324/9780429321597-6
32. Melnyk LH, Dehtyarova IB, Kubatko OV, Kharchenko MO (2020) Economic and social challenges of disruptive technologies in conditions of industries 4.0 and 5.0: the EU experience. Mech Econ Regul. https://doi.org/10.21272/mer.2019.86.04
33. Mosca F, Losè MT (2018) Evolution experiences in industry 4.0. Geoingegneria Ambientale e Mineraria
34. Onday O (2019) Japan's society 5.0: going beyond industry 4.0. Bus Econ J
35. Oztemel E, Gursev S (2020) Literature review of industry 4.0 and related technologies. J Intell Manuf. https://doi.org/10.1007/s10845-018-1433-8
36. SKobelev PO, Borovik SY (2017a) On the way from industry 4.0 to industry 5.0: from digital manufacturing to digital society. Industry 4.0
37. SKobelev PO, Borovik SY (2017b) On the way from industry 4.0 to industry 5. 0. Int Sci J "Industry 4.0."
38. Paschek D, Mocan A, Draghici A (2019) Industry 5.0—The expected impact of next industrial revolution. Manag Knowl Learn Int Conf
39. Paschek D, Luminosu CT, Draghici A (2017) Automated business process management—In times of digital transformation using machine learning or artificial intelligence. In: MATEC web of conferences, vol 121. EDP Sciences, p 04007. https://doi.org/10.1051/matecconf/201712104007
40. Pedro Serrasqueiro Martins J, Martins Rodrigues F, Paulo Ferreira Henriques N (2020) Modeling system based on machine learning approaches for predictive maintenance applications. KnE Engineering. https://doi.org/10.18502/keg.v5i6.7105
41. Polat L, Erkollar A (2021) Industry 4.0 versus society 5.0. In: Lecture notes in mechanical engineering. https://doi.org/10.1007/978-3-030-62784-3_28
42. Raharja HY (2019) Relevansi pancasila Era industry 4.0 dan society 5.0 di Pendidikan Tinggi Vokasi. J Dig Edu Commun Arts (DECA). https://doi.org/10.30871/deca.v2i1.1311
43. Rauch E, Linder C, Dallasega P (2020) Anthropocentric perspective of production before and within Industry 4.0. Comput Ind Eng. https://doi.org/10.1016/j.cie.2019.01.018
44. Roblek V, Meško M, Krapež A (2016) A complex view of industry 4.0. SAGE Open. https://doi.org/10.1177/2158244016653987
45. Rojko A (2017) Industry 4.0 concept: background and overview. Int J Interact Mob Technol. https://doi.org/10.3991/ijim.v11i5.7072
46. Salimova T, Vukovic N, Guskova N (2020) Towards sustainability through industry 4.0 and society 5.0. Int Rev. https://doi.org/10.5937/intrev2003048s
47. Salimova T, Guskova N, Krakovskaya I, Sirota E (2019) From industry 4.0 to society 5.0: challenges for sustainable competitiveness of Russian industry. IOP Conf Ser: Mater Sci Eng. https://doi.org/10.1088/1757-899X/497/1/012090
48. Sarfraz Z, Sarfraz A, Iftikar HM, Akhund R (2021) Is COVID-19 pushing us to the fifth industrial revolution (Society 5.0)? Pakistan J Med Sci. https://doi.org/10.12669/pjms.37.2.3387
49. Sołtysik-Piorunkiewicz A, Zdonek I (2021) How society 5.0 and industry 4.0 ideas shape the open data performance expectancy. Sustainability (Switzerland). https://doi.org/10.3390/su13020917
50. Świątek L (2019) From industry 4.0 to nature 4.0—Sustainable infrastructure evolution by design. Adv Intell Syst Comput. https://doi.org/10.1007/978-3-319-94199-8_42
51. Ungureanu AV (2020) The transition from industry 4.0 to industry 5.0. The 4Cs of the global economic change. In: 16th economic international conference NCOE 4.0 2020. https://doi.org/10.18662/lumproc/ncoe4.0.2020/07
52. Xu LD, Xu EL, Li L (2018) Industry 4.0: state of the art and future trends. Int J Prod Res. https://doi.org/10.1080/00207543.2018.1444806

53. Yin Y, Stecke KE, Li D (2018) The evolution of production systems from industry 2.0 through industry 4.0. Int J Prod Res. https://doi.org/10.1080/00207543.2017.1403664
54. Zengin Y, Naktiyok S, Kaygın E, Kavak O, Topçuoğlu E (2021) An investigation upon industry 4.0 and society 5.0 within the context of sustainable development goals. Sustainability (Switzerland). https://doi.org/10.3390/su13052682
55. Zhong RY, Xu X, Klotz E, Newman ST (2017) Intelligent manufacturing in the context of industry 4.0: a review. Engineering. https://doi.org/10.1016/J.ENG.2017.05.015

The manufacturer's authorised representative in the EU is Springer Nature Customer Service Centre GmbH, Europaplatz 3, 69115 Heidelberg, Germany. If you have any concerns regarding our products, please contact ProductSafety@springernature.com

Printed and bound by CPI Group (UK) Ltd, Croydon, CR0 4YY
30/09/2026
02225290-0012